Computer Networks, Architecture and Applications

IFIP – The International Federation for Information Processing

IFIP was founded in 1960 under the auspices of UNESCO, following the First World Computer Congress held in Paris the previous year. An umbrella organization for societies working in information processing, IFIP's aim is two-fold: to support information processing within its member countries and to encourage technology transfer to developing nations. As its mission statement clearly states,

> IFIP's mission is to be the leading, truly international, apolitical organization which encourages and assists in the development, exploitation and application of information technology for the benefit of all people.

IFIP is a non-profitmaking organization, run almost solely by 2500 volunteers. It operates through a number of technical committees, which organize events and publications. IFIP's events range from an international congress to local seminars, but the most important are:

- the IFIP World Computer Congress, held every second year;
- open conferences;
- working conferences.

The flagship event is the IFIP World Computer Congress, at which both invited and contributed papers are presented. Contributed papers are rigorously refereed and the rejection rate is high.

As with the Congress, participation in the open conferences is open to all and papers may be invited or submitted. Again, submitted papers are stringently refereed.

The working conferences are structured differently. They are usually run by a working group and attendance is small and by invitation only. Their purpose is to create an atmosphere conducive to innovation and development. Refereeing is less rigorous and papers are subjected to extensive group discussion.

Publications arising from IFIP events vary. The papers presented at the IFIP World Computer Congress and at open conferences are published as conference proceedings, while the results of the working conferences are often published as collections of selected and edited papers.

Any national society whose primary activity is in information may apply to become a full member of IFIP, although full membership is restricted to one society per country. Full members are entitled to vote at the annual General Assembly, National societies preferring a less committed involvement may apply for associate or corresponding membership. Associate members enjoy the same benefits as full members, but without voting rights. Corresponding members are not represented in IFIP bodies. Affiliated membership is open to non-national societies, and individual and honorary membership schemes are also offered.

Computer Networks, Architecture and Applications

Proceedings of the IFIP TC6 conference 1994

Edited by

S.V. Raghavan

Department of Computer Science and Engineering
Indian Institute of Technology
Madras
India

and

Bijendra N. Jain

Department of Computer Science and Engineering
Indian Institute of Technology
Delhi
India

SPRINGER-SCIENCE+BUSINESS MEDIA, B.V.

First edition 1995

Originally published by Chapman & Hall in 1995
MyCopy version of the original edition 1995

DOI 10.1007/978-0-387-34887-2

A catalogue record for this book is available from the British Library

∞ Printed on permanent acid-free text paper, manufactured in accordance with ANSI/NISO Z39.48-1992 and ANSI/NISO Z39.48-1984 (Permanence of Paper).
www.springer.com/mycopy

CONTENTS

PREFACE

NETWORKS' 94 is the fifth in the series of conferences, sponsored jointly by Computer Society of India and International Federation for Information Processing. The Conference emphasizes areas such as Architecture, Technology, Protocol and Applications. In line with the current trends we have selected papers which are in High Speed Networking, Multimedia, Distributed Systems and Performance. There are 19 contributed papers in all.

India, as a country, is going through a great change in business and financial sectors. As these two sectors are backbone of any economy, there is a perceptible impact in all spheres of activity. Education sector is no exception. While the global competitiveness exerts great pressure on performance of the system, the system in turn exerts a great pressure on the education of people. Everyone in the country without an exception sees networks and communication as the only possible cohesive force, in country as large, as populous and diverse as India.

NETWORKS' 94 addresses these expectations. The response is in the form of presentations based on research that is carried out in India and abroad and in terms of specialized Invited talks from key persons in the industry. Truly, the programme, as has been laid out, is a remarkable synthesis of exacting demands. In this volume, we have a complete and coherent picture of what went on at NETWORKS' 94.

The members of the programme committee did a great job in helping in reviewing the papers well in time. Each paper was carefully seen by at least two programme committee members, in addition to the chairmen.

The selected papers span the areas of High Speed, Multimedia, Distributed Systems, Performance and Protocols. We would like to thank all the reviewers of the papers for their timely reviews and support.

We hope the readers will enjoy the material presented in this book and find it useful in the course of their research or development or education.

S V Raghavan
B N Jain

Co-Chairs
Programme Committee

PROGRAMME COMMITTEE

	Raghavan S V	IIT Madras, India	Co Chairman
	Bijendra N. Jain	IIT Delhi, India	Co Chairman
1.	Aalvarez A	University of Madrid	Spain
2.	Agrawala A.K	University of Maryland	USA
3.	Casaca A	INESC	Portugal
4.	Gonsalves T.A	IIT Madras	India
5.	Haring G	University of Vienna	Austria
6.	Jhunjhunwala A	IIT Madras	India
7.	Janakiram D	IIT Madras	India
8.	Keshav S	AT & T Bell Labs	USA
9.	Kumar A	IISc Bangalore	India
10.	Mehndiratta S.L	IIT Bombay	India
11.	Pujolle G	University of Paris VI	France
12.	Ramakrishnan S	DOE Government of India	India
13.	Ramamurthy B	IIT Madras	India
14.	Ramani S	NCST Bombay	India
15.	Spaniol O	University of Aachen	Germany
16.	Srivatsan K.R	IIT Kanpur	India
17.	Tohme S	ENST	France
18.	Tripathi S.K	University of Maryland	USA
19.	VenkatRangan P	UCSD	USA

ADDITIONAL REFEREEES

20.	Bochmann G.v	University of Montreal	Canada
21.	Olli Martikainen	Telecom	Finland
22.	Pravin Bhagwat	University of Maryland	USA
23.	Ravi Mittal	IIT Madras	India
24.	Ravindran K	Kansas State University	USA
25.	Raymond Miller	University of Maryland	USA
26.	Saha D	Jadavpur University	India
27.	Sanjay K Bose	IIT Kanpur	India
28.	Venkataraman P	IISc Bangalore	India

Invited Talk

MOBILE COMPUTING

Mobile computing is attracting the attention of several researchers in recent times. As mobile networks will have to be built keeping the existing infrastructure in mind, one has to find solutions that can work within the existing framework and assume mobility. For example, issues such as routing, address translation, congestion control and flow control assume a new dimension. The heterogeneity of the existing infrastructure adds to the complexity.

The paper titled "Mobile Computing: A research perspective" by Pravin Bhagwat and Satish Tripathi, explains these issues and shows as to how the connectivity can be maintained at the transport level. The paper also discusses the problems in integrating new communication applications motivated by mobile environment with the existing networks and services.

1

Mobile Computing: A Research Perspective

Pravin Bhagwat Satish K. Tripathi
{ pravin, tripathi}@cs.umd.edu
Dept of Computer Science
University of Maryland
College Park, MD 20770

1 Introduction

The recent advances in wireless communication technology, coupled with the progress in hardware miniaturization, are taking us into a new era of computing, popularly known as nomadic computing. Battery powered, untethered computers are likely to become a pervasive part of our computing infrastructure [9]. There are, however, many technical challenges involved before the vision of ubiquitous computing can be realized. Paramount among these is the challenge of providing continuous, location independent network access to mobile computers. Traditionally, data networking protocols have been designed based on the assumption that network nodes are stationary. With the emergence of laptop computers this assumption no longer holds. From a network architecture point of view, a wireless network is another subnet attached to the backbone network. The key difference with respect to wired networks, however, lies in the mobility of hosts enabled by wireless access. Host mobility affects design considerations of communication systems at all layers of protocol stack. To name a few:

- At the *physical layer*, channel characteristics vary with the location of the user, and, because of mobility, vary in time. A mobile radio link is hindered by a number of propagation mechanisms, namely, multipath scattering from objects near the mobile transmitter, shadowing by dominant obstacles, and attenuation mechanisms on the propagation path between transmitter and receiver.

- At the *data link layer*, channel coding is to be chosen in accordance with the specific character of the fading radio channel. In contrast to wired channels with errors randomly distributed in time, bursts errors are experienced in mobile channels. A typical mobile channel allows relatively reliable communication during certain periods, interrupted by other periods of particularly poor communication known as fades.

- At the *network layer*, the network access point of a host may change from time to time as the user moves from one network to another. Networks must be capable of tracking the location of mobile hosts, and be able to route packets to mobile hosts at all time.

- At the *transport layer*, new mechanisms are needed so that active transport layer connections do not break as mobile hosts switch network access points. Wireless links coupled with host motion violate some of the basic design assumptions around which existing Transport Layer protocols were architected. For example, losses in wireless links are mistakenly interpreted by the transport protocol as the signs of congestion, and the corrective action taken by the transport protocol reduces the efficiency of the wireless link.

- At the *application layer*, new operating system primitives are required which provide support for disconnected and autonomous operation in face of intermittent network connectivity. Existing applications which depend on network support typically 'hang' when network connectivity is disrupted. Examples of such applications include NFS, X-windows and numerous other client-server applications.

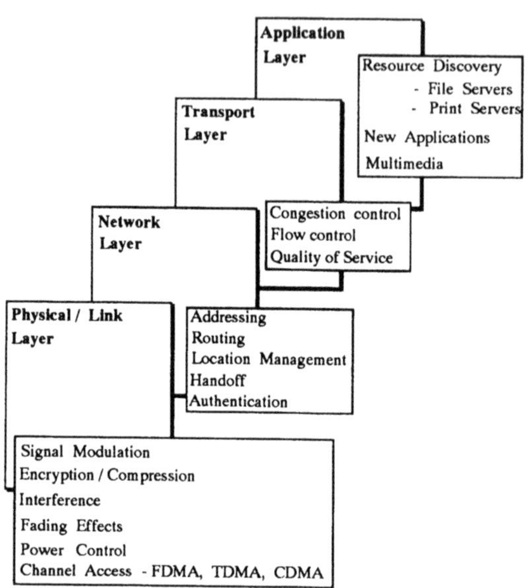

Figure 1: Mobility and the Layered Protocol Design

The issues summarized above capture the essence of all current research activity in the area of future generation mobile data communication systems. New system design constraints posed by host mobility require a careful re-evaluation of protocol design considerations at each layer in the protocol stack. In this paper, however, we limit our discussion to the impact of mobility on the design of the network and the transport layer protocols. First we present a discussion of the mobility problem from the Internet routing perspective, pointing out what makes this problem difficult to solve in practice. The next section is devoted to a short survey of all existing mobile-ip solutions. In section 4.1, we point out deficiencies in the existing transport protocols. Finally, we conclude with a short note on how host mobility affects some of the application design considerations.

2 Network Layer and Mobility

In this section we explore various network layer concepts that are of importance in the design of any mobile networking system. We also explore how these pertain to mobile systems, and what additional features are needed.

2.1 Internet Addressing and Routing

The Internet is a large collection of networks which share the same address space and inter-operate using a common set of protocols, such as TCP/IP [13, 14]. The Internet provides connection-less best effort delivery service to all users. Data are carried in the form of NPDUs[1] which contain source and destination addresses. Routers inspect destination address and forward NPDUs on to appropriate link based on the information contained in the routing table. The following three definitions will help clarify the subsequent discussion:

- Name: is a location independent identifier of a host. E.g. 'mimsy' is the name of the mail-server in out dept.
- Address: indicates where a given host is located. E.g. mimsy's address 128.8.128.8 indicates that it is connected to network 128.8.128
- Route: tells how to get to a destination

The mapping between *name* and *address* is maintained in a distributed data structure, which is know as Domain Name System (DNS). Finding the address of a host is essentially a directory lookup operation. In any large size network, addresses are allocated in a topologically significant manner. This has two advantages: First, it helps reduce the size of the routing tables maintained at various routes; second, the route to reach any host can be deduced from its address.

2.2 The Mobility Problem

In networks where hosts are static, *name* to *address* binding never changes. Host mobility makes this binding a function of time. Therefore, the solution to the mobility problem lies in constructing a directory lookup service, which, given a host name as input, would return the current address of the host. All network applications, prior to connection setup, acquire name to address binding through DNS lookup and then use the address of the peer entity for the entire duration of the connection. If name to address binding changes during an active transport session (as would be the case if the peer host moves) then some mechanisms are also required to notify affected hosts of this change.

Unfortunately, providing a solution for host mobility in the existing Internet environment is not so simple. The difficulty of providing a good solution stems from many sources. Internet has already grown over half a billion hosts in population, which makes any changes to host software almost impossible to achieve. Any changes to routing infrastructure, too, are not easy to achieve since routers in different addressing domains are managed by different administrative authorities. Difficulties are not always administrative in nature. There are some technical problems too. Over

[1] Network Protocol Data Unit

the years the Internet has evolved around the assumption that hosts are stationary. This assumption manifests itself in the following two forms:

1. Domain Name System has no provision to handle dynamic address changes.
2. Names and Addresses are used synonymously. E.g., Transport protocols such as TCP [14], and UDP [12] use host address as host identifier.

The second problem is more critical, and it is the sole reason which makes mobility a difficult problem to tackle. If it were only the first problem, someday DNS could be replaced by a newer version which would efficiently handle dynamic address changes. The second constraint requires that host addresses be preserved, regardless of a host's point to attachment to the network. Its implication is that the address of a mobile host bears no topological significance. As an immediate consequence, the existing routing infrastructure cannot route packets to mobile hosts. A solution might be to keep per-mobile-host routing information at all routers, but this completely breaks the hierarchical model of routing, causing the size of routing tables to grow unbounded. Thus, the problem of supporting mobile hosts within Internet is not just keeping track of where hosts are, rather, in addition, it has to do with designing a mechanism for packet forwarding to mobile hosts without modifying and sacrificing the scalable nature of the Internet routing mechanism.

2.3 The Solution Approach

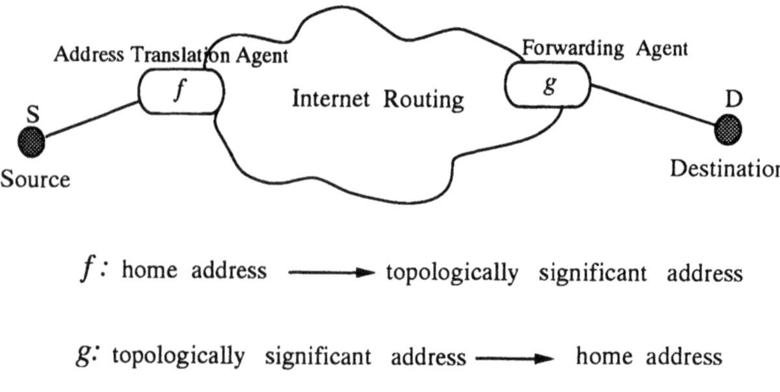

f: home address ⟶ topologically significant address

g: topologically significant address ⟶ home address

Figure 2: Packet Forwarding Model

The home address of a mobile host ceases to be topologically significant once it is not connected to its home subnet. An elegant solution can be designed by associating a forwarding address with each mobile host. A forwarding address is always topologically significant, and, therefore, it can be used within the existing routing system for the purpose of delivering packets to mobile hosts. Usually, the forwarding address is associated with a specialized entity known as a *forwarding agent.* This agent must be capable of directly delivering packets to mobile hosts in its domain. The agent which maintains mapping between a mobile host and its associated forwarding agent is called an *address translation agent.*

With the inclusion of address translation agents and forwarding agents, the operation of packet forwarding can be easily illustrated. Refer to figure 2 which illustrates how packets from a stationary host (S) are routed to a mobile host (D). S sends out packets which are addressed to the home address of D. These are intercepted by an address translation agent (f) which readdresses these packets towards the forwarding agent (g). Once these packets arrive at the forwarding agent, it strips readdressing information and delivers these packets to the mobile host.

Mobility in the Internet is essentially a re-addressing problem. *Encapsulation* and *Source Routing* are two ways of constructing the re-addressing function. In Encapsulation method a new packet header is appended at the beginning of the original packet. The outer header contains the address of the forwarding agent while the inner header contains the home address of the mobile host. Since the Internet routing system only looks at the outer packet header, it routes this packet to the forwarding agent. Forwarding agent strips the outer packet header and delivers the inner packet locally to the mobile host. Source Routing is another technique which can be used to perform re-addressing function. Using IP's source routing option, an address translation agent can cause packets addressed to a mobile host's home address to be routed via a forwarding agent.

A mobile host is always associated with some packet forwarding agent[2]. This association needs to be maintained, and updated at the address translation agent each time a mobile host moves from the coverage area of one forwarding agent to another. The protocol used for managing this information is called the *location update protocol.* All solutions proposed for the Internet mobility problem differ in terms of their choice of where to locate these agents, what the location update protocol is, and whether to use encapsulation or source routing. On surface these solutions may appear different, but they all conform to the model outlined above.

3 Mobile-IP

Mobile-IP is a generic term used to refer to all proposals aimed at solving the mobility problem in the Internet. Over the last two years several proposals, including one from us, have been made to address this problem [5, 1, 18, 19, 16]. Following is a brief outline of approach used in each one of them.

Columbia Scheme The scheme proposed by Ioannidis[5, 6] relies on a group of cooperating Mobile Support Routers, which advertise reachability to the same (sub)net. Each mobile host, regardless of its location within a campus, is always reachable via one of the Mobile Support Routers (MSR). When a host sends a packet to a mobile host, it first gets delivered to the MSR closest to the source host. This MSR encapsulates the packet and delivers it to the target MSR which strips the encapsulation header and relays the original packet to the mobile host. This approach is optimized to work within a campus environment and cannot be easily extended to support wide area mobility.

Sony Scheme In Sony's proposal [18, 17], a mobile host is assigned a new temporary address when it is attached to a new network. The mapping between the home address and the temporary address of a mobile host is kept in an Address Mapping Table (AMT), which is maintained at the routers. Packets transmitted to the home address of the mobile host get intercepted by some router

[2] Usually a wireless base-station functions as a packet forwarding agent

which holds an AMT entry for the mobile host. An address conversion is performed by the router before the packets are forwarded to the physical location of the mobile host. This method requires modifications to routers and host software and has problems inter-operating with the existing hosts unless so-called 'conversion gateways' are used.

MITL Scheme Another proposal to support mobile hosts is from Matsushita[19]. This method is also based on the encapsulation approach. A mobile host is assigned a temporary address when it visits a new network. The packets destined to the home address of the mobile host are intercepted by a Packet Forwarding Server(PFS). The PFS encapsulates the packet and forwards it using the temporary address of the target mobile host. The problem with this method is that routing is always sub-optimal unless the software on all stationary hosts is modified.

LSR Scheme The LSR approach [1, 11] is based on the use of an existing IP option and therefore, does not require any changes to the existing hosts and routers. The key idea is that each packet originating from a mobile host contains enough routing information that can be used by the remote host to send reply back to the source along an optimal path. The key feature of our design is that all functionality needed to manage and process location information is embedded in specialized entities, called Mobile Router and Mobile Access Station. This allows seamless integration of mobile hosts into current networking infrastructure without requiring any changes to existing routers and hosts. A summary of the notable features of our design is as follows:

- The system inter-operates with any TCP/IP compliant system.
- It achieves optimal routing.
- The system enables mobile hosts to roam without causing any disruption in active transport layer sessions.
- No modifications are required to routers and non-mobile hosts.
- Mobile hosts retain their address regardless of their location. This allows all networking applications to run on mobile hosts without any modifications.
- Support for mobility is provided completely within the network layer. This makes our solution very general and independent of any hardware technology used for wireless communication.

An implementation of the LSR scheme currently exists under AIX and MACH in our wireless lab facility at Maryland. The current implementation, in addition to providing mobile to mobile communication capability over wireless infrared link, also allows mobiles to move from the domain of one base-station to another without breaking any active transport layer connections.

4 Transport Layer Issues

4.1 Effect of wireless link and mobility on the Transport Layer

The network architecture providing ubiquitous access to mobile users will consist of a high speed broadband backbone and a set of wireless networks attached to the periphery. Solutions proposed for Mobile-IP would ensure continuous delivery of packets to mobile hosts even as they move from the range of one base-station to another. Enabling the packet forwarding function, as provided by any Mobile-IP system, would make a wireless system functional. However, the design of such a system would not be complete without a careful evaluation of how existing transport protocols and applications would perform in such an environment.

The integration of wireless and broadband networks will give rise to connections with routes encompassing both networks. Existing transport protocols, such as TCP and TP4 have been optimized to operate over wired links, where packet losses are rare and bandwidth is plenty. Wireless segments on the other hand are bandwidth limited and have error characteristics which are often bursty in nature. Moreover, user mobility can cause further packet losses during handoff [3]. It is evident that the underlying assumptions that guided the design of existing transport layer protocols are violated, and, as a result, a connection that encompasses a wireless link may encounter poor and unpredictable performance. To provide good and predictable performance in an integrated wireless and wired network, the transport layer issues need to be revisited. For any solution to be acceptable, it is important to provide inter-operability with the existing population of wired hosts, since they constitute the majority of the infrastructure. The challenge, therefore, is to architect changes in such a way that re-engineered protocols operate in the new environment with little or no changes to the existing infrastructure.

In the following we take a closer look at some of problems outlined above. The discussion is centered around dynamics of TCP, as it is the most widely used transport protocol in the Internet. Similar problems, however, are expected to arise in context of other protocols as well.

4.2 TCP's reaction to packet losses

TCP is designed around the assumption that links are reliable. It interprets packet losses to be resulting from congestion and consequently reacts by reducing its congestion window down to one and entering into *slow-start* [7, 15] phase. This control action is justified if underlying links are reliable in which case the true reason behind the packet loss is most likely to be congestion and the only remedy is to reduce input rate. However, triggering a congestion avoidance mechanism in response to each packet loss in wireless medium may lead to unacceptable level of performance in some cases [10]. Infact, if wireless link is indeed flaky then TCP's slow start mechanism may repeatedly get triggered, resulting in a a scenario where TCP's window size stays at the minimum possible value.

The source of this problem lies in inability to detect whether the losses are due to congestion or they are due to burst errors in the wireless link. Better control mechanisms can be designed if some form of feedback is available from wireless hardware about the quality of wireless link. Unfortunately, current generation wireless adapters do not have any provision to provide this information.

Another plausible solution to this problem is to use a reliable link layer protocol over the wireless segment. The intuition behind why this solution would work is that a reliable link layer would substantially reduce the effective packet loss rate see by the transport layer. This approach too does not warrant good performance due to reasons outlined in the next section; primary reason being the complex interaction of the link layer retransmission mechanism with the dynamics of the round trip time(RTT) and retransmission timeout(RTO) computation of TCP [4].

4.3 Problems originating from RTO computation

For each connection, TCP maintains a running average of RTT (referred to as τ_{rtt}). RTT is computed by measuring the time difference between the instant the packet was released in the network and the moment ACK for this packet arrives back at the source. In addition to τ_{rtt}, TCP also keeps a running average of variance in RTT (referred to as δ_{rtt}). Finally, the retransmission timeout(RTO) (i.e., the period beyond which a packet is retransmitted if its acknowledgement is not received) is computed as follows:

$$\tau^{i+1} = \alpha\tau^i + (1-\alpha)RTT$$

$$\delta^{i+1} = \beta\delta^i + (1-\beta) \mid \tau^{i+1} - RTT \mid$$

$$RTO^i = \tau^i + 2\delta^i$$

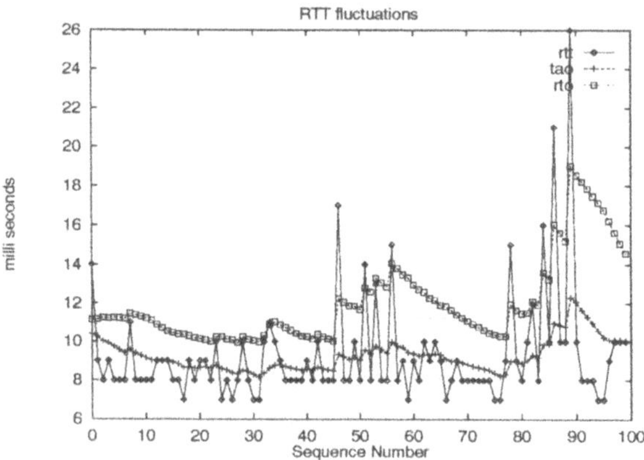

Figure 3: Round Trip Time and Retransmission Timeout Computation in TCP

Figure 3 shows a sample of TCP's τ_{rtt} and RTO computation. The data shown in the graph is taken from actual round trip time measurements collected over 3 hop Ethernet.

If delay variance in the wired segment is relatively small (as indeed would be the case if the backbone an ATM network), and losses in the wireless segment are bursty then it is easy to see that adding more reliability in the link layer segment may not always help. When a connection is in low error state, RTO would converge to the actual RTT of the session. Any packet loss in the wireless segment would not only trigger retransmission at the link layer, but it will also cause TCP's retransmission timer to expire resulting in another retransmission from the source This not only renders the retransmission at the link layer useless but it also wastes bandwidth of the wireless medium.

Burst errors in wireless medium may also adversely affect Karn's RTO adjustment algorithm [8]. According to Karn's algorithm, ACKs for retransmitted packets are never used for updating RTO estimate, rather RTO estimate is doubled each time TCP retransmits a packet[3]. If multiple retransmissions of a packet fail (which is very likely when the wireless link is in burst error mode) then the RTO estimate climbs at an exponential rate. As a result, even after the wireless link returns to a normal state, TCP does not transmit any packet until its timer expires. This behavior causes long periods of inactivity on part of transport protocol, resulting in substantial throughput degradation.

5 Summary

The rapid emergence of mobile computers as a popular, and increasingly powerful, computing device is presenting new challenges to the research community. A great deal of research is being carried out to support the needs of the growing mobile computer industry. A key challenge is to build the underlying infrastructure to support wireless communication. In this system, the main issue is not so much that the communication medium is wireless, but rather that the system must deliver information in the face of a constantly changing environment precisely because the end devices are on move. Intergrating this system with the existing wired infrastructure, applications and services is another important design consideration. The networking research attempts to solve this problem by providing a uniform *location-independent* networking interface to the mobile devices so that the end-users can use their desktop applications on their laptops without any modifications or reconfiguration. Most application designers, however, take a different approach. Their solution relies on making applications location- and situation-aware [2]. The idea here is to build enough intelligence into applications so that they can make use of the locally available resources to dynamically configure themselves in a distributed fashion. The issues discussed in this paper are only a few points in the vast territory of open problems that yet remain to be explored. Research in this area is only beginning, and it will take many more years of research effort before we develop a full understanding of all design tradeoffs.

References

[1] Pravin Bhagwat and Charles Perkins. A Mobile Networking System based on Internet Protocol(IP). In *Proceedings of USENIX Symposium on Mobile and Location Independent Computing*, pages 69–82, Cambridge, MA, Aug 1993.

[2] Pravin Bhagwat, Charles Perkins, and Satish K. Tripathi. Transparent resource discovery for mobile computers. In *Workshop on Mobile Computing Systems and Applications*, Dec 1994.

[3] Ramon Caceres and Liviu Lftode. The effects of mobility on reliable transport protocols. In *Proceedings of the International Conference on Distribnuted Computing Systems*, pages 12–20, Poznan, Poland, 1994. IEEE.

[4] Antonio DeSimone, Mooi Choo Chuah, and On-Ching Yue. Throughput performance of transport-layer protocols over wireless lans. In *Proceedings of the Conference on Global Communications (GLOBECOM)*, pages 542–549. IEEE, 1993.

[3] Reference [8] provides a detailed rationale behind why this approach has been adopted in TCP

[5] John Ioannidis, Dan Duchamp, and Gerald Q. Maguire Jr. IP-based Protocols for Mobile Internetworking. In *Proceedings of ACM SIGCOMM*, pages 235–245, 1991.

[6] John Ioannidis and Gerald Q. Maguire Jr. The Design and Implementation of a Mobile Internetworking Architecture. In *Proceedings of Winter USENIX*, pages 491–502, San Diego, CA, Jan 1993.

[7] Van Jacobson. Congestion avoidance and control. *ACM Computer Communication Review*, 18(4):314–329, August 1988. Proceedings of the Sigcomm '88 Symposium in Stanford, CA, August, 1988.

[8] Phil Karn and Craig Partridge. Improving round trip time estimates in reliable transport protocols. *ACM Transactions on Computer Systems*, 9:364–373, November 1991.

[9] Randy H. Katz. Adaptation and Mobility in Wireless Information Systems. *IEEE Personal Communication Magazine*, 1(1):6–17, Feb 1994.

[10] T. V. Laksman and U. Madhow. Performance Analysis of window-based flow control using TCP/IP: the effect of high bandwidth-delay products and random loss. In *High Performance Networking*, 1994.

[11] Charles Perkins and Pravin Bhagwat. A Mobile Networking System based on Internet Protocol. *IEEE Personal Communication Magazine*, 1(1):32–41, Feb 1994.

[12] J. Postel. User Datagram Protocol. RFC 768, Aug 1980.

[13] J. Postel. Internet Protocol. RFC 791, Sep 1981.

[14] J. Postel. Transmission Control Protocol. RFC 793, Sep 1981.

[15] Scott Shenker, Lixia Zhang, and Dave Clark. Some observations on the dynamics of a congestion control algorithm. *ACM Computer Communication Review*, pages 30–39, October 1990.

[16] Bill Simpson. draft-ietf-mobileip-protocol-03.txt. Draft RFC - work in progress, May 1994.

[17] Fumio Teraoka and Mario Tokoro. Host Migration Transparency in IP Networks. *Computer Communication Review*, pages 45–65, Jan 1993.

[18] Fumio Teraoka, Yasuhiko Yokote, and Mario Tokoro. A Network Architecture Providing Host Migration Transparency. In *Proceeding of ACM SIGCOMM*, Sept 1991.

[19] Hiromi Wada, Takashi Yozawa, Tatsuya Ohnishi, and Yasunori Tanaka. Mobile Computing Environment Based on Internet Packet Forwarding. In *proceeding of Winter USENIX*, pages 503–517, San Diego, CA, Jan 1993.

PART ONE

High Speed Networks

HIGH SPEED NETWORKS

High speed networks are gaining importance and their deployment for regular use is becoming widespread since late eighties. The availability of very high bandwidth alongwith very low latency times, rendered the earlier solutions such as handshake protocols, error recovery techniques and congestion control mechanisms irrelevant thereby forcing the researchers and developers to find more optimal solutions. There were Five papers presented in two sessions, one concentrating on network performance management and traffic control and the other focusing on analysis and implementation of DQDB media access control protocols.

The first paper titled "A study of bandwidth allocation strategies for interactive image applications" by Partho Mishra and Satish K. Tripathi, in the session High Speed Networks-I, basically compares the performance of Hop-by-Hop(HBH) control with rate adaptation with hierarchical round robin scheduling of multiplexers with bandwidth preservations. The study establishes that image transfer applications give better performance with the HBH adaptive rate control. The second paper titled "Stability and optimality of continuous time M-ary rate control mechanisms" by Ranasinghe, Gray and Davidson studies a model of M-ary rate control in a simple queuing network with information delays, using the method of delayed differential equations. The third paper titled "Mean end-to-end packet delays in hybrid integrated service networks" by Ganesh Babu and Anurag Kumar, considers networks in which CBR(Constant Bit Rate) traffic is carried via burst reservation and ABR(Available Bit Rate) traffic is carried in a packet queued best effort mode. The authors establish that the delay of ABR packets are better than that predicted using the poisson arrival assumption due to beneficial correlation between bandwidth availability and packet bursts when CBR traffic is light.

The fourth paper titled "A modified distributed queuing access protocol for dual unidirectional bus networks" by Radhakrishnan Pillai and Utpal Mukherji analyses the performance of the DQDB media access protocol. DQDB protocol is very complex for mathematical modeling and analysis. The authors model and analyse a modified version of the original DQDB protocol, by considering a probabilistic model for the access. A three node network was analysed with this model. The fifth paper titled "Implementation of an IEEE 802.6 compliant card for the ISA bus" by Lakshman Narayanswamy, Jasvinder Singh, Keerthi Mitra, Umesh Kulkarni, Anurag Kumar and Utpal Mukherji discusses the implementation details of an IEEE 802.6 complaint DQDB card for the PC-AT bus. Various issues involved in the design of the card are explained. A novel way of using the DQDB network as a distributed multiplexer for an ATM link is also discussed.

2

A Study of Bandwidth Allocation Strategies for Interactive Image Applications *

Partho P. Mishra
AT&T Bell Laboratories
Murray Bill, New Jersey

Satish K. Tripathi
Department of Computer Science
University of Maryland

Abstract

Image applications are expected to consume a significant portion of the bandwidth of future integrated service networks. In this paper, we examine the effectiveness of various bandwidth allocation mechanisms for this class of applications. First, we describe how a rate based hop by hop (HBH) congestion control scheme, proposed previously by us for controlling data traffic, may be modified to control the flow of image traffic. The performance of the modified HBH scheme as measured by the average image transfer times is shown to be within 15-30% of an unachievable lower bound. We also compare the effectiveness of the HBH scheme with a hierarchical round robin (HRR) scheme that reserves bandwidth before commencing each image transfer. Our results indicate that the likelihood of exceeding a desired transfer time requirement with the HBH scheme is *lower* than the blocking probability with the HRR scheme. Moreover, the average transfer times achieved with the HBH scheme are much lower. However, the variation of the transfer time is very small with the HRR scheme. We also show that the HBH scheme may be used to effectively trade transfer time vs image quality.

1 INTRODUCTION

Applications involving image transfers are expected to consume a significant portion of the bandwidth in integrated service networks. There are relatively few applications today

*This work was done when the first author was at the University of Maryland and was supported in part by TASC.

that use networks to transfer large images primarily because the required bandwidth is either unavailable or too expensive. For example, the transfer of a 200 Mbyte *cat scan* image over a 1.5 Mb/s leased line would take more than 25 minutes - an unacceptably large latency. Similarly, it is currently cheaper to transfer the high quality photographic images used in magazine publishing using courier services. However, the advent of high speed integrated service networks and greater accessibility to network resources promises to drastically alter this situation. There are an increasing number of applications, e.g. IMM [Dan], that provide various kinds of image access services and this trend is likely to accelerate.

The service quality requirements of image applications are in some ways similar to that of data traffic. For example, the major metric of performance is the *transfer time*: the time taken to transfer individual images or sets of images. This value is determined by the average throughput seen by a connection. Hence, as with bulk data transfer traffic, it is desirable to provide as much throughput as possible to individual connections. However, unlike data traffic the redundancy in information in each image implies that some amount of loss may be tolerable. Additionally, under circumstances where the bandwidth available is limited it is possible to transmit the information at a *coarser* level of granularity to trade image quality for transfer time latency.

There are two choices available for allocating bandwidth to image applications transported over a packet switched wide area network. One possibility is to allocate a fixed bandwidth to a connection, based on a quality of service contract between the application and the network. The other possibility is to adapt the bandwidth used by an application depending on available network capacity, as with data traffic applications [Jac88;RJ90]. However, the adaptive rate control mechanisms designed for data traffic applications need to be modified to reflect the differences in the service quality requirements of the two classes of traffic.

In this paper, we examine the effectiveness of various bandwidth allocation schemes for image traffic. First, we discuss how a rate based hop by hop (HBH) congestion control scheme, previously proposed by us to control data traffic [MK92;MKT93], may be modified to control the flow of image traffic. We compare the performance of the modified HBH scheme as measured by the average transfer times with an unachievable lower bound on the transfer time and show that the average transfer times achieved with the HBH mechanism are within 15-30% of the lower bound. Second, we show that allocating bandwidth to aggregates of connections results in superior performance specially in counteracting the degradation in the performance of adaptive control schemes as the transfer times become smaller vis-a-vis the end to end delay in the network. Third, we compare the effectiveness of adaptive and fixed bandwidth allocation using the HBH and Hierarchical Round Robin (HRR) scheduling schemes [KKK90] as examples. Our results indicate that the likelihood of exceeding a desired transfer time requirement with the HBH scheme is lower than the blocking probability with the HRR scheme. Moreover, the average transfer times are much lower when the HBH scheme is used. However, the variance of the transfer time is very low with the HRR scheme.

The rest of this paper is organized as follows. In section 2 we describe digital image representation and coding techniques as well as the traffic characteristics and the service qulaity requirements of typical applications. In Section 3 we describe the different congestion control mechanisms that we evaluate. In Section 4 we describe the experimental methodology we use to evaluate the performance of the control schemes and present the results. Section 5 concludes the paper.

2 APPLICATION CHARACTERISTICS

The digital representation of an image is a two dimensional array of pixels. The number of bits required to represent every pixel depends on the degree of resolution and whether or not color is used. The simplest way to represent an image is by explicitly storing the bits corresponding to every pixel in the image. However, this requires a huge number of bits to represent an image - for example a single 25000 x 25000 pixel image with 8 bits per pixel would require 5 Gigabits. The number of bits required to represent an image may be reduced by using compression techniques. Compression techniques achieve reduction in two ways. The first is by the use of a more efficient encoding scheme in which the representation for a symbol (a pixel value) is chosen based on how frequently it occurs in the image. Examples of these techniques include Huffman coding and arithmetic coding [NP85]. The second is by discarding information which is less relevant to image quality by quantization. The achievable compression and the loss in information varies depending on coding algorithm and the contents of an image. Existing compression algorithms can result in a reduction in the number of bits required to encode an image by a factor 10:1 to 50:1, without a significant reduction in the image quality. Due to the potentially large size of images, many applications also operate on smaller portions of an image, which we will refer to as *tiles*. In this paper we will assume that a tile is 256 x 256 pixels.

Most image transfer applications send or receive information in either a continuous or interactive mode. In a continuous mode transfer, a set of images is transferred from an image repository to a group of recipients. An example of such an application is the transmission of satellite weather images to meteorological bureaus. In an interactive mode transfer, image transfers proceed in cycles with a user requesting a few images or parts of an image, viewing/analyzing the images retrieved and then requesting more images. An example of such an application is image browsing where a user browses through an image database with images being transmitted on a demand basis. The main difference between the two categories of applications is that for continuous mode transfers, a much larger amount of information is transferred between idle periods and the duration of these idle periods is also much longer.

Applications belonging to the first category typically require a fixed level of quality and transmit large amounts of information. Hence, the traffic characteristics and service quality requirements of these applications is very similar to that of bulk data transfers. Thus the insights derived from the study of congestion control mechanisms for data traffic apply to this class of applications as well. However, for the second category of applications the quality may be changed dynamically depending on the service quality requirements of the user and the network load and the transfer times are relatively small. Hence, in this paper we have chosen to focus on the problem of how control mechanisms should be designed to handle this class of interactive application. We model the traffic characteristics of this class of applications using a client server communication model. We assume that clients issue image transfer request messages from time to time. In response to these messages, a server commences an image transfer. Thus a client is either idle, corresponding to the time during which the user analyzes the image, or waiting for an image transfer to finish. The start of the idle period starts from the instant at which the image transfer completes. This model is based on empirical studies of various interactive image applications such as image browsing [Bra].

The service quality requirements of image applications is dependent both on the nature of the applications as well as the specific operations being performed. As we have already indicated there are two principal metrics of service quality: transfer time and image quality. Desirable transfer times should be less than 2000-3000 ms for the image fetches. The desired image quality typically varies depending on the user's preferences.

3 CONTROL MECHANISMS

For the class of image applications modeled, an image server needs to transfer a certain number of tiles in response to client requests. These tiles could correspond to a part of an image, an entire image or a set of images. The number of bytes required to represent the image depends on the desired quality q and the information content of the image I: we denote this as $f(q, I)$ [1]. Let the minimum and maximum desired image quality be Q_{min} and Q_{max}, respectively, and let the maximum desired transfer time be T. How should the server decide at what rate to send packets to the client? There are two possibilities. The first possibility is to decide a fixed sending rate based on bandwidth availability and the transfer parameters: T, Q_{max}, Q_{min} and $f(q, I)$. In this case the quality stays constant for the duration of the transfer. Alternatively, the server can send packets at a variable rate with sending rate being updated periodically based on network conditions. The sending rate adaptation may be achieved by either increasing the transfer time or by reducing the image quality. In the second case, the quality may (potentially) be altered every time the

[1] Since there may be more than one way to adjust image quality, q is a composite measure.

sending rate is recomputed. For both the fixed and adaptive rate techniques it is possible to tradeoff the image quality with the time taken to complete the transfer.

3.1 Rate Reservation

The fixed rate approach to provide service quality guarantees is based on either explicit or implicit bandwidth reservation. With explicit reservation, a scheduling mechanism is used at each switch to ensure that individual connections get a certain guaranteed level of bandwidth [Gol91; KKK90; Zha91]. In the case of implicit reservation, admission control mechanisms are used to ensure that the arrival rate of packets for all connections does not exceed the available bandwidth [Bou92]. However, the scheduling mechanism used is simple, typically something like FCFS. Since the use of explicit reservation isolates connections from each other, it can provide better service quality guarantees. Hence, for the experiments described in this paper we choose a explicit reservation policy, using the HRR scheduling mechanism to allocate bandwidth at switches [KKK90].

The HRR scheduling discipline guarantees a minimum service rate to a connection independent of the traffic generated by other connections. In conjunction with an admission control mechanism, this discipline allows clients to get performance guarantees in terms of throughput, delay, delay jitter and loss rate. In this scheme, transmission on the output link of a switch occurs in *rounds* of a fixed time duration. The parameters used to define the rate for a connection are the *quantum* which defines the number of packets that can be sent per round and the *interval* which defines the number of rounds to skip when the rate of the connection is less than one packet per round. The reader is referred to [KKK90] for a more complete description of the mechanism and possible implementation strategies.

For the image applications that we model, the explicit reservation approach involves choosing a value of the desired bandwidth and reserving it at each switch on the route every time an image transfer takes place. The reserved bandwidth is freed as soon as the image transfer is complete. We assume that a client sends in an image transfer request over an unguaranteed connection. When the server receives such a request from a client, it sends out a bandwidth reservation request packet specifying a set of acceptable rates R. We assume that these rates are calculated as:

$$R = \frac{f(q, I)}{T}, \quad \forall q \in (Q_{min}, Q_{max}) \tag{1}$$

If any of the intermediate switches cannot provide even the minimum requested rate, the server waits for a fixed interval of time, t_w, and then generates a new request. The new rates requested are computed using $T - t_w$, rather than T, in Equation 1. A later request may succeed because of extra bandwidth becoming available in the network due to the termination of other image transfers. The server keeps sending requests until it succeeds or the transfer time requirements are exceeded. In the latter case the image transfer blocks, as a result of which the client is assumed to abort the request and send a new request after the usual idle period.

3.2 Rate adaptation

The adaptive rate based approach attempts to optimize the transfer time by adjusting the sending rate based on the network load. The control mechanisms used for adapting the flow of image traffic may be derived from those used for adapting the flow rate of data traffic with suitable modifications. The most commonly used congestion control mechanisms for adjusting the flow of data traffic are the JK and the DECbit schemes [Jac88; RJ90]. Our previous work [MK92; KKM93] suggests that the HBH control mechanism performs much better than either of these schemes in high bandwidth delay product networks. Hence, we choose the HBH control mechanism as a representative scheme to test the performance of adaptive control mechanisms for adjusting the sending rate. However, we modify the scheme in the following manner to allow the sender to reduce image quality to obtain lower transfer times.

The HBH control mechanism computes the sending rate for a connection at each host/switch on the path of the connection. The sending rates of connections are dynamically adjusted at a switch, using feedback information provided by the neighbouring switches. The desired service rate is computed using a control equation: the goal being to optimize the transfer time for individual connections by trying to get as much throughput as possible while avoiding excessive packet losses. The control equation attempts to keep the queue occupancy for each connection, at the downstream, at a target level. We refer to this value as the ***setpoint***. We refer the reader to [MK92; MKT93] for greater details about the control mechanism.

We modify the HBH mechanism to control the flow of image traffic in the following manner. The sender computes the estimated completion time of an image transfer every time it recomputes the sending rate. If we denote the estimate of the bandwidth at the host (sender) at time t as $R(t)$ and the number of bytes remaining to be transferred as $f'(q, I, t)$, the estimated completion time (F) at a particular quality level q, may be computed as

$$F = t + \frac{f'(q, I, t)}{R(t)} \tag{2}$$

Based on the above computation, the application picks the highest quality level which will enable the transfer to complete within the desired time. We assume that the changes in the quality level are done on a per-tile basis.

We also explore a version of the HBH scheme where the sending rate of an *aggregate* group of connections, rather than only a single connection, is adjusted. In this scheme it is assumed that all packets belonging to the same aggregated group are queued together at network switches and hosts and sent out in a First-In-First-Out manner. A rate control mechanism is enforced in the network layer protocol at the host and switches as described in [MK92; MKT93]. This control mechanism forces packets to be sent out from the common queue (at the host/switch) to the downstream switch at a particular rate. This rate is dynamically updated based on periodic feedback messages from the neighbouring switches. The bandwidth available to an individual connection depends on its share of the total output rate for the group to which it belongs. The transport layer protocol (for an individual connection) maintains an estimate of this bandwidth and picks the quality level using Equation 2.

In the experiments described in this paper, we assume completely reliable information transfer, although in practice it may be possible to tolerate some packet losses. To provide reliable transmission, we use a transport protocol which uses block based acknowledgements and selective retransmissions. A block consists of a fixed number of packets. We assume that a fixed number of blocks can be in transmission simultaneously. This is analogous to the usual concept of a window. Each packet is identified by a block number and an offset number defining its position within the block. The sender transmits all the packets within a block marking the last packet so as to force the receiver to send an acknowledgement. A timer is set when the last packet in a block is sent. A receiver maintains a B bit-vector, where B is the block size. When a packet is received the corresponding bit is set in the bit-vector. The acknowledgment packet sent by a receiver contains the number of the block being acknowledged and the bit vector for that block. By examining the acknowledgments from each receiver, the sender identifies lost packets and schedules them for retransmission. If no packets are lost, the sender advances the block window. If the timer goes off before an acknowledgment is received then all the packets in the block are retransmitted.

4 EXPERIMENTAL SETUP AND RESULTS

In this section, we evaluate the performance of the HBH and HRR schemes for image transport. We first study the performance of the HBH mechanism when used to control the sending rate of individual connections, assuming that quality changes are not allowed once a transfer commences. We next study the performance of the HBH mechanism when it used to control the sending rate of aggregates of connections. We then compare the performance of the HBH and HRR mechanisms. Finally we evaluate how well the HBH mechanism can be used to allow a transfer time vs image quality tradeoff.

We use discrete event simulations to evaluate the effectiveness of the control mechanisms. The traffic generated by a connection is created using the analytical models described in Section 2. We assume that the idle periods for an application are distributed exponentially with a mean of 30 seconds. We also assume that each of the applications switches on n secs from the start of the simulation, where n is an exponentially distributed random variable with a mean of 15 seconds. These values were chosed based on empirical measurements of image applications [Bra]. We use a very large number of connections to capture the effect of the aggregate traffic arrival process at a switch. We remark that it is not possible to achieve this type of aggregation by using an analytical model for the aggregate arrival process (created by a superposition of the arrival process for individual connections) because the traffic arrival process for a connection is partially dependent on the congestion control mechanism that is used. The experimental topology used is shown in Figure 1. This topology is chosen to capture a typical *fan-in* topology where a powerful image server handles multiple clients. The speed of the ingress links from each of the hosts to the switches (200 Mb/s) is deliberately chosed to be faster than the speed of the backbone link (50 Mb/s) between Switches 1 and 2 so that the backbone link is the bottleneck.

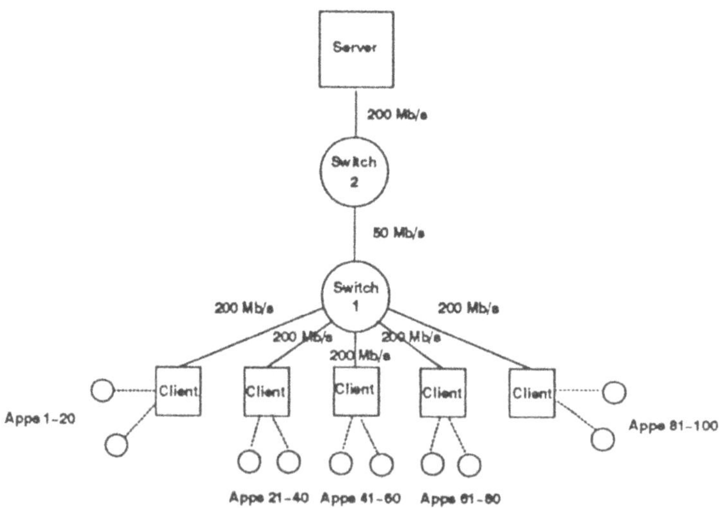

Figure 1: Simulation configuration

We used fixed size packets of 500 bytes. The duration of each simulation run is 2000

Table 1: Comparison of HBH and OPT

	Buffers	Number of conns	Transfer time	Utilization	Loss rate
HBH	400	100	1406 ms	.850	1.53%
HBH	1000	100	1237 ms	.864	0.35%
OPT	∞	100	1124 ms	.871	-
HBH	400	80	774 ms	.714	0.40%
HBH	1000	80	749 ms	.717	0.00%
OPT	∞	80	583 ms	.719	-

seconds[2] so that the confidence intervals for the mean measures are less than 10% of the sample mean.

The main goal of a rate adaptation mechanism, such as HBH, is to trade off the losses seen by a connection against the achievable throughput. A very conservative policy results in reduced utilization of the bottleneck bandwidth which leads to an increase in the overall transfer time. On the other hand, a very aggressive policy leads to excessive packet losses necessitating retransmissions with a concomitant increase in the overall transfer time. Hence the most useful metric that may be used to measure the performance of such control mechanisms is the transfer time[3]. This measure subsumes both transient and steady state performance. As we have pointed out, this is also one of the primary service quality metrics for image applications. To measure how well a rate adaptation mechanism performs, we compute a *lower bound* on the transfer time. This (unrealizable) lower bound corresponds to a situation where there is an infinite amount of buffering available at each node and each connection injects data into the network at the speed of the access link connecting the host to the first switch on the path of the connection. For simplicity, we will refer to this as the *optimal* scheme.

In the first set of experiments we compare the performance of HBH control with the optimal scheme. The topology used for the experiments is shown Figure 1 with either 400 or 1000 buffers at the switches. There are either 80 or 100 active connections and the traffic arrival process for each connection is generated as described previously. The time instant at which a new request is generated depends on when the previous transfer was completed. As a result, the average utilization depends both on the control scheme as well as the number of active connections. For the optimal scheme, the average utilization is about 72% and 87% with 80 and 100 connections respectively.

Table 1 shows the average transfer times, loss rates and the average utilization for this

[2]corresponding to about 2,500,000 packets going through the bottleneck link

[3]We will occasionally describe this as the *actual* transfer time to distinguish from the *target* transfer time

Table 2: Impact of aggregation

	Buffers	Number of conns	Transfer time	Loss rate
HBH	400	100	1406 ms	1.53%
HBH	1000	100	1237 ms	0.35%
Aggr	400	100	1146 ms	0.12%
HBH	400	80	774 ms	0.40%
HBH	1000	80	749 ms	0.001%
Aggr	400	80	712 ms	0.14%

set of experiments. The results indicate that the use of the HBH control mechanism results in transfer times that are about 10-30% greater than the optimal. The greater transfer times may be explained as follows. In the HBH control mechanism each connection linearly increases its sending rate when it starts a new transfer. For the application we model, the duration of a single transfer is relatively short lived. As a result the average throughput seen by a connection, when only a few connections are active, is not as high as it could be with a more aggressive policy.

Figures 2 and 3 which show the throughput for a single image transfer for a particular connection, controlled using either HBH or the optimal scheme, illustrates this point. Figure 3 shows the cumulative distribution of the average throughput. Note that the maximum throughput for a HBH controlled connection never exceeds 31 Mb/s while with the optimal policy a throughput of up to 50 Mb/s could be obtained[4].

The results also indicate (as might be expected) that a greater number of buffers leads to fewer losses and retransmissions. With the HBH scheme, it is possible to adjust the loss-throughput tradeoff seen by a connection by adjusting the parameters of the control mechanism. In our experiments we use an error control mechanism to guarantee totally reliable transmission. However, since totally reliable transmission may be avoided for certain types of image applications, it is possible to further reduce the average transfer time when using the HBH scheme using a less conservative error control policy.

The previous experiment indicated that sub-optimal performance is due to the relatively small transfer times. One way to reduce the average transfer time is to control aggregates of connections rather than individual connections. The idea here is that controlling aggregates of connections increase the effective lifetime of an aggregated connection, thereby reducing the impact of the linear probe-up phase on the average transfer time. Table 2 compares the effect of aggregated control with per connection control. For

[4]Even for optimal control a throughput of 50 Mb/s is never obtained because the transfer time used for computing the average throughput includes the time taken to send the request from the client to server.

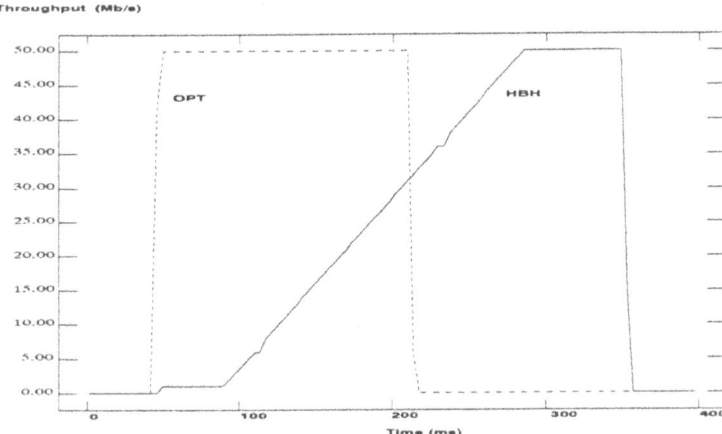

Figure 2: Instantaneous throughput for a single image transfer with HBH and OPT

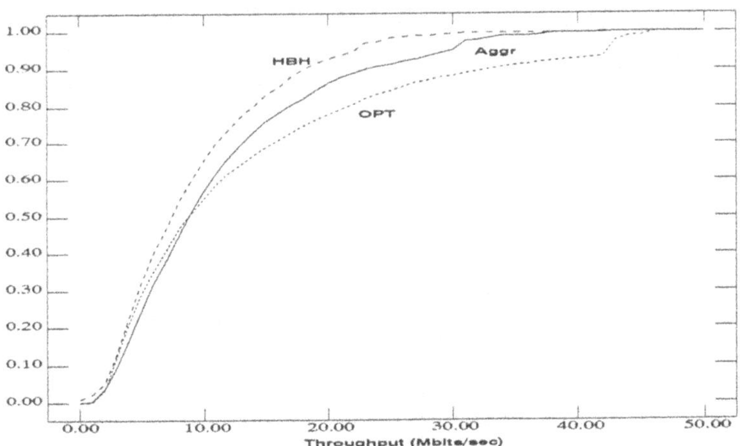

Figure 3: Cumulative distribution of throughput with HBH, Aggr and OPT

Table 3: Comparison of HBH and HRR, QBP: Quasi blocking probability percentage, BP: Blocking probability percentage.

	Conns	Transfer time	Utilization	QBP	BP
HBH(2s)	100	1268 ms	.839	14.37	-
HBH(3s)	100	1268 ms	.839	4.71	-
HRR(2s)	100	2062 ms	.676	-	14.68
HRR(3s)	100	3059 ms	.706	-	9.55
HBH(2s)	80	774 ms	.714	2.04	-
HBH(3s)	80	774 ms	.714	0.11	-
HRR(2s)	80	2062 ms	.610	-	7.30
HRR(3s)	80	3059 ms	.616	-	3.54

the same amount of buffering, aggregated control clearly results in much better performance than individual control.

4.1 Comparison with rate reservation

The main performance metric for reservation based approaches is the blocking probability which is the likelihood that a user requested transfer does not complete. For the modeled application, this value is dependent on the target transfer time and image quality requested for a particular transfer. For an adaptive bandwidth allocation scheme the main performance metric is the average value of the (actual) transfer time. To be able to compare the two types of schemes, we define a *quasi* blocking probability for an adaptive scheme as the likelihood that the actual transfer time will exceed the target transfer time.

Table 3 compares the blocking and quasi blocking probabilities for HRR and HBH for a target transfer time value of either 2 or 3 seconds and for a fixed level of quality. There are 400 buffers at the switch and either 80 or 100 connections. We also show the average utilization of the bottleneck link and the average transfer time. Figure 4 shows the cumulative distribution of the transfer time. The results show that the average transfer times with HBH are much lower [5] while the variation in the transfer time is much smaller with HRR. As the target transfer time is increased for the HBH scheme, the value of QBP decreases because the number of transfers whose duration exceeds the target transfer time value decreases. For example, Figure 4 shows that 2.04/only 0.11% take more than 3 seconds. More interestingly, the results show that the quasi blocking probability with HBH is *lower* than the blocking probability with HRR. This may be explained by the

[5] In general the ratio of the average time with HRR to that with HBH will tend towards n:1 as the load tends to zero.

following three factors. First, with HRR, connections are assigned rates in multiples of a set of basic rates. As a result, there may be an effect similar to fragmentation in memory allocation, with image transfers being blocked even though the utilization is less than 100%. This is why the average utilization with HRR is less than with HBH [6].

The second factor that contributes to a higher blocking probability for HRR is that the bandwidth at a switch can stay reserved but unutilized for up to 2 end to end round trip times due to the bandwidth reservation and release operations. Finally, with HBH control each connection finishes an image transfer faster, on the average, than with HRR. Consequently, when many connections commence image transfer over a relatively small time interval, the number of contending connections is lower with HBH control.

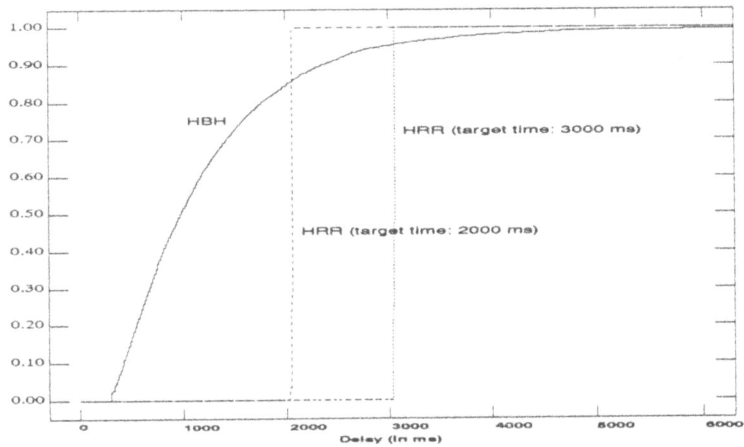

Figure 4: Comparison of HBH and HRR in terms of cumulative distribution of transfer times

4.2 Handling quality changes

For situations where an application needs the ability to trade image quality for lower transfer times, it is possible to use the HBH scheme with the sending rate being adjusted according to Equation 2. To investigate the effectiveness of this scheme, we look at the average quality, average transfer time and the quasi blocking probability for different values of the target transfer times at various loads. We assume that a set of 4 pre-recorded streams corresponding to four levels of quality (0,1,2 and 3, with 0 representing

[6] Another reason for the lower utilization is that the average transfer time with HRR is greater and as a result the average arrival rate is slightly lower.

best quality) is available. We assume that the average compression ratio at each of these levels is 1:1, 1:4, 1:10 and 1:50 respectively.

The results for this set of experiments is summarized in Table 4. Figure 5 shows the cumulative distribution of the transfer time for 100 connections for a target transfer time of 1, 2 and 3 seconds with 400 buffers. The quality value is computed as the mean of the quality of each individual tile that is transferred. The closer this value is to 0, the better the quality. These results show that the HBH scheme provides the ability to smoothly trade between the image quality and the transfer time. However, there is a non-negligible probability of exceeding the target transfer time which increases with the load. The main reason for this violation is due to the retransmission of lost packets, usually resulting from a timeout. Since the maximum timeout interval used in our experiments is 250 milliseconds, most of the transfers that exceed the target transfer time, exceed it by less than 250 milliseconds. This can be observed in Figure 5.

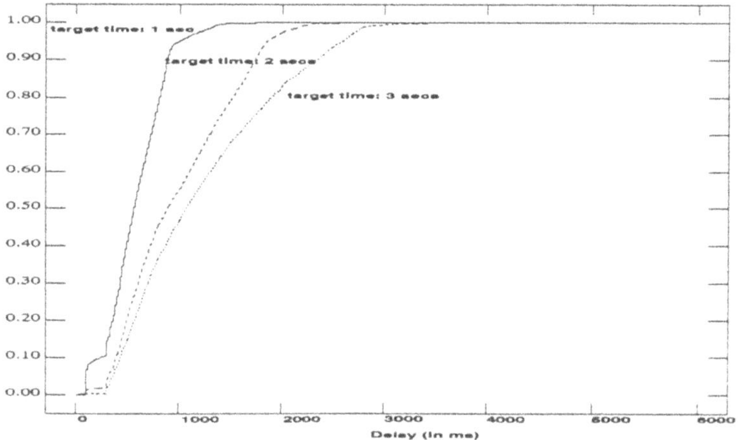

Figure 5: Cumulative distribution of the transfer time with HBH with variable quality

5 SUMMARY

In this paper, we have examined the effectiveness of various bandwidth allocation schemes for interactive image traffic applications. We have described how a hop by hop congestion control mechanism, proposed to control the flow of bursty data traffic, may be modified to control the flow of image traffic. The modified scheme allows a user to smoothly tradeoff the image quality against the transfer time. We have shown that for a fixed level of

Table 4: Effect of allowing quality changes

Conns	Transfer time		Quality	QBP
	Target	Actual		
120	1000 ms	556 ms	0.71	5.14
120	2000 ms	1275 ms	0.19	5.00
120	3000 ms	1761 ms	0.09	1.03
100	1000 ms	576 ms	0.45	4.94
100	2000 ms	989 ms	0.12	2.74
100	3000 ms	1228 ms	0.05	0.49
80	1000 ms	526 ms	0.33	2.19
80	2000 ms	745 ms	0.08	1.19
80	3000 ms	826 ms	0.003	0.05

QBP: Quasi blocking probability percentage.

image quality, the proposed scheme results in average transfer times which are close to optimal. We have observed that performance degrades as the image transfer times become smaller vis-a-vis the end to end delays. This may be countered by controlling aggregates of connections rather than individual connections; we have shown that this results in a significant improvement in performance. We have also studied the feasibility of using a rate reservation based approach for image transport. In our experiments, the blocking probability for a given target transfer time with rate reservation scheme is higher than the equivalent probability for the HBH scheme, even at high loads. These results suggest that adaptive rate schemes are more suitable for interactive image traffic applications.

In this paper, we have assumed that the quality is a linear function of the available bandwidth. However, this relationship is general non-linear. Moreover, the benefit derived by a user depends on the perceptual quality of the images transferred. As a result the choice of the control mechanism is likely to have a direct effect on the user perceived quality. For example, damping the changes in the quality so that there is gradual change in quality over contiguous tiles may result in better user perceived performance. We expect to examine this interaction in greater detail in future work.

The results of this paper demonstrate that aggregation is an effective way to control short bursts of traffic. In the simulations presented in this paper, we have assumed that aggregates always share the same end to end paths. It is important to study how aggregated control can be achieved in practice under conditions where connections are being set up and torn down rapidly and where it may not always be possible to have a large number of connections sharing the same end to end path. We intend to look at this issue in the future.

6 REFERENCES

[Bou92] J. Le Boudec. The Asynchronous Transfer Mode: a tutorial. Computer Networks and ISDN Systems, 24:279-309, December 1992.

[Bra] R. Braudes. Personal communication.

[Dan] W. Dang. Imm version 2.7. Announced on rem-conf mailing list 12-2-93. Available for anonymous ftp from ftp.hawaii.edu in /ftp/paccom/imm-2.7.

[Gol91] S.J. Golestani. A framing strategy for congestion management. IEEE Journal on Selected Areas in Communications,9(7):1064-1077, September 1991.

[Jac88] V. Jacobson. Congestion avoidance and control. In proceedings of ACM SIGCOMM, Stanford, California, August 1988.

[KKK90] C.R. Kalmanek, H. Kanakia, and S. Keshav. Rate controlled servers for very high-speed networks. In Proceedings of the Conference on Global Communications (IEEE GLOBECOM), San Diego, California, December 1990.

[KKM93] H. Kanakia, S. Keshav, and P.P. Mishra. A comparison of congestion control schemes. In Proceedings of the Fourth Annual Workshop on Very High Speed Networks, Baltimore, Maryland, March 1993.

[MK92] P.P. Mishra and H. Kanakia. A hop by hop rate based congestion control scheme. In Proceedings of ACM SIGCOMM, Baltimore, Maryland, August 1992.

[MKT93] P.P. Mishra, H. Kanakia, and S.K. Tripathi. Design and analysis of a hop by hop congestion control mechanism. Submitted to IEEE/ACM Transactions on Networking, 1993.

[NP85] A. Netravali and B. Prasada, editors. Visual communications Systems. IEEE Press, 1985.

[RJ90] K.K. Ramakrishnan and R. Jain. A binary feedback scheme for congestion avoidance in computer networks. ACM Transactions on Computer Systems, 8(2):158-181, May 1990.

[Zha91] L. Zhang. Virtual clock: A new traffic control algorithm for packet switching networks. ACM Transactions on Computer Systems, 9(2):101-125, May 1991.

3

Stability and Optimality of Continuous Time M-*ary* Rate Control Mechanisms

D.N. Ranasinghe,
*Department of Statistics and Computer Science, University of Colombo, Sri Lanka**

W.A. Gray,
Department of Computing Mathematics, University of Wales College of Cardiff, United Kingdom.

A.M. Davidson,
Department of Applied Mathematics, University of Wales College of Cardiff, United Kingdom.

Abstract

In feedback rate control mechanisms of both rate and window domains, M-ary control class can be considered as an extension of the widely implemented binary control class. In this paper, the stability properties of the class of continuos time M-ary rate controls are analytically established using dynamic theoretic tools. The dynamic behaviour is shown to exhibit a ***Hopf bifurcation*** where the stability of the system changes at a critical parameter value. The uniqueness of the optimal controller is established where the optimality criterion is defined as the ***convergence rate invariance*** property. This contrasts with the non unique and mutually exclusive controls of the discrete time M-ary rate control class. As a consequence of the rate-window duality, performance implications of this analysis for a recently proposed window control algorithm of the M-ary form are discussed. One such implication is that, the systemic stability of a partially distributed implementation of the window controller is guaranteed stable.

*This research was carried out while the author was at the Department of Computing Mathematics, University of Wales College of Cardiff, and was funded by a British Council TCT award.

1 INTRODUCTION

The class of feedback flow control mechanisms form a substantial part of overall flow control regimes in computer networks [Agarwala et al., 1992; Bolot et al., 1990; Chiu et al., 1989; Gerla et al., 1980; Jacobson, 1988; Jain, 1989; Keshav, 1991; Mitra et al., 1990; Mukherjee et al, 1991; Shenker, 1990]. The study of these mechanisms is prompted by their intrinsic analytical tractability, implementational simplicity and the relationships that exist between complementary classes of mechanisms. More recently, an attempt has been made to emphasise a precise analytical specification of the flow control problem compared to past efforts. In this respect a few complex flow controls with a sound theoretical basis have been proposed [Keshav, 191; Mitra et al., 1990].

Recently Fendick et al [Fendick et al., 1992] have analysed the window control algorithm of Mitra et al [Mitra et al., 1990], in an attempt to obtain certain performance characteristics of Mitra's algorithm. They have resorted to a direct analysis by mapping the original algorithm in a window domain to an equivalent piece-wise continuous rate control, thus obtaining results pertaining to unilateral stability, an optimal value of a control parameter and obtaining evidence for the existence of a large class of such controls. It is our view that, where a direct analysis of a complex algorithm is appropriate yet difficult, it would be more feasible to analyse a simpler complementary form of algorithm to that of the original, which may yield performance bounds and possibly implementational insight that could generally be valid for the original algorithm.

This study has its origins in two previous works. Bolkot et al [Bolot et al., 1990; Bolot et al., 1992] have studied binary rate control schemes and their dynamic behaviour, and have concluded that the absence of proper modelling tools has hindered the analytic establishment of stability properties. Shenker [Shenker, 1990] has studied a class of controls which we have identified as discrete time M-ary rate controls, where M-ary in our nomenclature refers to virtual M decision levels in the flow controller. The name is synonymous with the term M-ary in digital transmission schemes. The class of continuous time M-ray rate controls is hence identified as the natural counterpart to the discrete time class. We have also observed that Mitra's algorithm generally fits the definition of a M-ary control.

Our investigation in this paper centres on the analysis of the class of continuous time M-ary rate control mechanisms for its performance characteristics as an extension of Shenker's and Bolot et al's work, and as an analysis of Mitra's algorithm through a complementary yet a simpler approach. Our analysis makes the rigorous establishment of the stability of continuous time M-ary rate control possible by dynamic and queuing theoretic techniques. By implication of this stability for all deterministic continuous time M-ay controls, we predict the systemic stability of the partially distributed implementations, among other implementational insight [Ranasinghe, 1994].

In discrete time M-ary rate control, Shenker [Shenker, 1990] established that there are a number of non unique and mutually exclusive controls such as time scale invariant, time scale variant and logistic type forms. In contrast we show that the continuous time M-ary rate control class has a unique optimal controller for each specified equilibrium point. The optimality criterion is defined as the property of ***convergence rate invariance***, i.e., the property that the server utilisation remains constant for a range of convergence rates. This property was inherent in the discrete time M-ary rate class, yet it has to be explicitly specified in the optimisation process of the continuous time counterpart.

2 STABILITY OF A HYPOTHETICAL CONTROLLER

2.1 The Network Model

Consider a virtual circuit that traverses a single bottleneck node with a server rate μ (packets/sec). The service policy of the node is FIFO. The non-zero forward and reverse propagation delays are τ_f and τ_r (sec) respectively. The instantaneous source transmission rate is represented by $\lambda(t)$ (packets/sec), and a stochastic packet arrival and service process of general distribution may be assumed. The data flow is assumed to be unidirectional and it is assumed that there exists a reverse acknowledgement flow indicating the bottleneck node queue level $q_B(t)$ to the source. The access line has a rate $\mu_a(>> \mu)$. In order that feedback is enforceable, the source is modelled as an infinite data source. The pipe size, which is the delay-bandwidth product of the path following the bottleneck $\mu\tau_r$, is assumed to be $\mu\tau_r >> 1$. See Figure 1.

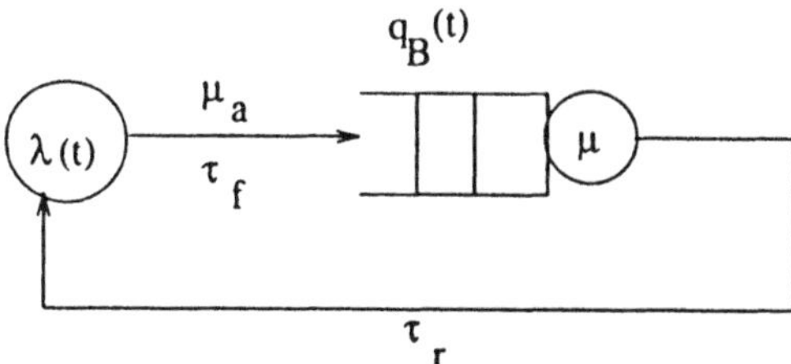

Figure 1: The Physical Model

2.2 Modelling and Analysis

The hypothetical controller we subject to analysis is given as:

$$\dot{\lambda}(t) = \alpha - \lambda(t)(R(t - \tau_r) - 1)/\beta \tag{1}$$

which, as can be seen is a generalised continuous time rate controller consisting of a linear rate increase policy with a convergence rate α (packets/sec) and an exponential rate decrease policy with a backoff time constant β(sec), reacting to a normalised round trip delay R. We shall call this an M-ary extension of the binary rate control of Bolot et al [Bolot et al, 1990]. It has been shown [Bolot et al, 1992] that a discrete space stochastic process at a node approaches a deterministic process in the presence of a large packet population under continuous time control and this enables R(t) to be substituted by $q_B(t)$ as:

$$R(t-\tau_r) = q_B(t-\tau_r)/\mu\tau + 1 \tag{2}$$

where, $\tau = \tau_f + \tau_r$.

The deterministic node process itself is described by the fluid mode [Bolot et al, 1990] as:

$$\dot{q}_B(t) = \left\{ \begin{array}{ll} 0 & \text{if } q_B(t) = 0 \\ & \text{and } \lambda(t-\tau_f) < \mu \\ \lambda(t-\tau_f) - \mu & \text{otherwise} \end{array} \right\} \tag{3}$$

By substitution of (2) into (1), we obtain the rate controller in the form:

$$f_1(\lambda, q_B) = \alpha - \lambda(t)q_B(t)/\beta\mu\tau \tag{4}$$

which in conjunction with (3) describes the dynamic system in delay-differential equation form. The desired queue equilibrium point of the system q_{Be}, can be any chosen value. We shall initially choose q_{Be} to be 1 in line with [Bolot et al., 1990]. The rate equilibrium point remains $\lambda_e = \mu$, as implied by the server rate of the node. The delay-differential model describing the queue and rate dynamics becomes analytically tractable if we convert it to a set of differential equations. Therefore, to obtain a R^n model of the dynamic system which can be subjected to a dynamic theoretic analysis, we make use of a queue model description. The node queue is modelled by an M/M/1 process whose fluid model description in (3) fits the deterministic process required [Filipiak 1988; Ohta, 1988]. The delay segments are modelled by infinite server queues such as $M/G/\infty$. We also choose to model either but not both of the pipe segments for simplicity. This should not affect the qualitative performance expected of the real system as what we essentially need is a non zero propagation delay anywhere in the closed loop. From the two possible equivalent queue model representations we select the following models as the other model introduces unknown functional dependency on state variables into the equations

[Ranasinghe,1994] . See Figure.2

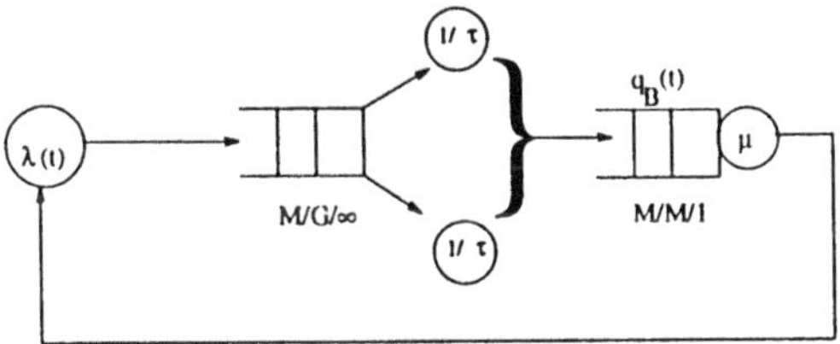

Figure 2: The Queueing Model

$$\dot{q}_p(t) = \lambda(t) - q_p(t)/\tau \tag{5}$$

$$\dot{q}_B(t) = q_p(t)/\tau - \mu \tag{6}$$

$$\dot{\lambda}(t) = \alpha - \lambda(t)q_B(t)/\beta\mu\tau \tag{7}$$

In these equations, $q_p(t)$ is the pipe equivalent queue, $q_B(t)$ is node queue and $\lambda(t)$ is the source rate. τ, μ, α and β are as defined in (1) and (2).

Now the dynamic system is in R^3 , where $\tau_f = \tau$ as $\tau_r = 0$. Given the state space description, the stability properties of the dynamic system at a specified equilibrium point (λ_e, q_{Be}) can be established. A similar proof to that of Fendick et al [Fendick et al, 1992] shows that the system is unstable in the vicinity of the equilibrium point $(\mu, 1)$ for example, and the interested reader is referred to [fendick et al, 1992]. Although the system is locally unstable, it can be proved that it exhibits a stable limit cycle behaviour for a range of parameter values of α (see Appendix) . We have found that there exists a critical convergence rate α_c such that for $0 < \alpha < \alpha_c$, there is a stable limit cycle surrounding the unstable equilibrium point and, for $\alpha > \alpha_c$, the system exhibits asymptotic stability. In dynamic systems theory, this is identified as a ***Hopf bifurcation*** where, as a parameter passes a critical value, the system changes from one state of stability to another [Beltrami 1987] . In contrast, for the discrete time counterpart as shown by Shenker [Shenker,1990] the rate iterations moved from an asymptotic stability position to chaotic instability as a critical convergence rate was passed. For the heuristic controller $f_1(\lambda, q_B)$ in (4) and for $(\lambda_e, q_{Be}) = (\mu, 1)$ this occurs for:

$$\alpha_c = \mu^2(1 - 1/\mu\tau)pkt/sec^2 \tag{8}$$

where $\mu\tau >> 1$. There is also a straightforward way to find the critical convergence rate. This involves the obtaining of the eigenvalue equation for the model in R^n and testing it for asymptotic stability or instability using Routh test. However in the present situation, it only indicates an instability in the region $0 < \alpha < \alpha_c$, which is not in itself a sufficient condition for the existence of a stable limit cycle.

2.3 Simulation Performance

The numerical solution of the delay-differential equation model or the model in R^3 provides an indication of the behaviour of the flow controlled system. In a study of continuous time binary rate control with direct implications to our analysis, Bolot et al [Bolat et al, 1992] have shown that there is a good agreement between the delay-differential model in the analytical domain and the discrete space stochastic model in the physical domain, i.e., the direct simulation of the packet transmission process, at large packet populations corresponding to $\mu\tau >> 1$. This validates the simulation results obtained by numerical solution as representing the packet transmission process. The figures corresponding to the model in R^n were obtained by numerical solution of (5) to (7) using the package. [ISIM 1987], which uses the standard 4th order Runge-Kutta procedure. The rate control form $f_1(\lambda, q_B)$ (9), with the values $\mu = 100$ and $\tau = 1$ have been used throughout. The simulation tests show that each of the models describing the forward or the reverse pipe segments have identical dynamic performance except for a phase shift between the rate variation and the queue variation, which is expected. Figures 3 and 4 show a large discrepancy between the delay-differential model and the model in R^3 at low q_{Be} of O(1). The delay-differential model behaviour further resembles the continuous time binary control of Bolot et al.

From Figures 5 and 6, it can be seen that the delay-differential model and the model in R^3 corresponding to $f_1(\lambda, q_B)$ agree well at large queue equilibria of $O(\mu\tau)$, with a substantial bias towards sinusoidal rate variation.

This observation confirms the applicability of a fluid approximation of he queue models used to describe the non-stationary stochastic behaviour at large packet populations. Therefore to retain equivalence between the model in R^3 and the delay-differential model, we shift the system equilibrium point to $(\lambda_c, q_{Be}) = (\mu, O(\mu\tau))$. However, our simulations also show that the delay-differential model and the model in R^3 differ in their prediction of asymptotic stability criterion, though not consistently. For example, at an $\alpha > \alpha_c$, the delay-differential model of $f_1(\lambda, q_B)$ displays a persistent backlog at the node queue, but the model in R^3 displays asymptotic stability (a.s) (Figures. 7 and 8). It is conjectured that, this discrepancy is the result of the residual modelling differences and provided the working convergence rate does not exceed the critical α by a wide margin, then the delay-differential model exhibits a backlog.

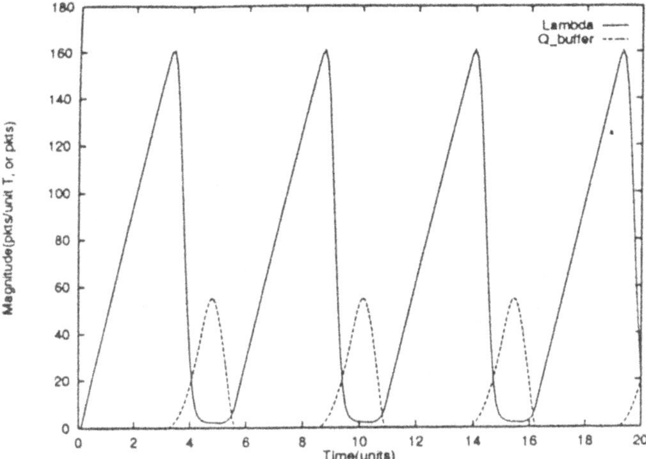

Figure 3: Delay-Differential model
($\alpha = \mu/2\tau, q_{Be} = 1$)

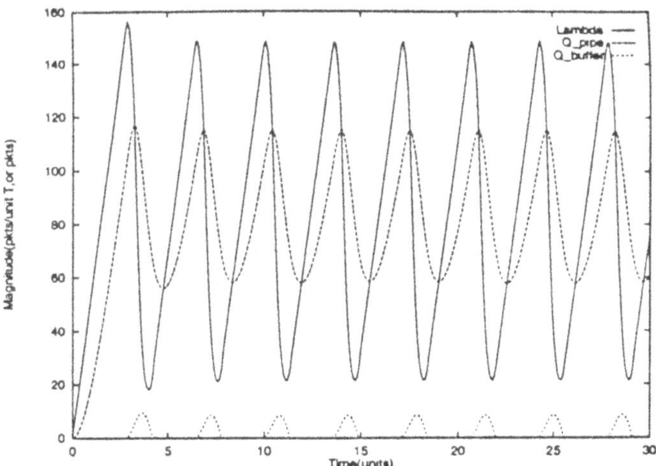

Figure 4: Differential model
($\alpha = \mu/2\tau, q_{Be} = 1$)

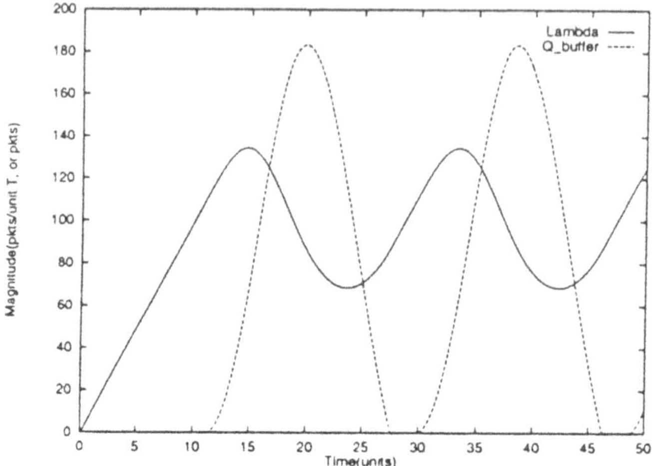

Figure 5: Delay-Differential model
($\alpha = \mu/10\tau, q_{Be} = 5\mu\tau/2\pi$)

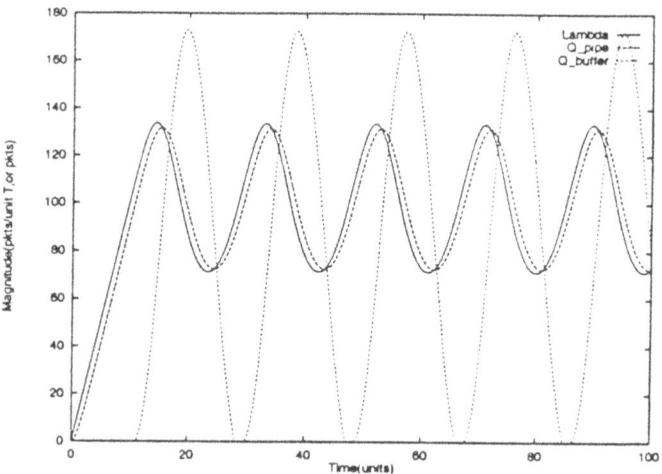

Figure 6: Differential model
($\alpha = \mu/10\tau, q_{Be} = 5\mu\tau/2\pi$)

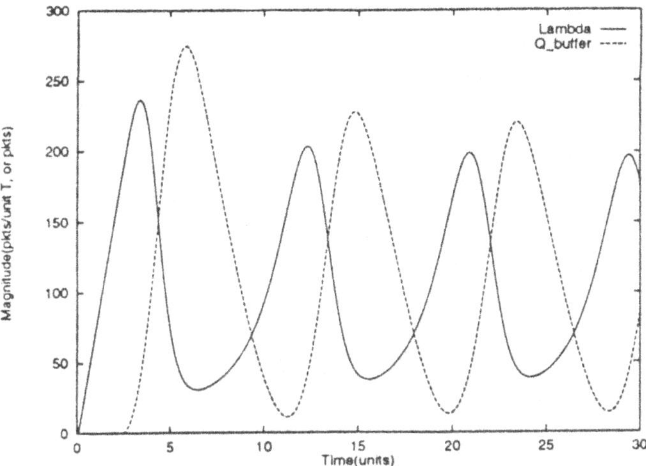

Figure 7: Delay-Differential model
($\alpha = 5\mu/2\pi\tau, q_{Be} = 5\mu\tau/2\pi$)

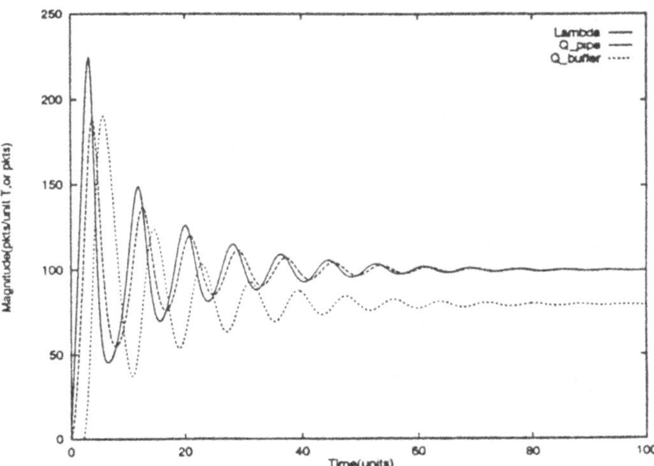

Figure 8: Differential model
($\alpha = 5\mu/2\pi\tau, q_{Be} = 5\mu\tau/2\pi$)

This is confirmed, where a slightly different form of heuristic control, $f_2(\lambda, q_B)$ (see next section) was used, which had a negative critical convergence rate at a specified q_{Be}, a positive α resulted in a.s. for the delay-differential model (Figure not shown). It is reasonable to conclude that a persistent backlog may be the practical reality at an $\alpha > \alpha_c$, for heuristic rate control.

3 OPTIMAL CONTROL

3.1 Optimality Criterion

We now consider two heuristic rate controllers (9) and (10) and compare their critical convergence rates at a specified queue equilibrium point, $q_{Be} = \mu\tau/k$, with k$\geq$ 1 would represent a range of q_{Be} values from $O(1)$ to $O(\mu\tau)$. Let:

$$f_1(\lambda, q_B) = \alpha - \lambda(t)q_B(t)/\gamma \tag{9}$$

which is 4 and

$$f_2(\lambda, q_B) = \alpha - \lambda^2(t)q_B(t)/\gamma \tag{10}$$

where $\gamma = \beta\mu\tau$

The corresponding critical convergence rates are:

$$\alpha_c = \mu(k-1)/\tau \tag{11}$$

for $f_1(\lambda, q_B)$, which is positive for $k > 1$ and

$$\alpha_c = \mu(k-2)/\tau \tag{12}$$

for $f_2(\lambda, q_B)$, which is positive for $k > 2$.

We note that a positive critical convergence rate for practical purposes is achieved at a lower queue equilibrium point for $f_2(\lambda, q_B)$ than for $f_1(\lambda, q_B)$. However , as q_{Be} approaches $O(1)$, and thus operating at a high convergence rate within the critical α_c, the continuous time M-ary rate controller's performance degrades to that of a continuous time binary rate control as observed in section 2.3, with the server utilisation being dependent on the convergence rate. Since our basis for the analysis of the M-ary class of rate controls is Shenker's work [Shenker, 1990], we have noted that the discrete time M-ary rate controls of all forms had the property of *convergence rate invariance* embedded in them. This property implies that for a range of convergence rates, the rate controller maintains a constant server utilisation $\eta(=1)$, in the case of deterministic continuous time control and, $\rho(<1)$ in case of stochastic discrete time controls. We shall define this property as the appropriate optimality criterion. We conclude therefore, any arbitrary heuristic M-ary rate controller is suboptimal in the sense that this property is not observed and, that there exists an optimal controller for any specified operating point (λ_e, q_{Be}).

3.2 On Optimisation

In the specification of the optimal controller, the position of the desired queue equilibrium point has to be defined. In open queuing networks with Markovian nodes, q_{Be} has been usually set to $O(1)$ which subsequently optimised the network power (Stidam, 1985). This approach is equally valid in selecting the operating point for discrete time M-ary rate controls. In continuous time rate controls of both binary and the proposed M-ary type, the network power argument cannot be effectively applied due to the deterministic nature of the node process An alternative approach is required and we raise the queue equilibrium point q_{Be}, to $O(\mu\tau)$ in order to have the widest possible range of convergence rates. For example, in bang-bang rate control, a unit server utilisation was reached at the highest possible convergence rate and at the minimum backoff time at a q_{Be} of $\mu\tau$ [Wang et al. 1991]. Since the idea behind optimal continuous time M-ary rate control is to reach the goal of convergence rate invariance, which would maintain a constant server utilisation η of 1 for $0 < \alpha < \alpha_c$, a sinusoidal rate variation would be the ideal description of such a behaviour due to it inherent symmetry. Therefore, consider the rate variation $\lambda(t)$ of the form:

$$\lambda(t) = \bar{\lambda} + a_0 sin\omega t \tag{13}$$

Where $\bar{\lambda} = \mu$ is the desired average rate, $\omega = 2\pi/T$ is the angular frequency of an oscillation with period T, and $(0 \leq a_0 \leq \mu)$ is the amplitude of the variation.

By application of the flow conversation rule(3) to the node queue and assuming a zero propagation delay between the source and the node, it follows that:

$$q_B(t) = \int_0^t (\lambda(t) - \mu)dt \tag{14}$$

$$= a_0(1 - cos\omega t)/\omega \tag{15}$$

by substitution of (13).

We have argued that the condition $\bar{q_B}$ be equal to q_{Be} , be satisfied when the controller is optimal and it is also seen that $\bar{\lambda} = \lambda_e$ is true [Ranasinghe, 1994] Figures.5 and 6, show $f_1(\lambda, q_B) at \lambda_e = \mu, q_{Be} = 5\mu\tau/(2\pi)$ and with an $\alpha = \mu/(10\tau)$, generating a rate variation that closely tracks the sinusoidal form expected, with average values of rate and node queue being approximately equal to respective equilibrium point values. This does not however mean that $f_1(\lambda, q_B)$ is optimal at the given operating point, as the "real" α_c for the specified q_{Be} may be less than or greater than the "apparent" α_c, that is obtained analytically for a heuristic control. For example, the negative critical α for $f_2(\lambda, q_B)$ clearly proves the case. We can state the optimal rate and node queue variations, taking round trip propagation delay into account as:

$$\lambda_{opt}(t) = \mu + a_0 sin\omega t \tag{16}$$

$$q_{B(opt)}(t) = a_0(1 - cos\omega(t - \tau))/\omega \tag{17}$$

In general, a heuristic controller $f_i(\lambda, q_B)$ can be optimised using established Lagrangian optimisation techniques, However, it has been shown that the closed form solution does not exist for the system in R^3, as it is a non linear servo mechanism problem [Sage er al., 1977] . A numerical solution alternative does exist, but results in an open loop controller , whereas a closed loop controller is desirable. Accordingly, the realisation of the optimal controller is not possible.

3.3 Uniqueness of the Optimal Control

To obtain the predicted variation of the critical convergence rate α_c with q_{Be} confirming the uniqueness of the optimal continuous time M-ary rate controller, we proceed as follows. Since all rate controls must contain a linear rate increase component of the form $\lambda(t) = \alpha$, that includes the convergence rate as a parameter see equation (1), by differentiating (16) and assigning $\omega t = 0$ to represent the highest rate increase effort, we obtain:

$$\alpha = a_0\omega \tag{18}$$

Since $\bar{q_B} = q_{Be}$ and :

$$\bar{q_B} = a_0/\omega \tag{19}$$

which follows from (17), we have:

$$\alpha = a_0^2/q_{Be} \tag{20}$$

As the peak amplitude of $\lambda_{opt}(t)$ is reached at the critical convergence rate α_c, we would then have, $\alpha_c q_{Be} = \mu^2$ as the form of the characteristic curve of $\alpha_c(q_{Be})$ as a function of q_{Be} for the region where a_0 remains constant at μ. The region $a_0 < \mu$ is predicted by reference to qualitative results obtained by simulations [Ranasinghe, 1994]. See Figure 9. These simulation results show that, as q_{Be} decreases, the effective critical convergence rate decreases as well, indicating that there should be an optimal operating point for the q_{Be} . Inferring from bang-bang rate control, which is a continuous time control that maintains a unit server utilisation, we note that the fundamental sinusoidal component of the source rate variation has a peak amplitude equal to μ corresponding to a q_{Be} of $\mu\tau$ [Wang et al., 1991]. It would not be incorrect to model the optimal continuous time M-ary rate control as the one that produces this sinusoidal variation, thus lowering the $\bar{q_B}$ of bang-bang control to the level of q_{Be} . Therefore we let, for the optimal control a value of $q_{Be(opt)} = \mu\tau$.

We reach the conclusion that the optimal controller is unique for a given operating point . In contrast, in the discrete time M-ary rate controls, there were time scale invariant, time scale variant and logistic type rate iterations, and therefore a wide class of controls in reaching and maintaining a stochastic server utilisation of less than one. Alternatively, as the deterministic server utilisation aimed by a controller approaches unity, it becomes increasingly difficult to maintain the time scale invariant property and hence the uniqueness of the continuous time controller.

4 STABILITY REVISITED

4.1 Oscillatory Behaviour

It has been shown that continuous time M-ary rate control exhibits an oscillatory behaviour for $0 < \alpha < \alpha_c(q_{Be})$. This is an artefact of reactive control in the presence of delayed feedback. In the implementation of heuristic controller $f_i(\lambda, q_{Be})$, it uses the a prior knowledge of μ the constant server rate, and τ the round trip propagation delay. Given these two pieces of information, it is possible to design a predictive controller [Keshav,1991], that would converge to the equilibrium state $\lambda_e = \mu$, in one round trip time. Basically, what the controller does is to inject and maintain a number of packets, known as the sliding window size $W(= \mu\tau)$ in the system, filling the pipe and thus reaching the unit server utilisation required.

Extending the analysis to more than one user (or connection), say N users, sharing the bottleneck node, the i-th user individual fair rate can be stated as $\lambda_i = \mu/N$, leading to the window size per user in a predictive controller of $W_i = \mu\tau/N$, Keshav [Keshav, 1991] describes a technique that uses a rate allocating server discipline, which enables the evaluation of μ/N adaptively by each user, which enables the predictive controller to be implemented . However, where a FIFO service discipline exists as in our case, this is not possible and, the way out of the problem is to use a partially distributed method, with an upper bound N_{max} on the number of users.

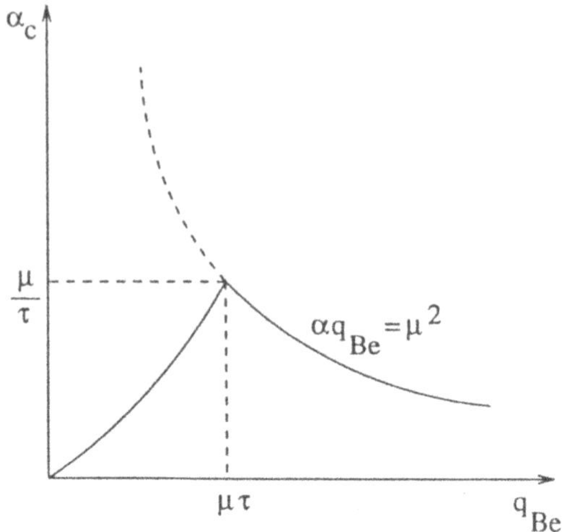

Figure 9: α_c variation with q_{Be} for the optimal controller (not to scale)

4.2 Systemic Stability in a Partially Distributed Implementation

Consider the physical model shown in Figure.1, but with N virtual circuits of equal round trip propagation delay, $\tau_i = \tau_j; \forall i \neq j$, sharing the single bottleneck node and employing the same hypothetical rate control $f_1(\lambda, q_B)$, it is a straightward task to write the model in R^5 by extending the equation as (5), (6) and (7) (Ranasinghe,1994). Further if $\gamma_i = \gamma_j; \forall i \neq j$, then the model reduces to R^3 , which makes the direct application of unilateral stability conditions to the systemic case possible.

Then by similar reasoning to that in section 2.2, and with the additional condition that $\alpha_i = \alpha_j; \forall i \neq j$ the systemic stability condition is obtained as:

$$N\alpha_i < \alpha_c \tag{21}$$

with equilibrium point condition,

$$N\alpha_i\gamma_i = \lambda_e q_{Be} \tag{22}$$

at the specified equilibrium point(λ_e, q_{Be})

Ideally, as the number of connections vary, each source control should hold its common β constant, and vary the individual convergence rate α_i based on a distributedly known value of N. Where this is not possible, partial distribution with an upper bound value, N_{max}, can be attempted [Jain, 1989 : Mitra et al. , 1990]. Suppose the active number of connections at a given instant N_{active} be greater than N_{max}. Then in satisfying the new equilibrium point condition.

$$N_{active}\alpha_i\gamma_i = \lambda_e q_{Be(active)} \tag{23}$$

where $\lambda_e = \mu$, it becomes $q_{Be(active)} > q_{Be}$, when compared to equation(22).

A controller which had been optimised for q_{Be} , when shifted to a higher $q_{Be(active)}$ without being re-optimised, will produce an "apparent" critical α that may be negative (see section 2.3). As a result, for a positive convergence rate, the system reaches a.s., or will exhibit a stable queue backlog. Similarly, it can be shown that in the case of $N_{active} < N_{max}$, the suboptimal behaviour is reflected in the form of a server under-utilisation which depends on the convergence rate.

Therefore, it can be argued that, the class of continuous time M-ary rate control is robust in the face of a partially distributed implementation. In contrast, a predictive control with $W_i = \mu\tau N_{active}/N_{max}$, is susceptible to a node buffer overflow. The complementariness of the continuous time M-ary rate control to that of Mitra's window algorithm [Mitra et al.. 1990] is further confirmed by the simulation observation that for $N_{active} > N_{max}$, a stable increase in the queue backlog has been observed in the later. This conclusion is valid to all continuous time M-ary controls.

5 CONCLUSIONS

In this paper we have considered the stability and optimality of continuous time M-ary rate control mechanisms. Following a theoretic analysis of the dynamic system, we have shown that the class of continuous time M-ary rate controls exhibit a stable limit cycle that surrounds an unstable equilibrium point which vanishes to asymptotic stability as the convergence rate is raised beyond a critical point. The critical point is a ***Hopf bifurcation*** point, in that the systems stability properties differ either side of this point.

Using a common optimality criterion for both the discrete and continuous time M-ary rate control forms, which is expressed as the ***convergence rate invariance*** property, it has been shown that the continuous time M-ary control produces a unique optimal control for specified operating point, whereas the discrete time counterpart consisted of a wider class of mutually exclusive controls satisfying the optimality criterion. The study of continuous time M-ary rate controls was found to have implications in the performance characterisation and implementational aspects of a complementary class of window controls.

Acknowledgements

The author is grateful for the encouragement provided by Prof.V.K. Samaranayake, Director, Institute of Computer Technology, University of Colombo , during this period.

Appendix

We present the proof of stability of the rate controlled system $f_1(\lambda, q_B)$, described by the model in R^3. We use the Hopf bifurcation theorem [Beltrami 1987] into prove the existence of a stable limit cycle surrounding an unstable equilibrium point.

Consider first, the nonlinear dynamic system described by:

$$\dot{x} = f(x, \xi)$$

where, $x(t) \in R^2$, and has an isolated equilibrium point $x_e(\xi)$, where ξ is a variable parameter. Let the linearised Jacobian of the system $A(\xi)$ have eigenvalues $\lambda_i(\xi) = \alpha(\xi) \pm j\beta(\xi)$ which are differentiable in ξ in a suitable range $|\xi| < \delta$. The Hopf Bifurcation theorem in R^2 states the following [Beltrami 1987].

Suppose that the equilibrium point $x_e(\xi)$ is asymptotically stable for $\xi < 0$ and unstable for $\xi > 0$, and that $\alpha(0) = 0$, i.e., $x_e(0)$ of the linearised system is neutrally stable. If $d\alpha(0)/d\xi > 0$ and $\beta(0) \neq 0$ then for all sufficiently small $|\xi|$, a closed orbit exists for ξ either positive or negative. In particular , if $x_e(0)$ is locally a.s, then there exists a stable limit cycle Γ about $x_e(\xi)$ for all small $\xi > 0$. Moreover, the amplitude of Γ grows as ξ increases.

Since the linearised system is neutrally stable at $x_e(0)$, we have to ensure that $x_e(0)$ is in fact locally a.s., which is a difficult procedure. However, extending the theorem to R^3 we remove this obstacle by observing that two of the three eigenvalues $\lambda_i(\xi); i = 1, 2, 3$ of Jacobian $A(\xi)$ are complex and one is real and negative. At $x_e(0)$ the complex eigenvalues become imaginary but, since $\lambda_3(0)$ is real and negative, the system is locally a.s.

Recall that the model in R^3 corresponding to $f_1(\lambda, q_B)$ as (see (5), (6) and (7)

$$\dot{q}_p(t) = \lambda(t) - q_p(t)/\tau$$

$$\dot{q}_B(t) = q_p(t)/\tau - \mu$$

$$\dot{\lambda}(t) = \alpha - \lambda(t)q_B(t)/\beta\mu\tau$$

with the state vector $x = (q_p, q_B, \lambda)$ and the equilibrium point $x_e = (\mu\tau, 1, \mu)$.

The Jacobian of the linearised system at x_e is obtainable as:

$$A = \begin{pmatrix} -1/\tau & 0 & 1 \\ 1/\tau & 0 & 1 \\ 0 & -\mu/\gamma & -1/\gamma \end{pmatrix}$$

The characteristic equation with eigenvalues λ_i as roots follows from the Jacobian and is given by:

$$\lambda^3 + (1/\tau + \alpha/\mu)\lambda^2 + \lambda\alpha/(\mu\tau) + \alpha/\tau = 0$$

Let there exist two complex conjugate and, one real and negative eigenvalue for this characteristic equation as:

$$\lambda_i(\xi) = \sigma(\xi) \pm j\rho(\xi)$$

for $i = 1, 2$ and λ_3, real.
Therefore from $\prod_{i=1}^{3}(\lambda - \lambda_i)$ we get:

$$\lambda^3 - (2\sigma + \lambda_3)\lambda^2 + (\sigma^2 + \rho^2 + 2\sigma\lambda_3)\lambda - \lambda_3(\rho^2 + \sigma^2) = 0$$

For linearised systems to be neutrally stable at a possible bifurcation point $\xi = 0$ it should be such that:

$$REAL(\lambda_i(0)) = \sigma(0) = 0$$

substitution of which simplifies the characteristic equation, and by comparison with the general equation we get:

$$\lambda_3 = -(1/\tau + \alpha/\mu)$$

$$\rho^2 = \alpha/(\mu\tau)$$

$$\lambda_3\rho^2 = -\alpha/\tau$$

Since $\lambda_3(0)$ is real and negative, the non-linear system is locally a.s. at $x_e(0)$, and since $IMAG(\lambda_i(0)) = \rho \neq 0$, we have satisfied two conditions required by the theorem. The bifurcation point can be determined now. Since;

$$\lambda_3\rho^2 = -\alpha/\tau = -\alpha(1/\tau + \alpha/\mu)/\mu\tau$$

by algebraic simplification we obtain the solution to α, as:

$$\alpha_c = (1 - 1/\mu\tau)\mu^2$$

or, $\xi = (\alpha_c - \alpha)$ with $\xi = 0$ as the possible bifurcation point.

Finally, we need to check that $REAL(d\lambda_i(0)/d\xi) > 0$. Noting that $d\xi = -d\alpha$, by differentiating the characteristic equation with respect to α, followed by the substitution $\xi = 0$, i.e., $\alpha = \alpha_c$ gives:

$$\frac{d\lambda(0)}{d\xi} = \frac{1/\tau + \lambda/\mu\tau + \lambda^2/\mu}{3\lambda^2 + 2\lambda(1/\tau + \alpha_c/\mu) + \alpha_c/\mu\tau}$$

Since $\lambda(0) = \pm j\rho(0)$ in the above equation, a complex fraction results, of which the real part is extractable as:

$$REAL(d\lambda_i(0)/d\xi) = \frac{(1/\tau - \rho^2/\mu)(\alpha_c/\mu\tau - 3\rho^2) + 2\rho^2/\mu\tau(1/\tau + \alpha_c/\mu)}{(\alpha_c/\mu\tau - 3\rho^2)^2 + 4\rho^2(1/\tau + \alpha_c/\mu)^2}$$

of which the denominator is positive, and the numerator further simplifies to give $2\alpha_c(\mu\tau - 1)/\mu\tau^3$ which is positive for $\mu\tau > 1$, thus satisfying the required condition.

Thus for small enough $\xi > 0$, there exists a stable limit cycle about $x_e(\xi)$, by implication of the theorem. Alternatively, from a dynamic theoretic point of view, for $0 < \alpha < \alpha_c$, a local instability gives way to a globally stable closed orbit, and for $\alpha > \alpha_c$, a globally unstable cycle degenerates into a stable point.

6 REFERENCES

[Agarwala et al., 1992] A.K. Agarwala, D.Sanghi. *Design and Evaluation of an Adaptive Flow Control Scheme.* IEEE INFOCOM, pp. 11.A.3.7 - 11.A.3.7, 1992.

[Beltrami 1987] E. Beltrami Mathematics for Dynamic Modelling *Academic Press*, 1987.

[Bolot et al., 1990] J.C. Bolot, A.U. Shankar. *Dynamical Behaviour of Rate Based Flow Control Mechanisms.* Computer Communication Review, Vol.20, No.2, pp.35-90, April 1990.

[Bolot et al., 1992] J.-C. Bolot, A.U.Shankar. *Analysis of a Fluid Approximation to Flow Control Dynamics.* IEEE INFOCOM, pp.11.A.4.1 - 11.A.4.10, 1992.

[Chiu et al., 1989] D.Chiu, R.Jain. *Analysis of the Increased and Decrease Algorithms for Congestion Avoidance in Computer Networks.* Computer Networks and ISDN Systems, Vol.17, No.1, pp.1-14, June 1989.

[Fendick et al., 1992] K.W.Fendick, M.A.Rodrigues, A.Weiss. *Analysis of a Rate Based Control with Delayed Feedback.* ACM SIGCOMM, Vol.22. No.4, pp.136-148, 1992.

[Fernow et al., 1985] j.P.Fernow, M.L.El-Sayed. *Stability of Adaptive Congestion Controls in Packet Networks.* Computer Networks and ISDN Systems, Vol.10, No.?. pp.7-18, 1985.

[Filipiak 1988]. J.Filipiak. Modelling and Control of Dynamic Flows in Computer Networks. *Springer-Verlag*, 1988.

[Gerla et al., 1980] M.Gerla, L. Klienrock. *Flow Control. A Comparative Survey*, IEEE Transaction on Communications, Vol. 28, No.4, pp.553-574, April 1980.

[ISIM 1987] Interactive Simulation Users Manual, *Manchester*, 1987.

[Jacobson, 1988] V.Jacobson. *Congestion Avoidance and Control.* ACM SIGCOMM, Vol.18, No.?, pp.314-329, August 1988.

[Jain, 1989] R.Jain. *A Delay Based Approach for Congestion Avoidance in Interconnected Heterogeneous Computer Networks.* Computer Communication Review, Vol.19, No.5, pp.56-71, October 1989.

[Keshav, 1991] S. Keshav. *A Control Theoretic Approach to Flow Control.* ACM SIGCOMM, Vol.21, No.4, pp.3-15, September 1991.

[Mitra et al., 1990] D. Mitra, J.B.Seery. *Dynamic Adaptive Windows for High Speed Data Networks: Theory and Simulations (extended abstract).* ACM SIGCOMM, Vol.20, No.4, pp.30-40, 1990.

[Mukherjee et al., 1991] A.Mukherjee, J.C.Strikwerda. *Analysis of Dynamic Congestion Control Protocols - A Fokker Plank Approximation.* ACM SIGCOMM, Vol.21, No.4, pp.159-161, 1991.

[Ohta, 1988] M. Ohta. *Effects of Control Delay and Transmission Delay on Congestion Controls.* Proceedings of the 9th International Conference on Communications (ICCC), pp.127-131, 1988.

[Ramakrishnan et al., 1990] K.K.Ramakrishnan, R.Jain. *A Binary Feedback Scheme for Congestion Avoidance in Computer Networks.* ACM Transactions on Computer Systems, Vol.8, No.2, pp.158-181, May 1990.

[Ranasinghe, 1994] D.N.Ranasinghe. On a class of Feedback Rate Control Mechanisms in Computer Networks - Ph.D. thesis (submitted). *University of Wales*, 1994.

[Sage et al., 1977] A.P.Sage, C.White. Optimum Systems Control. *Prentice Hall*, 1977.

[Shenker, 1990] S.Shenker. *A Theoretical Analysis of Feedback Flow Control.* ACM SIGCOMM, Vol.20, No.4, pp.156-165, 1990.

[Stidam, 1985] S. Stidam. *Optimal Control of Admission to a Queuing System.* IEEE Transactions on Automatic Control, Vol.30, No.8, pp.705-713, August 1985.

[Wang et al., 1991]Y.T.Wang, B.T.Sengupta. *Performance Analysis of a Feedback Congestion Control policy Under Non-negligible Propagation Delay.* ACM SIGCOMM, Vo.21, No.4, pp.149-157, 1991.

4

Mean End-to-End Packet Delays in Hybrid Integrated Services Networks

T.V.J. Ganesh Babu and A. Kumar
Department of Electrical Communication Engineering
Indian Institute of Science
Bangalore 560 012, INDIA.
e-mail: anurag@ece.iisc.ernet.in

Abstract

We study a scheme for approximate calculation of mean *end-to-end* packet delay in a network that integrates circuit switched and packet switched traffic, using the movable boundary traffic integration scheme on each link. Numerical results obtained from the approximation are compared with simulation results. We analyse the discrepancies between the simulation and the approximation. The analysis reveals an important phenomenon that occurs when packets and circuits are integrated in a *network*. When a link is shared between circuit switched and packet switched traffic, using the movable boundary scheme, then the packet departure process is burstier than Poisson. In a network of such links, at light circuit switched loads, the active calls occupy multiple hops, hence bursts of packets departing from a link are likely to find the next link available. Thus at light circuit loads there is a positive correlation between packet bursts and link availability, yielding packet delays less than what would be expected on the grounds of burstiness, and causing the Poisson process based approximation to overestimate mean packet delay. At higher circuit loads the active circuits occupy fewer hops, and the correlation is diminished; the Poisson approximation then underestimates the delay of the bursty packet arrivals.

Keywords

Circuit switching, packet switching, burst reservation, queueing models, Poisson arrival approximation, correlation effects in queuing networks

1 INTRODUCTION

In this paper we study end-to-end packet delays in integrated services networks that multiplex circuit switched and packet switched traffic using the movable boundary scheme [Fischer and Harris, 1976] on each link. Although much work has appeared in the literature, on the performance of the movable boundary scheme on a single link (e.g., [Atmaca and Pujolle, 1992], [Kraimeche and Schwartz, 1985], [Rosenberg and Le Bon, 1993]), little is known about the end-to-end performance when such links are interconnected to form networks.

The objective of this paper is to study the quality of and the errors in a natural approximation scheme for estimating the mean end-to-end packet delays. Calculations are carried out for the linear network topology (i.e., a linear array of nodes and full-duplex links) with single and multiple channels on each link. Path counting techniques, or well known approximation techniques, are used for calculating link blocking probabilities due to circuit switched traffic. The approximation uses these blocking probabilities to replace each link with an equivalent isolated link model. Mean packet delays from these link models are summed to obtain mean end-to-end packet delays.

Careful simulations, using the regenerative technique, are used to understand the accuracy of the approximation. Detailed stochastic analysis is performed to understand the discrepancies between the approximation results and the simulation results. Apart from explaining the trends in the errors in the approximation, this analysis reveals an important effect that will arise only in the network context. The movable boundary scheme, for sharing a link between circuits and packets, yields bursty packet departures. We show that the presence of multiple link circuit switched calls in the network tends to reduce packet delays below what would be expected on grounds of burstiness. As the circuit arrival rate of each source-destination pair increases, circuit switched calls tend to become shorter in length. It turns out that shorter circuit switched calls in the network result in larger packet delays.

Models of integration of circuit switched traffic and packet switched traffic on a TDM link have been analysed in [Fischer and Harris, 1976] [Sriram et al, 1983] [Li and Mark, 1985] [Yang and Mark, 1991] in the discrete time domain, and in [Atmaca and Pujolle, 1992] [Kraimeche and Schwartz, 1984] [Kraimeche and Schwartz, 1985] [Weinstein et al, 1980] [Williams and Garcia, 1984] in continuous time domain. Continuous time analysis is adopted here, i.e., if on a link there are C slots per frame, then these C servers are assumed to be operating as if they are C independent channels in parallel.

Two notable references on the end-to-end performance analysis of integrated circuit/ packet switching networks are [Li and Mark, 1988] and [Leon-Garcia and Aboul-Magd, 1986]. Our work differs significantly from these two references, in that we are concerned with the dynamics of the occupancy of network links by circuit switched traffic and the effect of this on the packet switched traffic. In particular, our work reveals the effect of the correlation between packet arrivals at a link and the availability of free channels on the link. Under lower circuit arrival rates for each source-destination pair, there is positive correlation, hence packet delays are less than one might expect based just on the burstiness of the packet arrival process. At higher circuit arrival rates for each source-destination pair, this correlation decreases and packet delays increase.

While the work reported in this paper has been motivated by channelised narrowband ISDN, very similar models will arise if in B-ISDN/ATM isochronous connections are allocated peak bandwidth, and data services are given "best-effort" service (as in the Internet). Further, in this scenario, data services could be allowed to use bandwidth temporarily left vacant by variable-bit-rate isochronous connections in between bursts (movable boundary).

The outline of this paper is as follows. In Section 2 we describe the model. In Section 3 we explain how the approximate analysis obtains mean end-to-end packet delays. In Section 4 we discuss the numerical results obtained from the approximation and from simulations. In Section 5 we make an extensive study to understand the discrepancies. In Section 6 we see how the approximation technique is applied to networks with many channels per link, and discuss simulation results for a particular type of network. In Section 7 we present our conclusions and discuss possible extensions of this work.

2 THE NETWORK MODEL

We consider a communication network, consisting of N nodes (switches) and L links. Link $j, 1 \leq j \leq L$ has c_j full-duplex channels; m_j channels are used by circuit traffic and $(c_j - m_j)$ channels are dedicated to packet traffic. Packets can use free circuit channels (i.e., the movable boundary scheme), but circuits have preemptive priority over packets. Hence, as far as circuit traffic is concerned, we just have a circuit switching network with m_j channels on link $j, \quad 1 \leq j \leq L$.

External arrival processes are assumed to be Poisson. We assume that for every source-destination pair $i, 1 \leq i \leq N(N-1)$, the circuit arrival rate is λ_c and the packet arrival rate is λ_p. The holding times of all circuit switched calls is exponentially distributed with mean μ_c^{-1}. All packet service times are exponentially distributed with mean μ_p^{-1}.

We assume fixed routing for circuits and packets. Routing for circuit traffic and packet traffic is defined by the same routing matrix R, where R is an $N(N-1) \times L$ matrix whose element $r_{ij} = 1$ if the route for the source-destination pair i uses link j, and $r_{ij} = 0$ otherwise.

3 THE APPROXIMATE ANALYSIS METHOD

We calculate the probability at each link that all the circuit channels are busy for a given λ_c and μ_c. Next we calculate the composite arrival rate of packets at each link, from the routing matrix R and the packet arrival rates for all the source-destination pairs. Then we consider each link in isolation, and assume that the blocking probability of circuit channels on the link is due to an independent Poisson circuit arrival process to this link. Further, we assume that the composite arrival process of packets to the same link is Poisson with arrival rate as calculated above. We now have an analysable model that can be used to obtain the approximate mean delay for packets at that link in the network. The mean end-to-end packet delay is obtained by summing the mean packet delays over the various links in the route of a packet.

3.1 Link blocking probability calculation

Let $a = \frac{\lambda_c}{\mu_c}$. In steady state, let $\underline{X} = (X_1, X_2, \ldots, X_{N(N-1)})$ denote the random vector whose j^{th} component X_j denotes number of calls active on route j. Let

$$\pi(\underline{x}) := P(\underline{X} = \underline{x}), \quad \underline{0} \leq \underline{x}R \leq \underline{m}$$

denote the joint distribution of the random vector $\underline{X}$. Here, $\underline{0}$ is the L-vector of all 0's and $\underline{m}$ is the L-vector of m_j's. It follows from the product form solution [Burman et al, 1984] [Kelly, 1986] that

$$\pi(\underline{x}) = \frac{1}{G(a)} a^{\sum_{i=1}^{N(N-1)} x_i} \cdot \prod_{i=1}^{N(N-1)} \frac{1}{x_i!} \quad \text{for } \underline{x}, \quad \text{such that} \quad \underline{0} \leq \underline{x}R \leq \underline{m}$$

where $G(a)$ is for normalisation of $\pi(\cdot)$ to 1. Letting $M = \sum_{j=1}^{L} m_j$, it follows that there exist constants $A_K^{(N)}, 1 \leq K \leq M$, such that

$$G(a) = 1 + A_1^{(N)} \cdot a + \ldots + A_M^{(N)} \cdot a^M$$

For networks with simple topologies and only one channel per link, with uniform traffic for circuits, it is possible to calculate the blocking probability of each link directly from the product form solution [Babu, 1991]. For networks with multiple channels on each link, it becomes extremely difficult to calculate these link blocking probabilities directly from the product form solution. So in these cases we resort to an approximate iterative method to calculate the link blocking probabilities (see Section 6.1).

3.2 Model for packet delay on a single link in the network: single channel per link

In the approximate link model, the circuit arrival rate γ_c is obtained as the Poisson arrival rate that yields the blocking probability β for this link in the network, and the packet arrival rate γ_p is obtained from λ_p and the routing matrix R. Thus, γ_c is obtained by Erlang's blocking formula. Since we are interested in packet delays, when the server is working on circuits it is equivalent to the server being on "vacation". The preemptive vacation model with service order independent of service times was discussed by Gaver in [Gaver, 1962]. Denote by S the stationary sojourn time of packets in this model. For, $\frac{\gamma_c}{\mu_c}(1 - \beta) + \frac{\gamma_p}{\mu_p} < 1$, using Laplace transform technique and using regenerative analysis [Babu, 1991], we can write,

$$E(S) = \frac{1}{\mu_p} \left[\frac{(1 + \rho_c)^2 + \alpha \rho_c}{(1 - \rho_p(1 + \rho_c))(1 + \rho_c)} \right]$$

where $\rho_c = \frac{\gamma_c}{\mu_c}$, $\rho_p = \frac{\gamma_p}{\mu_p}$ and $\alpha = \frac{\mu_p}{\mu_c}$. This result can also be derived using Markov chain analysis and the z-transform technique as shown in [Schwartz, 1987].

4 NUMERICAL RESULTS (APPROXIMATION AND SIMULATION): LINEAR NETWORKS, SINGLE CHANNEL PER LINK

4.1 Approximate analysis

In the single channel per link case, each link has two servers one for either direction. A circuit switched call is bidirectional and hence seizes both servers of each link in its path. Packets traverse the network unidirectionally, and hence require service from only one server of a link in the direction of traversal. The busy probability due to circuits on each link of an N-node linear network, with a single channel per link, can be obtained directly from the product form solution [Babu, 1991]. In the case of linear networks, the total packet arrival rate on link j (we use the convention of labelling the links in the linear network from left to right), in both directions, based on uniform packet traffic assumption, is $\gamma_p = j(N-j)\lambda_p$. We calculate the approximate mean packet delay at each link using the isolated link model discussed in the Section 3.2. This calculated approximate mean delay will be the same for packets in both directions of a link.

We fix the mean holding time for circuits and mean service time for packets in our entire calculations and simulations. We have chosen $\mu_c = 0.01$ and $\mu_p = 0.1$. Observe that these parameter values are quite reasonable for a voice-data integrated network that does fast circuit switching of voice bursts and packet switching of data. This combination of parameter values has led to some very interesting phenomena.

4.2 Comparison with simulation

The simulation was carried out using the SIMSCRIPT II.5 simulation language. Our main focus is on estimating the mean delay experienced by packets. We use regenerative simulation to obtain the estimates and confidence intervals [Lavenberg, 1983], [Wolff, 1989]. We use as regeneration epochs those departure epochs of packets that leave the network empty.

The central link in a linear network with an even number of nodes, or the two central links in a linear network with an odd number of nodes, experience the maximum blocking probability and a maximum arrival rate of packets. Since a linear network is not symmetric, the mean packet delay on each link has to be estimated individually, in order to estimate mean end-to-end packet delay. Since the load on the central link/s is the maximum, however, we only maintain regeneration statistics for the central link/s. The simulation is run until an adequately narrow confidence interval is obtained for the central link/s. It is then expected that the confidence interval for the estimated mean packet delays on the other links, are at least as narrow.

The simulation results and the corresponding values obtained from the approximate analysis, for various values of λ_c and keeping λ_p constant, are shown in Tables 1, 2 and 3 for the second link in a 3-node (2 link) linear network. In each of these tables (and in the tables to follow) two sets of simulation results

Table 1: Approximate mean packet delay and its corresponding simulation result for link 2 of a 3-node linear network with $\mu_c = 0.01$ and $\mu_p = 0.1$. Packet load on link 2 = 0.1

Circuit occupancy on Link 2	Mean delay on Link 2 With 95% confidence interval (Sampling of packet size done only once at network entry)	Mean delay on Link 2 with 95% confidence interval (Sampling of packet size done at each link)	Mean delay from approximate analysis
0.1	21.35 ± 0.19	21.17 ± 0.19	23.75
0.2	32.69 ± 0.28	32.48 ± 0.28	37.14
0.3	45.49 ± 0.36	45.24 ± 0.36	51.67
0.4	60.99 ± 0.46	60.84 ± 0.46	68.00
0.5	81.66 ± 0.59	81.02 ± 0.59	87.50
0.6	109.81 ± 1.57	108.87 ± 1.59	113.33
0.7	159.64 ± 2.49	158.79 ± 2.54	155.00
0.8	279.01 ± 5.19	279.58 ± 5.51	260.09

are shown. In one case the packet length is sampled from the packet length distribution when the packet enters the network, and the packet retains this length as it traverses the links of the network. In the other set of simulations the packet length is sampled each time it is queued at a new link. Thus the latter results incorporate Kleinrock's independence assumption, and are included to show that the errors in the approximation under study are *not* due to the independence assumption, and that in any case there is enough mixing in these example networks that Kleinrock's independence assumption is quite good.

In Tables 1, 2 and 3, the packet load on link 2 is 0.1, 0.3, and 0.6, respectively. As packet load increases, the maximum circuit occupancy must, of course, decrease. We make the following observations from these tables:
(i) The maximum error is less than 10%.
(ii) For a particular packet load, as the circuit occupancy increases from a low value to a high value, the approximation overestimates at low circuit occupancies and underestimates at high circuit occupancies.
(iii) The circuit load at which the crossover occurs decreases as packet load on link 2 increases (≈ 0.65 when packet load is 0.1, ≈ 0.55 when the packet load is 0.3, and ≈ 0.325 when the packet load is 0.6).

Tables 4 and 5 show simulation results along with the corresponding results from the approximate analysis for fixed packet load, and various circuit loads for a 6-node linear network. The delays are shown at each link for packets that are entering the link from the left.

From Tables 4 and 5 we make the following observations:
(i) For each link, and each level of packet load, the approximation overestimates at light circuit loads and underestimates at circuit loads that drive the link close to saturation.

Table 2: Approximate mean packet delay and its corresponding simulation result for link 2 of a 3-node linear network with $\mu_c = 0.01$ and $\mu_p = 0.1$. Packet load on link 2 = 0.3

Circuit occupancy on Link 2	Mean delay on Link 2 With 95% confidence interval (Sampling of packet size done only once at network entry)	Mean delay on Link 2 with 95% confidence interval (Sampling of packet size done at each link)	Mean delay from approximat analysis
0.1	28.94 ± 0.25	28.67 ± 0.77	31.667
0.2	46.94 ± 1.25	46.52 ± 1.23	51.99
0.3	69.54 ± 1.83	69.41 ± 1.83	77.50
0.4	107.12 ± 3.08	106.77 ± 3.01	113.33
0.5	174.32 ± 5.57	173.96 ± 5.42	175.00
0.6	360.06 ± 8.13	364.09 ± 7.49	340.00

Table 3: Approximate mean packet delay and its corresponding simulation result for link 2 of a 3-node linear network with $\mu_c = 0.01$ and $\mu_p = 0.1$. Packet load on link 2 = 0.6

Circuit occupancy on Link 2	Mean delay on Link 2 With 95% confidence interval (Sampling of packet size done only once at network entry)	Mean delay on Link 2 with 95% confidence interval (Sampling of packet size done at each link)	Mean delay from approximat analysis
0.1	58.02 ± 2.03	57.95 ± 1.97	63.34
0.15	82.22 ± 1.42	83.82 ± 1.45	90.99
0.2	116.42 ± 4.45	117.47 ± 4.42	129.99
0.25	176.61 ± 4.29	181.64 ± 4.36	191.67
0.3	295.99 ± 9.01	302.94 ± 8.62	310.04
0.35	671.82 ± 11.38	676.21 ± 12.45	656.477

Table 4: Mean packet delays (moving from left to right) for various links of a 6-node linear network with $\mu_c = 0.01$ and $\mu_p = 0.1$. For link 3 the 95% confidence interval obtained from regenerative simulation is also shown. Packet load is as follows: link 1: 0.11, link 2: 0.18, link 3: 0.2, link 4: 0.18, link 5: 0.11

l	Circuit occupancy on link l.	Mean packet delay (simulation) (Sampling of packet size done only once at network entry.) at link l	Mean packet delay (simulation) (Sampling of packet size done at each link.) at link l	Approximate mean packet delay at link l
1	0.058	18.366	18.34	18.55
1	0.116	26.00	26.03	26.17
1	0.181	35.40	35.37	35.06
1	0.251	45.56	45.51	45.03
1	0.335	58.66	58.64	58.22
1	0.437	78.72	78.69	76.32
1	0.588	118.49	118.90	113.35
2	0.091	23.107	22.82	24.96
2	0.180	33.42	33.34	38.50
2	0.276	47.96	47.71	54.62
2	0.369	66.28	65.83	73.19
2	0.470	91.45	91.21	98.69
2	0.570	138.06	137.03	136.01
2	0.687	256.76	256.19	230.39
3	0.102	23.36 ± 0.62	22.56 ± 0.60	27.39
3	0.201	34.10 ± 0.60	33.19 ± 0.59	43.35
3	0.304	49.32 ± 0.73	48.42 ± 0.71	62.69
3	0.404	69.64 ± 0.91	68.74 ± 0.89	85.59
3	0.506	106.75 ± 1.16	105.26 ± 1.13	118.31
3	0.600	174.56 ± 1.96	173.64 ± 2.04	168.36
3	0.701	375.93 ± 5.38	377.09 ± 4.11	307.49
4	0.091	20.48	19.15	24.96
4	0.180	28.55	27.42	38.50
4	0.276	40.42	39.05	54.62
4	0.369	55.55	54.16	73.19
4	0.470	84.53	83.35	98.69
4	0.570	135.89	134.39	136.01
4	0.687	282.45	283.63	230.39
5	0.058	15.59	14.57	18.55
5	0.116	19.84	18.73	26.17
5	0.181	25.33	24.09	35.06
5	0.251	32.97	31.48	45.03
5	0.335	45.60	44.18	58.62
5	0.437	67.48	65.83	76.32
5	0.588	119.56	117.92	113.35

Table 5: Mean packet delays (moving from left to right) for various links of a 6-node linear network with $\mu_c = 0.01$ and $\mu_p = 0.1$. For link 3 the 95% confidence interval obtained from regenerative simulation is also shown. Packet load is as follows: link 1: 0.223, link 2: 0.356, link 3: 0.400, link 4: 0.356, link 5: 0.223.

l	Circuit occupancy on link l.	Mean packet delay (simulation) (Sampling of packet size done only once at network entry.) at link l	Mean packet delay (simulation) (Sampling of packet size done at each link.) at link l	Approximate mean packet delay at link l
1	0.0577	21.01	20.98	21.45
1	0.1160	30.45	30.36	30.63
1	0.1813	42.02	41.87	41.68
1	0.2507	55.60	55.66	54.65
1	0.3354	73.96	73.98	73.03
1	0.3788	87.65	87.80	84.09
2	0.0912	31.00	30.58	33.09
2	0.1804	47.55	47.26	53.45
2	0.2755	74.24	73.68	81.29
2	0.3692	115.64	115.35	121.12
2	0.4703	201.93	201.01	200.93
2	0.5158	293.14	291.56	272.92
3	0.1022	33.21 ± 0.875	32.47 ± 0.938	38.55
3	0.2009	52.77 ± 1.052	52.04 ± 1.063	65.38
3	0.3046	88.72 ± 1.737	88.37 ± 1.699	105.73
3	0.4039	155.33 ± 3.240	158.22 ± 3.480	174.22
3	0.5062	394.68 ± 10.31	399.73 ± 9.82	375.08
3	0.5500	789.35 ± 23.74	804.67 ± 23.49	702.25
4	0.0912	27.24	25.86	33.09
4	0.1804	39.97	39.36	53.45
4	0.2755	61.13	61.63	81.29
4	0.3692	96.26	99.55	121.12
4	0.4703	188.02	198.98	200.93
4	0.5158	280.503	299.63	272.92
5	0.0577	18.69	17.17	21.45
5	0.1160	23.29	22.16	30.63
5	0.1813	30.28	29.38	41.68
5	0.2507	40.42	40.04	54.65
5	0.3354	59.14	59.71	73.03
5	0.3788	72.18	73.10	84.09

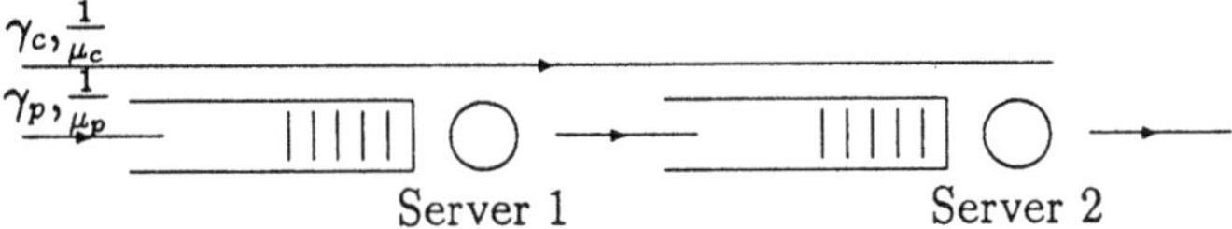

Figure 1: A two link integrated services model with two link circuits (Model-I)

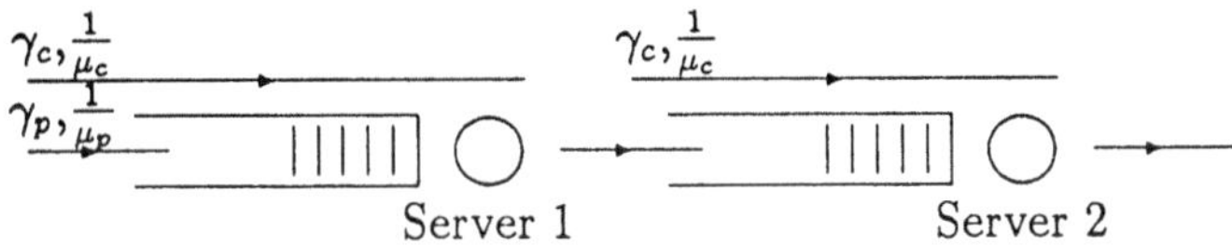

Figure 2: A two link integrated services model with single link circuits (Model-II)

(ii) On each link, the circuit load at which crossover occurs decreases as the packet load increases.

These two observations are the same as for Tables 1, 2 and 3. Further we observe:

(iii) There is load symmetry between links 2 and 4. For a given packet load, however, underestimation starts at a higher circuit load on link 4 than on link 2. As we will argue later (Section 5), this has to do with the fact that, as compared with link 2, more of the arrivals to link 4 are bursty departures from the previous link, rather than "fresh" Poisson arrivals. A similar observation holds for links 1 and 5.

We explain these observations in the next section.

5 A DETAILED STUDY OF THE APPROXIMATION

We begin by considering the two simple models shown in Figure 1 and Figure 2.

In both models, there are two single server queues in tandem. Packets entering the queue of Server 1, after finishing their service will enter the queue of Server 2. In the first model (Model I), circuits arriving will seize both servers simultaneously. In the second model (Model II), the servers will be seized by independent circuit arrivals and no arriving circuit can seize both servers. In other words, in the first model we allow only double link circuits, and in the second model we allow only single link circuits. The approximation we are studying in this paper will yield the same approximate model for link 2 in both models, i.e., the model (Model III) shown in Figure 3.

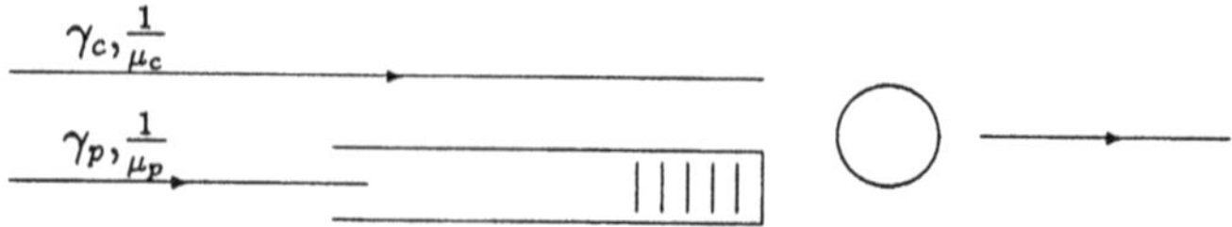

Figure 3: Approximate model for the second link of both models in Figures 1 and 2. (Model-III)

Table 6: Comparison of simulation estimates of mean delay experienced by packets at Server 2 of model-I and model-II with the results obtained using the approximate model for link 2 (model-III) with $\mu_c = 0.01$ and $\mu_p = 0.1$.

γ_c	γ_p	Delay on Link 2(simulation) (model-I)	Delay on Link 2(simulation) (model-II)	Mean Delay on Link 2(approximation) (model-III)
0.001	0.01	12.71± 0.21	22.78 ± 0.716	22.57
0.002	0.02	17.58± 0.29	39.63 ± 0.97	37.72
0.003	0.03	25.37± 0.85	66.13 ± 3.73	59.14
0.004	0.04	43.41± 1.6	116.1 ± 4.82	96.75
0.005	0.05	92.94± 5.515	237.61± 3.84	193.33
0.006	0.05	127.68±10.65	330.73 ± 6.50	267.5
0.007	0.05	174.05±21.46	491.79± 13.22	387.84
0.008	0.05	281.93±39.35	803.45 ± 39.49	624.44
0.009	0.05	507.58± 87.1	1629.28± 107.15	1327.37

In Table 6, we compare the simulation estimates of mean delay experienced by packets at Server 2 of model-I and model-II with the results obtained by analysing, model-III.

Let S_I and S_{II} be the random variables representing the stationary sojourn times of packets at Server 2 of model-I and model-II respectively. Let S_{III} be the random variable representing the stationary sojourn time of packets at the single server of model-III. Then these results suggest that

$$E(S_I) \leq E(S_{III}) \leq E(S_{II})$$

Note that the first link has the same stochastic behaviour in model-I and model-II. The observed inequality thus suggests that if there are *multiple link circuits* then the approximation tends to *overestimate* the packet delays, whereas if there are only *single link circuits* then the approximation tends to *underestimate* the packet delays at the second link of the 2 link model.

We now present some arguments and results that explain the observations.

Observe that the packet departure process from node 1, is the same in models I and II. Yet in model-I the mean packet delay at the second server is less than in model-II. It can be shown ([Babu, 1991]) that the interdeparture times of packets from the first link have a coefficient of variation greater than 1, and that the successive interdeparture times are correlated. Thus one would expect that the Poisson process approximation in model-III should yield smaller packet delays. Yet our observation is that the approximation yields larger packet delays than model-I.

Note, however, that the bursts in the packet departure process from node 1 occur because packets accumulate while a circuit holds the server, and are released in a burst when the circuit terminates. In model-I, this burst of packets will find the second server free of any circuit, whereas in model-II there is a probability that this burst of packets will again encounter a server occupied by a circuit.

Thus in model-I, there is a correlation between server availability and bursts in the arrival stream, which helps in reducing delay even below that in model-III. On the other hand the combined effect of a bursty packet arrival process and uncorrelated circuit arrivals makes the delay at node 2 in model - II exceed that in model - III.

It is interesting to note the similarity between our observations and arguments and those in [Sriram and Lucantoni, 1989]. In this paper the effect of bit dropping congestion control on the performance of a packet voice multiplexer is studied. It is observed that bit dropping during overload periods reduces the effect of correlation between packet interarrival times, thus causing a bursty arrival process to yield delays a Poisson process would yield.

It is, in fact, possible to create a mixture of 1 link and 2 link circuits with total circuit arrival rate γ_c at Server 2, such that the mean packet delay at the second node is close to that obtained via model-III.

The above arguments suggest that if the network is more likely to have multiple link circuits then the approximation will tend to overestimate the packet delays. Whereas if single link or "short" circuits are more likely then the approximation will tend to underestimate the packet delays.

It seems intuitively clear that, as circuit load increases, the length of active circuits will tend to become shorter. If this can indeed be established, then it would explain the fact that the approximation overestimates packet delays at light circuit loads and underestimates at high circuit loads. In fact, we have proved the following two theorems [Babu, 1991].

Theorem 1 *The mean number of active circuits in the network is an increasing function of the offered circuit switched load per route.*

Theorem 2 *If all single link circuits are routes, then the average length of the active circuits approaches 1 as the load goes to infinity.*

It may also be conjectured that the average length of active circuits is a decreasing function of the offered circuit load. In [Babu, 1991] we present numerical results to support this conjecture.

From the above analysis, we come to the conclusion that the length of active circuits tends to become shorter as circuit load increases. At low circuit loads, the isolated single link model overestimates the mean delay experienced by packets at downstream links, even though the packet traffic is bursty. This is because, the correlation between channel activity on neighbouring links actually overcomes the effect of burstiness and improves the queueing performance, since a burst of packets can traverse several links in the network as soon as a multiple link circuit releases its set of servers. But as circuit load increases, the effect of single link circuits becomes predominant. This means that the internode circuit correlation decreases. This combined with the fact that packet traffic is bursty, results in an underestimation by the approximation.

At low circuit loads, the overestimation at downstream links of the linear network is higher. This is because at downstream links there is more internodal bursty packet traffic than "fresh" Poisson packet traffic, and bursts in the servers' availability correlate with bursts of arrivals. If the arrivals are uncorrelated Poisson arrivals then the approximate model underestimates the packet delays. This explains the observation (iii) following Tables 4 and 5. At circuit loads that drive the links close to saturation, the packet departure process is very bursty. If the packet load is large then this situation would correspond to a relatively low circuit load; thus there will be a high probability of multiple link circuits, and in spite of the highly bursty packet traffic the approximation can underestimate.

6 NETWORKS WITH MULTIPLE CHANNELS PER LINK

In this section we apply the approximation to linear networks with multiple channels per link.

Calculation of link blocking probabilities directly from the product form distribution becomes extremely difficult as the size of the network increases and as the link capacity vector becomes large. Hence we resort to the Erlang fixed-point approximation procedure (outlined in [Kelly, 1986]).

We approximate each link of the network by an isolated link with the same number of circuit channels and packet channels and a Poisson circuit arrival process and a Poisson packet arrival process. Assuming that the link blocking probability calculated by the fixed point method [Kelly, 1986] is due to a Poisson circuit arrival process, we calculate the equivalent Poisson circuit arrival rate, using Erlang's formula. Further, given the packet arrival rate for each source-destination pair, and the routing matrix, we find the packet arrival rate at each link. This yields an approximate model for each link, whose analysis we have adopted from [Kraimeche and Schwartz, 1985].

We have done extensive simulation studies on a 6-node linear network in which each link has 4 full duplex channels, 3 circuit channels and 1 packet channel. We draw our conclusions by comparing delays experienced by packets at link 3 and their corresponding approximate estimates. We do this because link 3 experiences maximum blocking for circuits and maximum packet arrival rate.

As before it is found that the simulation that samples packet size at each link (i.e., simulates Kleinrock's Independence Assumption) yields values that are quite close to those obtained when the packet size is sampled upon entry into the network. Hence, as before, the effect of the assumption of independent packet sizes is insignificant. The approximate mean packet delays are calculated in two ways. One uses the approximate link blocking probability obtained from the Erlang fixed-point method, and the other uses the "exact" link blocking probability obtained from the simulation. The latter values were obtained to study the behaviour of the basic idea of the approximation, i.e., replacing a link in the network with an "equivalent" isolated link. These latter results are the ones that are plotted in Figures 4, 5, 6 and 7.

In Figure 4 the packet load on link 3 is 0.7 which is less than the number of channels dedicated to packet service (i.e., 1 channel). This is referred to as the "underload" situation [Schwartz, 1987]. The packet queue is stable without help from the circuit channels, and the departure process from a link is close to Poisson. In fact, for light circuit loads the network of packet queues is close to a product form network, and hence the approximation is very accurate. As circuit load increases, however, the interaction between the circuits and packets at a link increases. While the packet departure process is close to Poisson, correlation in channel availability due to multiple link circuits reduces packet delay below that estimated by the approximation. Hence (see Figure 4) the approximation overestimates for high circuit loads. In Figures 5, 6 and 7 the packet load on link 3 is in the "overload" region. Packets queues now need circuit channels for stability and the packet departure process becomes burstier than Poisson as circuit load increases. The packet load goes deeper into the overload region from Figure 5 to 6 to 7; hence the packet departure process burstiness correspondingly increases. In order to keep the packet queues *stable*, however, the circuit load cannot be taken too high, hence, even at the maximum circuit load shown in these figures there will be multiple link circuits with a high probability. Thus the approximation overestimates in almost all cases except when the packet queues are close to instability and the packet departures are very bursty (see Figures 6 and 7).

When the approximation uses link blocking probability obtained from the Erlang fixed point method, errors in the estimates of link blocking probability seem to dominate the final result. Either the approximate mean packet delay is close to the simulation estimates, or is a considerable under estimation.

7 CONCLUSION

Our study, that began as an attempt to evaluate a simple approximation for packet delays, ended up revealing some very interesting characteristics of packet delays when packets are integrated with circuits in a *network* context. In the movable boundary scheme for sharing channels on a link between circuits and packets, it can be shown that the packet departure process is bursty. The coefficient of variation of packet interdeparture times is > 1 and the interdeparture times are correlated. Yet it is found that in some cases the Poisson assumption at a downstream link yields an *overestimate* of packet delays.

This is explained by observing that in these cases the active circuits span

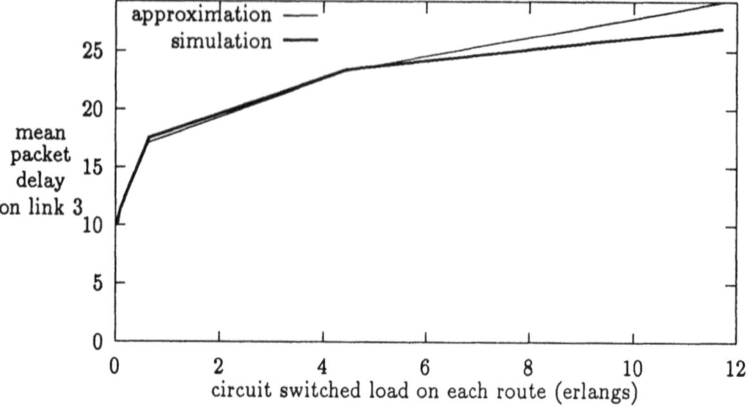

Figure 4: Mean packet delay on link 3 vs. circuit switched load on each route, in a 6-node linear network with 3 channels for circuits and 1 channel for packets on each link. Estimates from approximation are based on link blocking probability obtained from simulation. Packet load on link 3 = 0.7

multiple links, consequently when a circuit terminates it releases several channels simultaneously. Packets waiting for use of the freed circuit channels are now released in a burst but can flow through several links with low delay. In these cases the Poisson arrivals assumption overestimates packet delays. On the other hand, when active circuits are short then packets departing one link may again find channels on the next link occupied by circuits. Clearly this phenomenon is more significant at loads for which packets require the use of circuit channels (see Section 6). We also show that the average number of active circuits decreases as circuit load increases, and at heavy circuit loads the active circuit load approaches 1 (Section 5, Theorems 1 and 2). Thus the correlation effect is pronounced at low circuit loads leading to less than expected packet delays, and the correlation effect decreases with high circuit loads.

As mentioned in the Introduction, models similar to those discussed here will also arise in the B-ISDN/ATM context, if peak rate allocation is done for isochronous connections. Hence these models will continue to be of interest. Future work would include establishing the truth, or otherwise, of the conjecture in Section 5. Further, clearly the net arrival process of circuit switched calls to a link is not Poisson. The approximation may be improved by modelling these arrivals with a Markov modulated Poisson process.

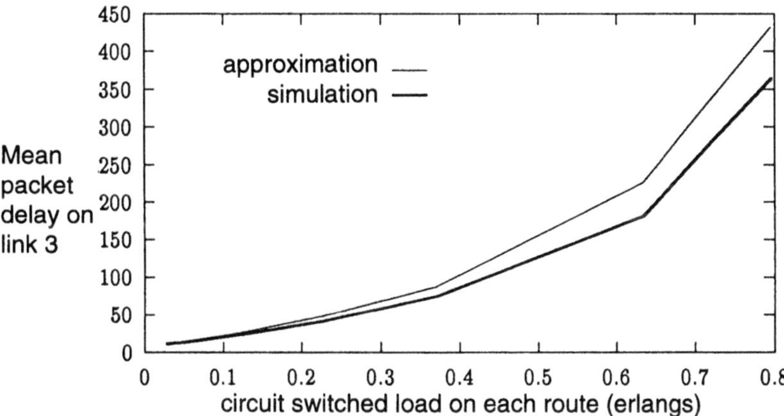

Figure 5: Mean packet delay on link 3 vs. circuit switched load on each route, in a 6-node linear network with 3 channels for circuits and 1 channel for packets on each link. Estimates from approximation are based on link blocking probability obtained from simulation. Packet load on link 3 = 1.5

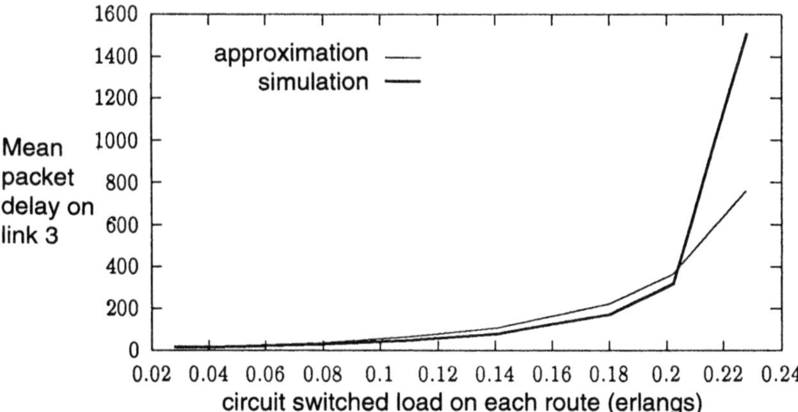

Figure 6: Mean packet delay on link 3 vs. circuit switched load on each route, in a 6-node linear network with 3 channels for circuits and 1 channel for packets on each link. Estimates from approximation are based on link blocking probability obtained from simulation. Packet load on link 3 = 2.0

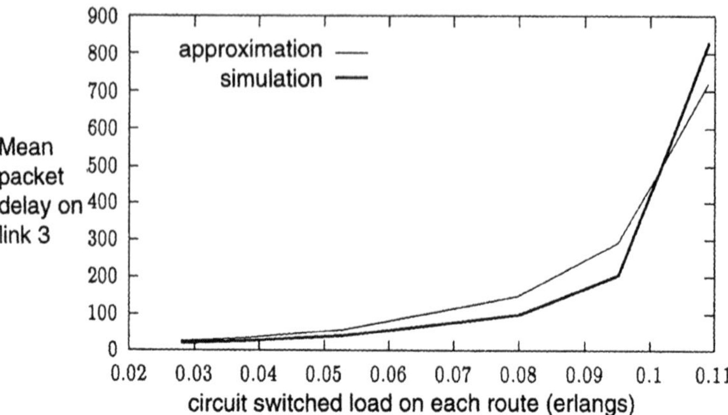

Figure 7: Mean packet delay on link 3 vs. circuit switched load on each route, in a 6-node linear network with 3 channels for circuits and 1 channel for packets on each link. Estimates from approximation are based on link blocking probability obtained from simulation. Packet load on link 3 = 2.5

REFERENCES

[Atmaca and Pujolle, 1992] T. Atmaca and G. Pujolle, "Performance Analysis of an Access Control Strategy in Integrated Networks", *Computer Network and ISDN Systems*, Vo.24, 1992, pp421-434.

[Bertsekas and Gallager, 1987] D. Bertsekas and R. Gallager, *Data Networks*, Prentice-Hall Inc., Englewood Cliffs,N.J, 1987.

[Burke, 1956] P.J.Burke, "Output of a Queueing System", *The Journal of the Operations Research Society of America*, 4, 1956, pp. 699-704.

[Burman et al, 1984] D.Y.Burman,J.P.Lehoczky and Y. Lim, "Insensitivity of Blocking Probabilities in a Circuit-Switching Network", *Journal of Applied Probability*, Vol.21, No.4, December 1984, pp. 850-859.

[Fischer and Harris, 1976] M.J.Fischer and T. C. Harris, "A Model for Evaluating the Performance of an Integrated Circuit- and Packet-Switched Multiplex Structure", *IEEE Transactions on Communications*, Vol.COM-24, No.2, February 1976, pp 195-202.

[Babu, 1991] T.V.J. Ganesh Babu, "Analysis of the End-to-end Performance of Integrated Services Networks", M.Sc(Engg) thesis, Indian Institute of Science, Bangalore, India, 1991.

[Leon-Garcia and Aboul-Magd, 1986] A.Leon-Garcia and O.S.Aboul-Magd, "Performance Analysis of Integrated Voice and Data Tandem Link Network", *Proc.IEEE Globecom*, 1986, pp. 944-948

[Gaver, 1962] D.P.Gaver,Jr, "Waiting Line with Interrupted Service Including Priorities", *Journal of Royal Statistical Soceity*, Vol.b24, 1962, p73-90.

[Kelly, 1986] F.P. Kelly, "Blocking Probabilities in Large Circuit-Switched Networks", *Adavances in Applied Probability*, Vol.18, No.2, June 1986, pp. 473-505.

[Kraimeche and Schwartz, 1984] B. Kraimeche and M. Schwartz, "Traffic Access Control Strategies in Digital Integrated Networks", *INFOCOM' 84, San Francisco*, CA, April 1984.

[Kraimeche and Schwartz, 1985] B. Kraimeche and M. Schwartz, "Analysis of Traffic Access Control Strategies in Integrated Services Networks", *IEEE Transactions on Communications*, Vol.Com-33, No.10, October 1985, pp. 1085-1093.

[Lavenberg, 1983] S.S.Lavenberg, Editor, *Computer Performance Modelling Handbook*, Academic Press INC, 1983.

[Li and Mark, 1985] S.Q.Li and J.W.Mark, "Performance Of Voice/Data Integration on a TDM System", *IEEE Transactions on Communications*, Vol.Com-33, No.12, December 1985, pp. 1265-1273.

[Li and Mark, 1988] S.Q.Li and J.W.Mark, "Simulation Study of a Network of Voice/Data Integrated TDM's", *IEEE Transactions on Communications*, Vol.36, No.1, January 1988, pp. 126-132.

[Rosenberg and Le Bon, 1993] C. Rosenberg and A. Le Bon, "Performance Models for Hybrid Broadband Networks", *Computer Networks and ISDN Systems*, Vol 25, 1993, pp1155-1163.

[Ross, 1983] S.M.Ross, *Stochastic Processes*, John Wiley & Sons INC, 1983.

[Schwartz, 1987] M.Schwartz, *Telecommunication Networks: Protocols, Modeling and Analysis*, Addison Wesley Publishing Company, 1987.

[Sriram et al, 1983] K.Sriram, P.K.Varshney and J.G.Shanthikumar, "Discrete-time Analysis of Integrated Voice/Data Multiplexers with and without Speech Activity Detectors", *IEEE Journal on Selected Areas in Communications*, Vol. SAC-1, No.6, December 1983, pp 1124-1132.

[Sriram and Lucantoni, 1989] K.Sriram and D.M.Lucantoni, "Traffic Smoothing Effects of Bit Dropping in a Packet Voice Mutiplexer", *IEEE Transactions on Communications*, Vol.37, No.7, July 1989, pp. 703-712.

[Weinstein et al, 1980] C.J.Weinstein,M.L.Malpass and M.J.Fischer, "Data Traffic Performance of an Integrated Circuit- and Packet-Switched Multiplex Structure", *IEEE Transactions on Communications*, Vol.COM-28, No.6, June 1980, pp 873-877.

[Williams and Garcia, 1984] G.F.Williams and A.L.Garcia, "Performance Analysis Of Integrated Voice and Data Hybrid-Switched Links", *IEEE Transactions on Communications*, Vol.COM-32, No.6, June 1984, pp. 532-539.

[Wolff, 1989] R.W. Wolff, *Stochastic Modeling and the Theory of Queues*, Prentice Hall, Englewood Cliffs, N.J, 1989.

[Yang and Mark, 1991] O.W.Yang and J.W.Mark, "Queueing Analysis of an Integrated Services TDM System Using a Matrix-Analytic Method", *IEEE Journal on Selected Areas in Communications*, Vol.9, No.1, January 1991, pp. 88-94.

BIOGRAPHIES

T.V.J. Ganesh Babu has a B.E. from Madurai University, and a M.S. in Electrical Communication Engg. from the Indian Insititute of Science. He is currently pursuing a Ph.D. in Concordia University, Montreal, Canada.
Anurag Kumar has a B.Tech. in E.E. from the Indian Institute of Technology at Kanpur, and a PhD from Cornell University. He was a Member of Technical Staff at AT&T Bell Labs, Holmdel for over 6 years. Since 1988 he has been with the Indian Institute of Science (IISc), Bangalore, in the Dept. of Electrical Communication Engineering, where he is now Associate Professor. He is also the Coordinator at IISc of the Education and Research Network Project, which has set up a country-wide computer network for academic and research institutions, and conducts R&D in the area of communication networks. His own research interests are in the area of modelling, analysis, control and optimisation problems arising in communication networks and distributed systems.

5

A Modified Distributed Queueing Access Protocol for Dual Unidirectional Bus Networks

R. R. Pillai and U. Mukherji
Department of Electrical Communication Engineering
Indian Institute of Science
Bangalore - 560 012, India
(Tel : (+91) (080) 3340855; Fax : (+91) (080) 3347991
e-mail : {pillai, utpal}@ece.iisc.ernet.in)

Abstract

The protocol is described. Packet access delay is the sum of request bus access delay and data bus access delay. An upper bound on data bus access delay is obtained that is equal to twice the propagation delay from the node to the most upstream node. The expected request bus access delay for Poisson offered traffic is the expected waiting time in a slotted $M/D/1$ non-preemptive priority queue, for which an expression is obtained. The upper bound on data bus access delay is smaller and the expected request bus access delay is larger at a more upstream node. A bandwidth balanced version of the protocol is also described, for which the upper bound continues to hold.

Keywords

Distributed queueing, performance modelling and analysis

1 INTRODUCTION

The Distributed Queue Dual Bus (DQDB) Metropolitan Area Network (MAN) is a dual unidirectional slotted bus network standardized by the IEEE Project 802 Working Group on Metropolitan Area Networks (IEEE 802.6) (IEEE, 1990). The unique feature of this network is the Distributed Queueing algorithm for Medium Access Control (MAC) which regulates access to the slots on the two buses. A description of the MAC protocol is given in (Newman, 1988).

The principal components of a DQDB MAN are two slotted unidirectional buses, each with a frame generator, and a number of nodes which access the network. Each bus is made up of unidirectional point-to-point links between the nodes. The architecture is as shown in Figure 1. The nodes are attached to both the buses by read and write connections.

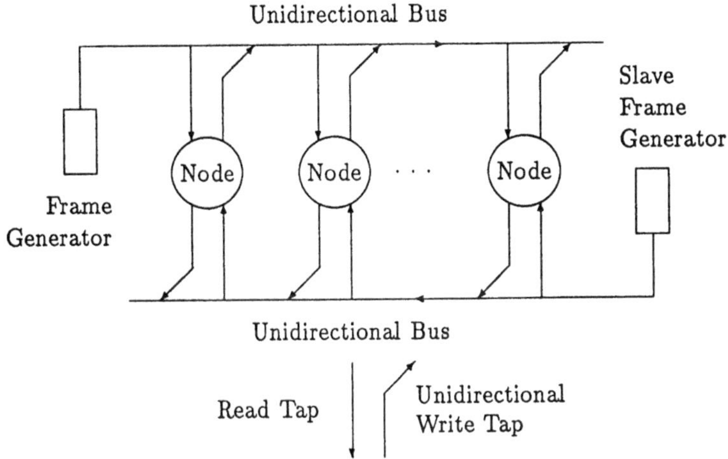

Figure 1 A dual bus network.

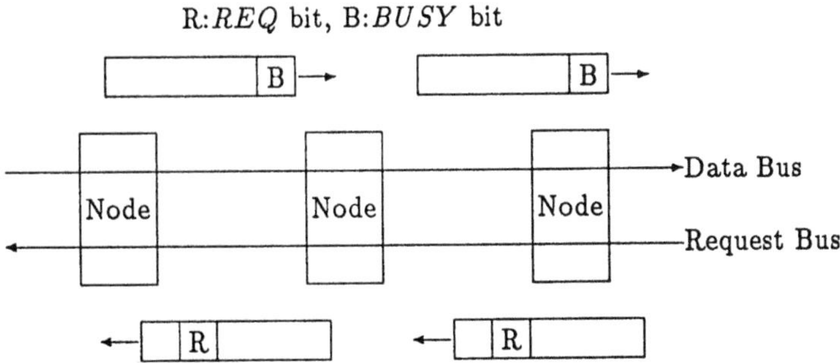

Figure 2 *BUSY* and *REQ* bits in a dual bus network.

The network architecture considered in this paper is the same as that of a DQDB network (Figure 2). Nodes are attached to two slotted unidirectional buses. Each slot contains two control bits, the REQ bit and the $BUSY$ bit (Figure 2). The REQ bit of a slot on a bus is used by a node to reserve a slot on the other bus, and the $BUSY$ bit of a slot is used to indicate that the slot contains a packet. Since the access mechanism on both buses is identical, access on only one bus is analysed. This bus is called the data bus, and the other bus is called the request bus. Propagation on the data bus is assumed to be in the downstream direction, while that on the request bus is in the upstream direction. Slots on the request bus are called *request slots* and those on the data bus are called *data slots.* If the REQ bit of a request slot is set to 0, then it is called a *free request slot.* Similarly, if the $BUSY$ bit of a data slot is set to 0, then it is called a *free data slot.* Each packet fits exactly into a data slot. The terms packet and segment will be used interchangeably, as also the terms bandwidth and capacity, and the terms request and reservation.

Several studies have been made of the performance of the Distributed Queueing MAC protocol. For infinite backlog, if the network size is small, i.e., propagation delays are insignificant, the Distributed Queueing MAC protocol results in a fair allocation of the bus capacity (Zukerman, 1989). But, the throughput division is unfair when two or more nodes in a large DQDB network are overloaded. This problem is overcome in a DQDB network through a mechanism called Bandwidth Balancing (Hahne, 1992).

The effect of various network and traffic parameters on the medium access delay has been studied in the literature through simulation and approximate analytical models (Potter, 1991), (Stock, 1990), (Tran-Gia, 1989) (Tran-Gia, 1990). Medium access delay analysis involves the analysis of a system of interacting queues, which is in general a difficult problem. The nature of interaction depends on the size of the network in addition to the MAC protocol. In a large DQDB network, the interaction between the queues at various nodes is complex and hence an exact delay analysis is quite difficult.

A variation of the Distributed Queueing protocol is proposed in this paper, and is described in Section 2. In this variation it is possible to analyse certain aspects of the performance of the protocol exactly. Data segment access delay in this variation is the sum of request bus access delay and data bus access delay. The protocol has an upper bound on data bus access delay that is equal to twice the propagation delay from the node to the most upstream node on the data bus. The expected request bus access delay for Poisson offered traffic is the expected waiting time in a slotted $M/D/1$ non-preemptive priority queue, for which an expression is obtained. A version of the protocol with rate control similar to Bandwidth Balancing is also described, for which the upper bound continues to hold.

Section 3 introduces the notation used in the following section. Some delay properties of the protocol are discussed in Section 4. Section 5 discusses the throughput performance. The request bus access is analysed in Section 6 and conclusions are drawn in Section 7.

2 THE MODIFIED MAC PROTOCOL

Each node maintains two first-come first-served queues: (i) local queue and (ii) combined queue. When a node wants to transmit a packet on the data bus, it first queues a local request in its local queue of requests. Each request in the local queue requires a service time of one request slot duration. When a request slot appears on the request bus with REQ bit set to zero, one request from the local queue is served by setting the REQ bit of that slot to 1 provided that the local queue is not empty. Simultaneously, a local segment is queued in the combined queue. If the REQ bit of the slot on the request bus is already set to 1 by a downstream node, a remote request is queued in the combined queue. Also, the node repeats that request bit in the upstream direction. The combined queue has both remote requests and local segments, and each such entry requires a service time of one data slot duration. There can be at most one arrival to the combined queue per request slot. For simplicity, local segments in the combined queue will also be referred to as remote requests.

Request and data buses are assumed to have the same capacity and, for purposes of analysis, the slot boundaries on these buses are assumed to coincide. The access delay of a packet/segment consists of two components: (i) request bus access delay and (ii) data bus access delay. Request bus access delay is the queueing delay in the local queue, i.e., the time elapsed since the arrival of the segment till the beginning of the request bus slot in which the local request is transmitted. Data bus access delay is the queueing delay in the combined queue, i.e., the time elapsed since the arrival of the segment in the combined queue till the beginning of the data bus slot in which it is transmitted. If a remote request or a local segment arrives in the slot beginning at time k to the combined queue of a node which is empty at time k, and the data bus slot beginning at time k is free, then the same remote request can be served in the data slot beginning at time k. Therefore, the data bus access delay of a segment can be zero.

3 NOTATION

M denotes the number of nodes in the network, N_i denotes the ith node from the beginning of the data bus, and $d_{i,j}$ denotes the value of propagation delay between nodes N_i and N_j normalised so that one slot duration is taken as the time unit. It is assumed that

$$\begin{aligned} d_{i,j} &= d_{j,i} \in \{0,1,2,\ldots\} \ \forall\, i,j \in \{1,2,\ldots,M\}, \\ d_{i,j} &= 0 \quad \forall i = j, \\ d_{i,k} + d_{k,j} &= d_{i,j}, \quad i \le k \le j, \\ d_{1,M} = d_{M,1} &= d. \end{aligned}$$

Let T_n be the time at which the nth ($n \ge 1$) busy request slot is observed at node N_1. Thus,

$$T_1 < T_2 < T_3 < \ldots.$$

The requests are labeled based on their arrival time at node N_1. The local request that has set the REQ bit of the nth busy request slot arriving at N_1 is denoted as $REQ(n)$. Let r_n denote the index of the node that queued a local segment corresponding to $REQ(n)$, and $SEG(n)$ denote that local segment. $REQ(n)$ appears as a remote request at node N_m, $1 \le m < r_n$, and as $SEG(n)$ at node N_{r_n}. $REQ(m)$ is said to be a future request with respect to $REQ(n)$ if and only if $m > n$, independent of the times at which these requests are queued in the local queues and of the times that they set the REQ bit of the request slots at the respective nodes. Every node except N_{r_n} observes either $REQ(n)$ or $SEG(n)$ and not both. The data bus access delay of $SEG(n)$ refers to the data bus access delay of $SEG(n)$ at node N_{r_n}.

4 SOME PROPERTIES OF THE ACCESS PROTOCOL

All remote requests have zero data bus access delay at node N_1.

Lemma 1 *If the remote request corresponding to $REQ(n)$, queued at node N_j is served at node N_i $(i \leq j)$ at time T, then the remote request corresponding to $REQ(n)$ will be served at node N_j by time $T + d_{i,j}$.*

Proof: (By induction)
Let R_i^n be the set of remote requests from downstream nodes served at N_i till time T with the last one (corresponding to $REQ(n)$) served exactly at time T. The time at which the last free data slot released by N_i reaches N_{i+1} is $T + d_{i,i+1}$. For each remote request, from the time instant it is queued at N_{i+1}, at least one free data slot reaches N_{i+1} by time $T + d_{i,i+1}$.

Any remote request m in R_i^n $(1 \leq m \leq |R_i^n|)$ gets queued at N_i after getting queued at N_{i+1}. N_i serves this remote request queued by releasing a free data slot. This free slot arrives at N_{i+1} after the corresponding remote request is queued at N_{i+1}. Also, this arrives at N_{i+1} by time $T+d_{i,i+1}$. Therefore, each remote request m in R_i^n $(1 \leq m \leq |R_i^n|)$ will be served at N_{i+1} by time $T + d_{i,i+1}$. If $j = i + 1$, then this proves the lemma.

Assume that the last remote request (corresponding to $REQ(n)$) served at N_i by time T is served at N_k $(i < k < j)$ by time $T + d_{i,k}$. Let R_k^n be the subset of R_i^n consisting of remote requests served at N_k. The remote request corresponding to $REQ(n)$ is the last remote request in the set R_k^n. Therefore, it follows that the last free data slot released by N_k for the set R_k^n is so released by time $T + d_{i,k}$. This last free data slot arrives at N_{k+1} by time $T + d_{i,k} + d_{k,k+1} = T + d_{i,k+1}$. Therefore, for each remote request, from the time instant it is queued at N_{k+1}, at least one free data slot reaches N_{k+1} by time $T + d_{i,k+1}$.

Any remote request m in R_k^n $(1 \leq m \leq |R_k^n|)$ gets queued at N_k after getting queued at N_{k+1}. N_k serves this remote request queued by releasing a free data slot. This free slot arrives at N_{k+1} after the corresponding remote request is queued at N_{k+1}. Also, this arrives at N_{k+1} by time $T + d_{i,k+1}$. Therefore, each remote request m in R_k^n $(1 \leq m \leq |R_k^n|)$ will be served at N_{k+1} by time $T + d_{i,k+1}$. If $j = k + 1$, then this proves the lemma.

□

Theorem 1 *The maximum value of the data bus access delay experienced by $SEG(n)$ at node N_{r_n} is $2d_{1,r_n}$.*

Proof:
Node N_1 observes the remote request corresponding to $REQ(n)$ at time T_n and it serves that request with zero delay. By Lemma 1, the remote request corresponding to $REQ(n)$ will be served at node N_{r_n} by time $T_n + d_{1,r_n}$. $SEG(n)$ is queued in the combined queue at node N_{r_n} at time $T_n - d_{1,r_n}$. Therefore, the maximum value of the data bus access delay for $SEG(n)$ is $2d_{1,r_n}$.

□

Table 1 Average data bus access delay performance of the modified distributed queueing protocol as a function of the propagation delay between adjacent nodes

Prop.	*Nodes*								
delay	1	2	3	4	5	6	7	8	9
1	0.00	1.14	2.18	3.11	3.98	4.85	5.73	6.65	7.61
2	0.00	1.77	3.03	4.13	5.16	6.26	7.44	8.73	10.20
3	0.00	2.16	3.50	4.64	5.79	7.04	8.44	10.00	11.60
4	0.00	2.41	3.83	5.04	6.26	7.59	9.01	10.74	12.96
5	0.00	2.61	4.08	5.36	6.61	8.02	9.58	11.56	13.81
6	0.00	2.78	4.28	5.63	6.96	8.42	9.99	11.94	14.55
7	0.00	2.91	4.45	5.84	7.20	8.75	10.32	12.38	15.08
8	0.00	3.04	4.63	6.01	7.43	8.96	10.56	12.87	15.81
9	0.00	3.19	4.79	6.14	7.60	9.14	10.90	13.13	16.06
10	0.00	3.27	4.88	6.26	7.79	9.33	11.14	13.54	16.89

Corollary 1 *The maximum value of the combined queue length at node N_i is $2d_{1,i}+1$.*

Proof:
The arrival rate to the combined queue is at most one remote request per request bus slot. By Theorem 1, the waiting time (including the transmission time on the data bus) of a remote request in that queue is upper bounded by $2d_{1,i}+1$. Therefore, it follows that the maximum value of the combined queue length at node N_i is $2d_{1,i}+1$.

□

The upper bound on the data bus access delay decreases in the upstream direction. In a small network the data bus access delay of a segment is zero. The properties of the modified MAC protocol discussed so far do not depend on the offered traffic. However, the distribution of the data bus access delay depends on the offered traffic. To find the expected data bus access delay of a packet for Poisson offered traffic, simulations are carried out. These results are shown in Table 1 for different values of the propagation delay between adjacent nodes. A 10 node network is considered and the propagation delay between adjacent nodes is assumed to be equal. The total offered load is 0.9 packets/slot and the individual packet arrival rates to the nodes are $\lambda_1 = 0.18$, $\lambda_2 = 0.16$, $\lambda_3 = 0.14$, $\lambda_4 = 0.12$, $\lambda_5 = 0.10$, $\lambda_6 = 0.08$, $\lambda_7 = 0.06$, $\lambda_8 = 0.04$, and $\lambda_9 = 0.02$. Simulation results show that the average data bus access delay can be much less than the upper bound even for high bus utilisation.

In a small network, the protocol results in a priority system where the most downstream node has the highest priority. The data bus access delay of a segment becomes zero and hence the access delay is equal to the request bus access delay.

5 THROUGHPUT PERFORMANCE

Assume that the capacity of the bus is equal to 1. When the offered load for a bus at all nodes together is less than the capacity of the bus, i.e., $\Sigma_{i=1}^{M-1}\lambda_i \leq 1$, the long term throughput of node N_i is equal to its offered load λ_i. But, when the offered load exceeds the capacity of the bus, i.e., $\Sigma_{i=1}^{M-1}\lambda_i > 1$, the long term throughput of a node can be different from its offered load, depending on its position. The throughputs of nodes $N_i, N_{i+1}, \ldots, N_{M-1}$ are equal to their offered loads where $i = \min j$ satisfying $\Sigma_{k=j}^{M-1}\lambda_k \leq 1$, and the throughputs of other nodes are less than their offered loads. However, this throughput unfairness does not depend on the time at which the nodes become active as in the case of the distributed queueing protocol. Therefore, under such overload conditions, the protocol has to be suitably modified to remove the throughput unfairness.

In a small network, the distributed queueing protocol performs as a perfect scheduler and results in throughput fairness under heavy load (Zukerman, 1989). Throughput unfairness of the distributed queueing protocol under heavy load in a large network has been studied in the literature (Hahne, 1992). A scheme called *Bandwidth Balancing* is used along with the MAC protocol to overcome the throughput unfairness that can occur in large DQDB networks (Hahne, 1992). This scheme achieves throughput fairness by wasting a small amount of bus bandwidth. Throughput fairness can be achieved with the modified distributed queueing protocol too by applying a rate control similar to Bandwidth Balancing as described below.

Assume that all nodes that access the data bus have sufficient traffic to warrant rate control. Under rate control, each node uses only a fraction α of the spare capacity left over by the remaining nodes. Let γ_i be the carried throughput of node N_i after applying the rate control. Then,

$$\gamma_i = \alpha[1 - \Sigma_{j \neq i}\gamma_j] \tag{1}$$

subject to the constraint

$$\Sigma_{j=1}^{M-1}\gamma_j < 1.$$

The unique solution to the set of Equations 1 and satisfying the constraint is

$$\gamma_1 = \gamma_2 = \ldots = \gamma_{M-1} = \frac{\alpha}{1 + (M-2)\alpha}.$$

The rate $\Sigma_{j=1}^{i-1}\gamma_j$ of arrival of busy data slots at node N_i is the fraction of the capacity used by the nodes upstream to node N_i. Since every node transmits one request corresponding to every packet to be transmitted, the rate $\Sigma_{j=i+1}^{M-1}\gamma_j$ of arrival at node N_i of request slots with *REQ* bit set to 1 is the fraction of the capacity used by the nodes downstream to node N_i. Hence, node N_i can control its carried throughput to

$$\gamma_i = \alpha[1 - \Sigma_{j \neq i}\gamma_j].$$

Thus, under heavy load, the carried throughputs of all nodes are equal and are independent of the times at which they become active and of the positions of the nodes. The convergence properties of the the rate control depend on the value of α.

To implement this rate control for the modified MAC protocol, each node N_i needs a counter. At the beginning of a slot, the counter is incremented by 1 when the node finds either a data slot with *BUSY* bit set to 1 or a request slot with *REQ* bit set to 1, and by 2 when the node finds both a data slot with *BUSY* bit set to 1 and a request slot with *REQ* bit set to 1. At the end of a slot, the counter is decremented by one provided that the counter value is not zero. Thus, the rate at which the counter is incremented is $\Sigma_{j \neq i}\gamma_j$ and the rate at which it is decremented is 1. Therefore, the fraction of slots during which the value of the counter is zero is $1 - \Sigma_{j \neq i}\gamma_j$. If the counter value is zero immediately after the beginning of a slot, then if the local queue of the node is not empty, then the node transmits a request with probability α. Thus, the fraction of slots during which the node transmits requests is $\alpha[1 - \Sigma_{j \neq i}\gamma_j]$.

Queueing of remote requests, local segments, and operation of combined queue are as described earlier. Under rate control, all properties of the modified DQDB protocol described in Section 4 hold good. Thus, with rate control, the modified MAC protocol results in throughput fairness at heavy load with bounded access delay for the data bus. This kind of bounded delay along with throughput fairness is important in networks that have to support a wide variety of traffic with different performance requirements. For example, an interactive application generates short messages with stringent delay requirements, while a file transfer application generates long messages for which the carried throughput is more important than the delay.

6 REQUEST BUS ACCESS ANALYSIS

The behaviour of the local queue at a node depends largely on the nature of the traffic offered to the node. The performance of the local queue is discussed in this section assuming that the segment arrival process to node N_i ($1 \leq i \leq M - 1$) for the data bus is Poisson with arrival rate λ_i segments/slot, independent of the arrival processes to the other nodes. It is assumed that $\Sigma_{i=1}^{M-1}\lambda_i < 1$. All segments are of equal length and each segment requires a transmission time of one slot duration. Since a node has priority over its upstream nodes in accessing the request bus, the request bus access mechanism can be modelled as a single server slotted priority queueing system with $M - 1$ priority classes.

Since the arrival processes are Poisson and the service time of each request is deterministic (equal to unity), the request bus access mechanism is modelled as a slotted $M/D/1$ non-preemptive priority queueing system with $M - 1$ priority classes. Requests from node i ($1 \leq i \leq M - 1$) belong to class $M - i$ and class i requests have priority over class $i + 1$ requests. The superposition of a number of independent Poisson processes is a Poisson process with rate equal to the sum of rates of the individual processes.

A slotted $M/D/1$ non-preemptive priority queue with two classes of customers is analysed in Section 6.1, and the results are used in Section 6.2 to understand the performance of local queues.

6.1 Slotted $M/D/1$ Non-preemptive Priority Queue

Consider a slotted $M/D/1$ non-preemptive priority queue with two classes of arrivals. The arrival processes are Poisson with rates λ_1 and λ_2 customers per slot respectively. Assume that $\lambda_1 + \lambda_2 < 1$. Customers of each class require a service time of one slot duration, and service begins only at slot boundaries. Class 1 customers have non-preemptive priority over class 2 customers. Let $E[W_i]$ be the average queueing delay of a class i customer and $E[Q_i]$ be the average number of class i customers in the queue (this does not include the customer in service). The slotted $M/D/1$ queue is a queue with vacations, where each vacation period is equal to one slot duration.

When there is only one class of customers, the average queueing delay of a customer in an $M/G/1$ queue with vacations is (Bertsekas, 1992) (p. 192)

$$E[W] = \frac{R'}{1-\rho}$$

where R' is the mean residual time for completion of service or vacation in progress when a customer arrives, and ρ is the utilisation.

Therefore, in the slotted $M/D/1$ priority queue, $R' = 1/2$ and hence the average values of the queueing delay of a class 1 customer and the number of class 1 customers in the queue are

$$E[W_1] = \frac{1}{2(1-\lambda_1)} \quad and \quad E[Q_1] = \frac{\lambda_1}{2(1-\lambda_1)}.$$

The queueing delay experienced by a customer of class 2 has the following three components:

(i) The residual time for completion of service or vacation in progress when the customer arrives.

(ii) Delay due to the customers of its own priority and above that arrived earlier.

(iii) Delay due to the arrivals of higher priority that arrive while the class 2 customer is waiting.

Thus,

$$E[W_2] = R' + E[Q_1] + E[Q_2] + E[W_2]\lambda_1.$$

From Little's law, $E[Q_i] = \lambda_i E[W_i]$. Therefore,

$$E[W_2] = \frac{1}{2(1-\lambda_1)(1-\lambda_1-\lambda_2)} \quad and \quad E[Q_2] = \frac{\lambda_2}{2(1-\lambda_1)(1-\lambda_1-\lambda_2)}.$$

The z-transform of the number of class 1 customers in the system is (Mukherji, 1986)

$$G_1(z) = (1-\lambda_1)[\frac{z-1}{ze^{-\lambda_1(z-1)}-1}].$$

From the superposition property of Poisson processes, the average queueing delay of any customer in the queue is given by $E[W_1]$ with λ_1 replaced by $\lambda_1 + \lambda_2$. Again, the z-transform of the total number of customers in the system is given by $G_1(z)$ with λ_1 replaced by $\lambda_1 + \lambda_2$.

6.1.1 Inter-Departure Time Distribution for Class 1 Customers

Let D_1 be the random variable denoting the inter-departure time in slots of class 1 customers. Let r_1 be the probability that a customer of class 1 entering for service leaves behind at least one class 1 customer in the queue. Then,

$$\begin{aligned} Prob\{D_1 = 1\} &= r_1 + (1-r_1)(1-e^{-\lambda_1}), \\ and\ for\ d \geq 2, & \\ Prob\{D_1 = d\} &= (1-r_1)(1-e^{-\lambda_1})e^{-(d-1)\lambda_1}. \end{aligned}$$

In steady state $E[D_1] = 1/\lambda_1$. From this, the value of r_1 is obtained as

$$r_1 = \frac{\lambda_1 - (1 - e^{-\lambda_1})}{\lambda_1 e^{-\lambda_1}}$$

and hence

$$\begin{aligned} Prob\{D_1 = 1\} &= \frac{\lambda_1 - (1-\lambda_1)(1 - e^{-\lambda_1})}{\lambda_1}, \\ \textit{and for } d \geq 2, & \\ Prob\{D_1 = d\} &= \frac{(1 - e^{-\lambda_1})^2 e^{-(d-2)\lambda_1}(1-\lambda_1)}{\lambda_1}. \end{aligned}$$

The distribution of the inter-departure time between any two consecutive departures has the same distribution as D_1 with λ_1 replaced by $\lambda_1 + \lambda_2$.

6.1.2 Distribution of Server Idle Period With Respect to Class 1 Customers

Let I_1 be the random variable denoting the length of this idle period in slots. For $i \geq 1$,

$$\begin{aligned} Prob\{I_1 = i\} &= Prob\{D_1 = i + 1/D_1 \geq 2\} \\ &= \frac{Prob\{D_1 = i+1\}}{Prob\{D_1 \geq 2\}} \\ &= (1 - e^{-\lambda_1})e^{-(i-1)\lambda_1}. \end{aligned}$$

Therefore, I_1 is geometrically distributed with mean $1/(1 - e^{-\lambda_1})$ and variance $1/(1 - e^{-\lambda_1})^2$. The distribution of server idle period irrespective of customer class is the same as that of I_1 with λ_1 replaced by $\lambda_1 + \lambda_2$.

6.1.3 Distribution of Server Busy Period With Respect to Class 1 Customers

Let B_1 be the random variable denoting the length of this busy period in slots. Let T_1 be the time elapsed between two successive slots in which the server does not serve any class 1 customer. Then,

$$\begin{aligned} Prob\{T_1 = 1\} &= e^{-\lambda_1}, \\ \textit{and for } u \geq 2, & \\ Prob\{T_1 = u\} &= Prob\{u-1 \textit{ class } 1 \textit{ arrivals in } u-1 \textit{ consecutive} \\ &\quad \textit{slots with at least } i \textit{ arrivals in the} \\ &\quad \textit{first } i \textit{ slots together } \forall i,\ 1 \leq i \leq u-1, \\ &\quad \textit{and no class } 1 \textit{ arrivals in the } u\textit{th slot}\} \\ &= P(\lambda_1, u-1)e^{-\lambda_1} \end{aligned}$$

where $P(\lambda_1, u-1)$ is the probability of $u-1$ Poisson arrivals of rate λ_1 in $u-1$ consecutive slots with at least i arrivals in the first i slots together $\forall i \in \{1, 2, \ldots, u-1\}$. Thus, for

$b \geq 1$,

$$\begin{aligned} Prob\{B_1 = b\} &= Prob\{T_1 = b + 1/T_1 \geq 2\} \\ &= \frac{Prob\{T_1 = b+1\}}{Prob\{T_1 \geq 2\}} \\ &= \frac{P(\lambda_1, b)}{e^{\lambda_1} - 1}. \end{aligned} \tag{2}$$

Lemma 2 *$P(\lambda, N) = e^{-N\lambda}\lambda^N C(N)$ where*

$$C(N) = \Sigma_{\{(a_1,a_2,\ldots,a_N):\ i \leq \Sigma_{k=1}^{i} a_k \leq N,\ \ i=1,2,\ldots,N\}} \frac{1}{a_1!\ a_2!\ \ldots\ a_N!}.$$

Further, $C(N)$ can be obtained from the recursion

$$C(N') = \frac{1}{N'!} - [\Sigma_{j=1}^{N'-1} \frac{(-j)^{N'-j}C(j)}{(N'-j)!}], \quad N' \geq 2,$$

with $C(1) = 1$.

Proof:
Let A_k be the random variable that denotes the number of Poisson arrivals in the interval $(k-1, k]$. Let $S = \{(a_1, a_2, \ldots, a_N) :\ i \leq \Sigma_{k=1}^{i} a_k \leq N,\ \ i = 1, 2, \ldots, N\}$.

$$\begin{aligned} P(\lambda, N) &= Prob\{i \leq \Sigma_{k=1}^{i} A_k \leq N,\ \ i = 1, 2, \ldots, N\} \\ &= \Sigma_{\{(a_1,a_2,\ldots,a_N)\in S\}} Prob\{A_1 = a_1, A_2 = a_2, \ldots, A_N = a_N\} \\ &= \Sigma_{\{(a_1,a_2,\ldots,a_N)\in S\}} Prob\{A_1 = a_1\}Prob\{A_2 = a_2\}\ldots Prob\{A_N = a_N\} \\ &= \Sigma_{\{(a_1,a_2,\ldots,a_N)\in S\}} \frac{e^{-\lambda}\lambda^{a_1}}{a_1!}\frac{e^{-\lambda}\lambda^{a_2}}{a_2!}\cdots\frac{e^{-\lambda}\lambda^{a_N}}{a_N!} \\ &= e^{-N\lambda}\lambda^N \Sigma_{\{(a_1,a_2,\ldots,a_N)\in S\}} \frac{1}{a_1!\ a_2!\ \ldots\ a_N!} \end{aligned}$$

From Equation 2,

$$\Sigma_{N=1}^{\infty} \frac{P(\lambda, N)}{e^{\lambda} - 1} = 1.$$

Hence,

$$\begin{aligned} \Sigma_{N=1}^{\infty} P(\lambda, N) &= e^{\lambda} - 1, \\ i.e.,\ \lambda e^{-\lambda}C(1) + \lambda^2 e^{-2\lambda}C(2) + \ldots &= \frac{\lambda}{1!} + \frac{\lambda^2}{2!} + \frac{\lambda^3}{3!} + \ldots. \end{aligned}$$

Equating the coefficient of $\lambda^{N'}$ on both sides,

$$\begin{aligned} C(N') &= \frac{1}{N'!} - [\Sigma_{j=1}^{N'-1} \frac{(-j)^{N'-j}C(j)}{(N'-j)!}], \\ C(1) &= 1. \end{aligned}$$

□

The distribution of server busy period irrespective of customer class is the same as that of B_1 with λ_1 replaced by $\lambda_1 + \lambda_2$.

Since $E[I_1] = 1/(1 - e^{-\lambda_1})$ and $E[B_1]/E[I_1] = \lambda_1/(1 - \lambda_1)$, $E[B_1] = \lambda_1/(1 - \lambda_1)(1 - e^{-\lambda_1})$.

6.2 Request Bus Access

Requests from node i ($1 \leq i \leq M-1$) belong to class $M-i$ and class i requests have priority over class $i+1$ requests. Let $E[W_i]$ be the average request bus access delay of a request from node i and $E[Q_i]$ be the average number of requests in the local queue of node i (this does not include the request in service). The request bus access delay is measured in units of slots.

Since the superposition of a number of independent Poisson processes is a Poisson process with rate equal to the sum of rates of the individual processes, the effect of the requests from the downstream nodes on the requests from node i is that of a single class of Poisson arrivals with arrival rate $\Sigma_{j=i+1}^{M-1}\lambda_j$ and with higher priority.

6.2.1 Request Bus Access Delay

From the analysis of a slotted $M/D/1$ non-preemptive priority queue (Section 6.1), the average values of the queueing delay of a request from node N_i and of the number of requests in the queue at node N_i are

$$E[W_i] = \frac{1}{2(1-\Sigma_{j=i}^{M-1}\lambda_j)(1-\Sigma_{j=i+1}^{M-1}\lambda_j)} \quad \textit{and} \quad E[Q_i] = \frac{\lambda_i}{2(1-\Sigma_{j=i+1}^{M-1}\lambda_j)(1-\Sigma_{j=i}^{M-1}\lambda_j)}.$$

The average request bus access delay at a node is higher compared to that at any downstream node. In the example considered for Table 1, calculations show that though the data bus access delay decreases in the upstream direction, it does not compensate for the increase in request bus access delay. The z-transform of the total number of requests in the local queues of nodes $N_i, N_{i+1}, \ldots, N_{M-1}$ together is

$$G_i(z) = (1-\Sigma_{j=i}^{M-1}\lambda_j)[\frac{z-1}{ze^{-\Sigma_{j=i}^{M-1}\lambda_j(z-1)}-1}].$$

6.2.2 Inter-departure Time Distribution of Requests

Let D_{i+1} be the random variable denoting the inter-departure time of requests from nodes $N_{i+1}, \ldots, N_{M-1}$ seen by N_i. Then, from the analysis of a slotted $M/D/1$ non-preemptive priority queue (Section 6.1),

$$\begin{aligned} Prob\{D_{i+1}=1\} &= \frac{\lambda-(1-\lambda)(1-e^{-\lambda})}{\lambda}, \\ &\textit{and for } d \geq 2, \\ Prob\{D_{i+1}=d\} &= \frac{(1-e^{-\lambda})^2 e^{-(d-2)\lambda}(1-\lambda)}{\lambda} \end{aligned}$$

where $\lambda = \Sigma_{j=i+1}^{M-1}\lambda_j$.

6.2.3 Distribution of Idle Period on the Request Bus

Let I_{i+1} be the random variable denoting the period during which N_i does not find requests from nodes downstream of it. For $a \geq 1$,

$$\begin{aligned} Prob\{I_{i+1}=a\} &= Prob\{D_{i+1}=a+1/D_{i+1}\geq 2\} \\ &= \frac{Prob\{D_{i+1}=a+1\}}{Prob\{D_{i+1}\geq 2\}} \\ &= (1-e^{-\lambda})e^{-(a-1)\lambda} \end{aligned}$$

where $\lambda = \Sigma_{j=i+1}^{M-1}\lambda_j$. I_{i+1} is geometrically distributed with mean $1/(1-e^{-\lambda})$ and variance $1/(1-e^{-\lambda})^2$.

6.2.4 Distribution of Busy Period on the Request Bus

Let B_{i+1} be the random variable denoting the length of the period in which N_i finds requests from nodes downstream of it. From the analysis of a slotted $M/D/1$ non-preemptive priority queue (Section 2.6.1.3), for $b \geq 1$,

$$Prob\{B_{i+1} = b\} = \frac{P(\lambda, b)}{e^{\lambda} - 1}$$

where $\lambda = \Sigma_{j=i+1}^{M-1}\lambda_j$, and $P(\lambda, b)$ can be obtained using Lemma 2.

Since $E[I_{i+1}] = 1/(1 - e^{-\lambda})$ and $E[B_{i+1}]/E[I_{i+1}] = \lambda/(1-\lambda)$, $E[B_{i+1}] = \lambda/(1-\lambda)(1-e^{-\lambda})$.

7 CONCLUSION

A modified distributed queueing MAC protocol is studied in this paper. The access delay of a packet at a node consists of two components (i) request bus access delay and (ii) data bus access delay. An upper bound on the data bus access delay is obtained that is independent of the offered traffic. The data bus access delay bound decreases in the upstream direction. Simulations show that the average data bus access delay can be much less than the upper bound even for high bus utilisation. At heavy load, by applying a rate control similar to the DQDB bandwidth balancing mechanism, it is possible to achieve equal throughputs with bounded data bus access delay. This is useful when the network has to support different traffic with guaranteed delay and throughput requirements. For Poisson offered traffic, an analytical expression is obtained for the expected request bus access delay, the value of which increases in the upstream direction.

ACKNOWLEDGEMENTS

This work was carried out in the ERNET Laboratory. The first author was supported by a CSIR Senior Research Fellowship.

REFERENCES

Bertsekas, D. and Gallager R. (1992) Data Networks (2nd. ed.). Prentice-Hall International.

Hahne, E.L., Choudhury, A.K. and Maxemchuk N.F. (1992) DQDB Networks with and without Bandwidth Balancing *IEEE Trans. Commn.*, **40**, 1192-1204.

IEEE 802.6 Working Group (1990), IEEE standard 802.6: Distributed Queue Dual Bus (DQDB) Subnetwork of a Metropolitan Area Network (MAN), *Final Draft D15, approved by the IEEE Standards Board on Dec. 6.*

Mukherji, U. (1986) A Schedule-based Approach for Flow-control in Data Communication Networks, *Doctor of Science Thesis, Massachusetts Institute of Technology.*

Newman, R.M., Budrikis, Z.L. and Hullet, J.L. (1988), The QPSX MAN, *IEEE Commn.*, **26**, 20-8.

Pillai, R.R. (1993) Performance of Multiaccess Dual Slotted Unidirectional Bus Networks, *Ph.D dissertation, Indian Institute of Science, Bangalore.*

Potter, P.G. and Zukerman M. (1991) Analysis of a Discrete Multipriority Queueing System Involving a Central Shared Processor Serving Many Local Queues, *IEEE Jl. on Selected Areas in Commn.*, **9**, 194-202.

Stock, Th. and Tran-Gia (1990) A Discrete-time Analysis of the DQDB Access Protocol with General Input Traffic *Proc. IEEE Int. Telecom. Symposium, ITS'90*, Rio deJaneiro.

Tran-Gia, P. and Stock T. (1989) Approximate Performance Analysis of the DQDB Access Protocol, *ITC Specialist Seminar*, Adelaide, Paper No. 16.1.

Tran-Gia, P. and Stock T. (1990) Modelling of the DQDB Access Protocol and Closed-Form Approximation, *Proc. NATO Advanced Research Workshop on Architecture and Performance of High-Capacity Local and Metropolitan Area* Networks, Sophia Antipolis, France, 253-65.

Zukerman, M. and Potter P. (1989) The DQDB Protocol and its Performance Under Overload Conditions, *ITC Specialist Seminar*, Adelaide, Paper No. 16.4.

Dr. R. Radhakrishna Pillai received the B.Tech degree in Electronics and Communication Engineering from the University of Kerala in 1987 and the M.E degree in Electrical Communication Engineering from the Indian Institute of Science, Bangalore, in 1989. He was awarded the Ph.D degree by the Indian Institute of Science, Bangalore, in 1993. Since July 1993 he has been working in the R&D group of Tata Elxsi (India) Ltd., Bangalore, and is involved in the development of networking and multimedia products. His area of research is communication networks.

Dr. Utpal Mukherji received the B.Tech. degree in electrical engineering from the Indian Institute of Technology, Bombay, in 1980, and the S.M., E.E., and Sc.D. degrees in electrical engineering and computer science from the Massachusetts Institute of Technology in 1982, 1984, and 1986, respectively. He was a Member of Technical Staff at AT&T Bell Laboratories, Murray Hill, from 1986 to 1989. Since 1989, he has been an Assistant Professor in the Department of Electrical Communication Engineering at the Indian Institute of Science, Bangalore. His research interests are in communication networks.

6

Implementation of an IEEE 802.6 Compliant Card for the ISA Bus

Lakshman Narayanswamy, Jasvinder Singh, Keerthi Mitra
Umesh Kulkarni, Anurag Kumar, Utpal Mukherji
ERNET Project
Department of Electrical Communication Engineering
Indian Institute of Science
e-mail:{laks,jassi,umesh,anurag,utpal}@ece.iisc.ernet.in

Abstract

This paper describes an implementation of an IEEE 802.6 compliant, DQDB card for the PC-AT bus. We discuss the various issues involved in design of the card and describe the various functional blocks of the card in detail. We also discuss how a DQDB network can be used as a distributed multiplexer for an ATM link.

Keywords

IEEE802.6, DQDB, MAN, ISA, ATM multiplexer, high-speed network

1 INTRODUCTION

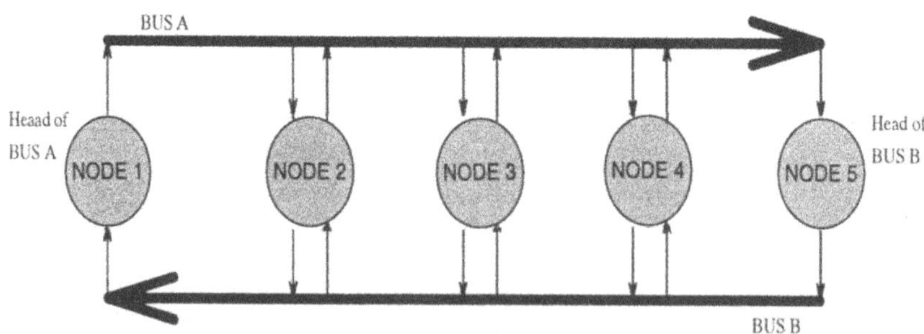

Figure 1: DQDB Network Topology.

Rapid growth of local area networks (LAN) which span a limited distance have given rise to the need for interconnecting LANs. Metropolitan area networks (MAN) that span distances greater than that of LAN and can meet the bandwidth requirements for the interconnection of many LANs are ideally suited to meet this need. The IEEE 802.6 MAN standard for a Distributed Queue Dual Bus (DQDB) network defines a high speed common channel access protocol over dual unidirectional buses. The topology of the DQDB network is as shown in Fig. 1 . We briefly describe the media access protocol. With reference to Bus A and any of the intermediate nodes in Fig. 1, nodes to the left are called upstream nodes and nodes to the right are called downstream nodes. With respect to Bus B, it is vice-versa. The structure of the DQDB cell is as shown in Fig. 2. The head of bus node for both buses generates cells at regular intervals that are sized to transport fifty-three octets. The MAC protocol uses the Busy bit and the Request bit of the Access Control Field to gain access to the bus. We describe the procedure for a node to transmit data on Bus A. When a node has cells to transmit on bus A, it sets the request bit of a cell passing on Bus B. This set request bit is noted by all downstream nodes by which the cell passes and a counter which indicates the number of requests that have passed by on bus B is incremented. This is the *request counter*; at a given node the count value held by this counter indicates the number of outstanding requests made by upstream nodes. To transmit a cell, the node copies the value of the *request counter* into a *countdown counter*. The value held by the *countdown counter* indicates the number of outstanding requests that have to be serviced before the node can transmit its cell. For every cell that passes by on Bus A without the busy bit set, the value of the *countdown counter* is decremented. When the *countdown counter* value reaches zero, the node transmits in the next available free cell. A node at any given time can have only one outstanding request per bus. Detailed description of the DQDB protocol are presented in [Newman *et al.*,1988], [Newman *et al.*,1986], [DQBD Draft]. Reference [Pillai *et al.*] reviews the performance of the DQDB MAC protocol.

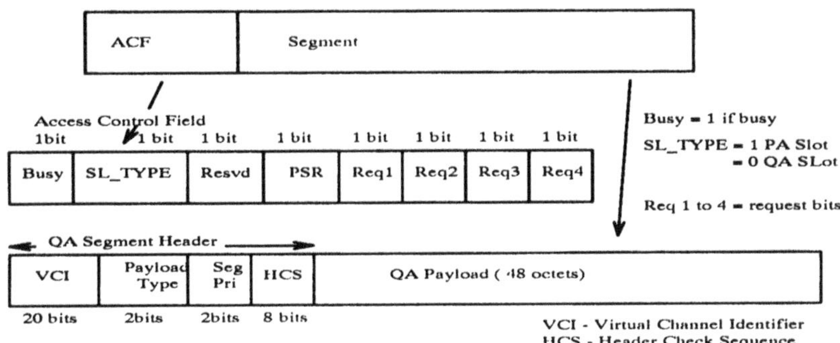

Figure 2: DQDB cell structure.

DQDB provides for a range of services that include connectionless data transfer, connection oriented data transfer and isochronous services. DQDB provides these services by using the Queued Arbitrated (QA) and Pre-Arbitrated (PA) access methods. The QA access provides for the asynchronous transfer of cells using the distributed queue access protocol. The QA function supports four priority queues for the connectionless and connection-oriented data services. The PA function provides for connection oriented transfer over a guaranteed bandwidth. The PA functionality requires the establishment of a connection, for which a *Virtual Connection Identifier* (VCI) value is assigned.

We have undertaken to prototype a DQDB network for attractive reasons. This exercise is part of an overall effort at building a broadband networking testbed. The DQDB protocol provides a degree of fairness in medium access by use of a distributed queue, and offers a mechanism for supporting isochronous services. The second attractive feature of DQDB is the close alignment of its cell structure with that of the Asynchronous Transfer Mode (ATM) cell structure. The goal of accommodating diverse source characteristics in B-ISDN is to be met using ATM. DQDB can be used to provide the function of multipoint access to an ATM link.

1.1 Related Work

Reference [IEEE JASC] is devoted to high-speed computer-network interfaces. Reference [Davie] and [Traw *et al.*] discuss the design of an ATM network interface for high speed workstations, the lessons are relevant to the design of the DQDB network interface, as the cell structure of DQDB and ATM are similar. Paper [Davie] describes a network interface for a 622 Mb/s ATM link using microprocessors to implement transmit and receive functions. This card is designed for a DEC workstation whose I/O bus supports a peak bandwidth of 800 Mb/s. Reference [Traw *et al.*] describes a network interface for connecting to a SONET based ATM link at the rate of 155 Mb/s. This interface is designed for a RISC 6000 workstation which supports a peak I/O bus bandwidth rate of 320 Mb/s. This implementation is done using dedicated hardware. Segmentation and reassembly is done by the network interface card so that the number of buffer copies required is reduced and the high I/O bus bandwidth is effectively utilized.

This paper is organized as follows. Section 2 discusses the issues that influenced the design of the DQDB add-on card. Section 3 describes the various blocks that make up the DQDB card implementation and Section 4 charts the future direction that is planned for the project. Section 5 concludes the paper.

2 DESIGN ISSUES

Within the framework of the OSI reference model, the IEEE 802.6 standard [DQBD Draft] defines the distributed queue protocol and a convergence function (also referred to as DQDB layer) that provides connectionless MAC service access point to the Logical Link Control sublayer. The VCI value provides a means to identify the service to which the cell belongs. The VCI value of all bits set to one is the default value for the connectionless MAC service provided by the convergence function. The DQDB layer receives the Mac Service Data Unit (MSDU) for transmission from the LLC sublayer and forms an Intermediate Mac Protocol Data Unit (IMPDU) by the addition of a header and trailer. The important fields of the header are the source address, destination address, length and a unique tag. The trailer is made up of the same length and the unique tag. The IMPDU is segmented into fifty-two octet cells which is made up of four octets of QA header, four octets of DMPDU header and trailer and forty-four octets of payload. Fig. 3 diagrams the segmentation process which is described in section 3. On the receiving node the DQDB layer receives fifty-two octets from the bus, checks to see if the destination address matches and reassembles the forty-four octet payloads to form an IMPDU which is handed to the LLC sublayer.

The service offered by a network interface adapter to the host is dictated by the host architecture. For maximum flexibility of the network adapter card, the obvious design choice is to use a microprocessor to implement the card functionalities. After having carefully studied this option, we find that none of the available CISC processor or microcontrollers satisfied the timing requirements. In a typical microprocessor based design, the processor is informed of external events (e.g., arrival of a new cell) by means of a hardware interrupt. In such a design, the interrupt latency time of the microprocessor should be less than the interarrival time of the external event (in a 43.776Mbps DQDB network, the cells interarrival time is 10.42 μs) in order that all events be processed. We find locally available microprocessors and microcontrollers are not suitable as their interrupt latency time is in the order of milliseconds. Hence we have implemented all time critical functionalities in hardware.

The issue of whether the DQDB layer should be implemented on the card or on the host can be resolved by considering the capabilities of the host. The DQDB card that we describe is designed for the PC-AT bus. The PC-AT platform was chosen for its ease of availability at affordable cost. The main constraints while designing for the PC-AT bus are the maximum data rate of the bus, mapping of the card memory to the host memory, and the physical dimension of the card that can be accommodated by the PC-AT platform. We take the worst case condition of having to receive all the cells on both the buses and calculate the time available to process each cell. The DQDB network that we have designed operates at 43.776 Mbps. Every 125μs the head of bus node generates twelve, fifty-seven octet slots. The first four octets are management octets, the fifth octet is the Access Control Field used by the DQDB MAC. The rest of the fifty-two octets form the segment payload. At 43.776 Mbps the per slot timing is 10.41667 μs. Considering both the buses, the time available to process a slot is 5.21 μs. For the DQDB layer to be implemented on the host, the host should be able to copy the incoming cells into host memory at the maximum input rate. This requires a host bus transfer rate of (52/5.26μs)=9.886 MBps. The maximum bus transfer rate using the byte DMA mode of the PC-AT bus is 1 MByte/sec [Solari]. [1] Hence, the host is not capable of copying the cells at the maximum input rate from the buses. The other alternative is to provide buffering on the card. The incoming cells destined to the node are copied into

[1]The exact DMA throughput is a function of: the length of the instruction cycle that is interspersed with the DMA cycle, higher priority DMA request that are queued to be serviced and the DRAM refresh cycle every 15 μs. We have assumed the DMA rate to be 1 MB/s given that the maximum input clock to the DMA controller can be 5 MHz and that we use the single byte DMA transfer mode.

the buffer and the host interrupted. The host then copies the fifty-two octet cells from the buffer on the card for further processing. The disadvantage of interrupting the host for every fifty-two octet cell is the high overhead associated with the processing of a hardware interrupt.

Our design overcomes the low bus data transfer rate and high overhead associated with interrupt processing by having a buffer on the card and reassembling the received cells into IMPDU's in the buffer. The host is not interrupted for every cell, but only after the complete IMPDU has been reassembled. Our design supports a maximum IMPDU size of 6 KBytes. An important consideration in implementing the receive function of the DQDB layer is the number of IMPDU's that can be concurrently assembled. Our design provides for concurrent reassembly of five MID's and maximum IMPDU size of 6 KBytes. We feel that support for five concurrent reassemblies are sufficient for a PC-AT host running a multi-tasking operating system (e.g. UNIX) to support popular networking services (e.g. file transfer, remote login ..). Support of every additional reassembly requires additional logic and hence board space, which is a scarce resource in our implementation.

The DQDB layer transmit function accepts MSDU's from the LLC sublayer, segments them into forty-four octet payload and adds the header to form a fifty-two octet cell which is then queued for transmission. To resolve the issue of whether to implement the transmit function of the DQDB layer on the card or on the host, we examine the times involved in transmission of a IMPDU. They are, the processing time required to calculate the header and the trailer to form a fifty-two octet cell, the time required to copy data from the host memory to the card and the time the cell remains queued before it is transmitted onto the network. The third component is an attribute of the network load and not a function of the host-card architecture. The other two times are the processing delay and the time taken to copy data from the host memory to the card. The minimum time required to copy data from the host memory to the card is calculable from the maximum data rate of the PC-AT bus. Assuming a maximum I/O channel rate of 1 MB/sec, 52 octets can be transferred from the host memory to the card in 49.5 μs. A 80386 processor with a clock of 25 MHz, can execute 309 instructions in 49.5 μs if the average instruction cycle is taken to be 4 machine cycles. We argue that an optimized implementation of the segmentation process will take lesser time than that required for copy of the cell from the host memory to the card. Following the above, we can say that there is little gain in implementing the processing required for segmentation of a IMPDU on the card in hardware, since the bottleneck in using the complete bandwidth of the DQDB bus is not the CPU processing speed of the host, but its limited I/O bus speed. Our design assumes that the segmentation is done by software on the host. A nineteen cell FIFO buffer is provided on the card into which the host writes the cells to be transmitted. A significant part of the trailer calculation for every cell is the CRC. To cut down on the processing time required to calculate the CRC, we have implemented CRC calculation in hardware.

Calculation of the 8-bit and 10-bit CRC requires the 3 octet QA header and the 46.75 octet (46 octets and 6 bits) payload to be serialized and fed to the 8-

bit and 10-bit CRC generator. The transmit CRC calculates the remainder and appends 8 bits to the QA header and 10 bits to the QA payload. The receiver calculates the CRCs of the incoming cell and checks for zero remainder. CRC calculation on the transmit and the receive path are ideally suited to be done parallely as the octets are read into the transmit and receive FIFO from the host and network respectively. On the receiver, as the incoming octets are read from the network and copied into the receive FIFO, the octets are also copied to a shift register. The clock to the shift register should be at least eight times as fast as the incoming octet rate of 5.472 MBytes/s so that all eight bits of the octet are shifted to the CRC calculator before the next octet arrives from the network. On the transmitter, since the host transfers 16 bits over the I/O bus, the clock to the shift register should be at least 16 times as fast as the data transfer rate from the host to the transmit FIFO.

To reassemble five IMPDUs of 6 KBytes each, a 30 KByte buffer is required. Further, buffer for copying SSM segments has to be provided. To satisfy the above requirement we have chosen to use 32Kx8bit dualport RAM as the buffer which can accomodate five IMPDUs and forty-two SSM segments. The choice of dualport RAM is obvious since there are two functional blocks that need access to the RAM. One is the Receive block that writes the incoming cells into the RAM. The other is the host, which reads the assembled packet from the DQDB card buffer into the host memory. One option would have been to use a single port RAM and give priority to the receive block to write the incoming cells into the RAM. This strategy requires the host to wait till the receive block gives up access to the RAM and then transfer the packet from the RAM to the host memory. This method is not acceptable as it increases the host-to-card data transfer time, which is the bottleneck in our design. Use of a dualport RAM solves this problem by providing simultaneous and independent access to the RAM for both the host and the receive block.

A node on the DQDB network has access to the optical fiber bus by means of an active or passive read and write connection. Writing on to the bus then is a logical OR of data from the node and data from upstream. We have implemented the bus as segmented cables between successive nodes. The option of implementing a bus using optical couplers is not flexible for addition of nodes to the network. Each DQDB interface card has two optical receivers and two optical transmitters, one transmit and receive pair for each bus. For each bus, each node on the DQDB network has an optical fiber coming in to a receiver and and an outgoing fiber connected to the corresponding transmiter.

To optimise on the board space and to meet our timing and power requirements, we have implemented most of our hardware logic using ALTERA's Programmable Logic Devices [Altera Databook].

3 DESIGN DESCRIPTION

We have implemented the connectionless service using the QA convergence function provided by the DQDB MAC to the logical link sublayer. The functionality of the DQDB layer for the connectionless protocol is described in detail with reference to Fig. 3.

Fig. 3 shows the construction of the IMPDU from the MSDU by the addition of header and trailer. The IMPDU is then segmented into fifty-two octet DMPDUs and queued for transmission. The QA segment transports the 48 octet DMPDU as its segment payload. The receive block does the reassembly of DMPDUs to get back the IMPDU. The reassembly is done by associating one MID (message identifier) value to the entire IMPDU. This MID value is contained in the DMPDU header of every cell that contains the IMPDU's data as its payload. Hence for the interval of time that all the DMPDUs from an IMPDU are received, the MID value associated with that IMPDU is unique across the DQDB network. The first DMPDU derived from the IMPDU is the Beginning of Message (BOM). The BOM has both the destination node address and the MID value associated with the IMPDU being transferred. On the arrival of a BOM, the receive block checks for a destination address match, if there is a match, the MID value contained in the BOM cell is stored. Subsequent DMPDU constructed from the IMPDU being transferred are Continuation of Message (COM) which contain the MID value as part of the DMPDU header. If the received COM cell's MID value matches, then the cell is copied into the RAM. The last DMPDU derived out of the IMPDU is the End of Message (EOM). If the received EOM cell's MID matches with one of those being assembled, the cell payload is copied into RAM and the MID value removed from the list of MIDs being reassembled. Further, no cell containing the MID value will be copied until the receiver gets a fresh BOM with the same MID value. To maintain uniqueness of the MID value, the DQDB standard defines allocation of MID values through the management octets in the overhead part of the DS3 frame. Our design statically allocates a set of usable MID values to each node.

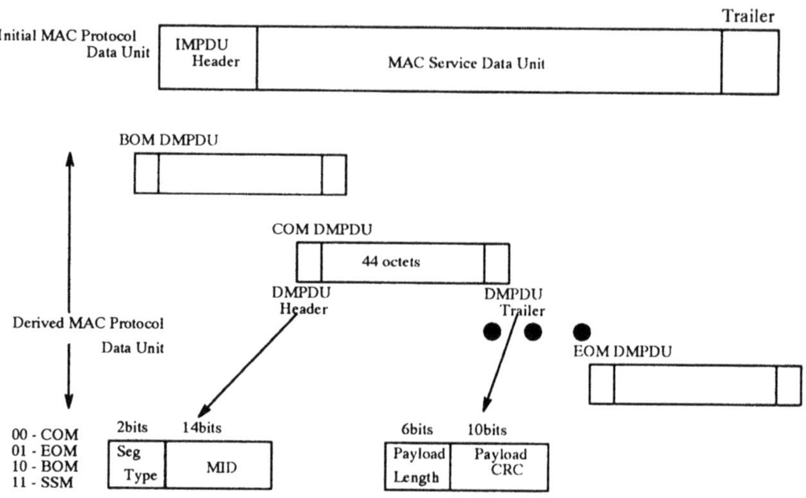

Figure 3: DMPDUs from IMPDU.

3.1 Physical Layer Block Design

This section describes the physical protocol block implementation. The block diagram of this layer for one of the buses is as shown in Fig. 4.

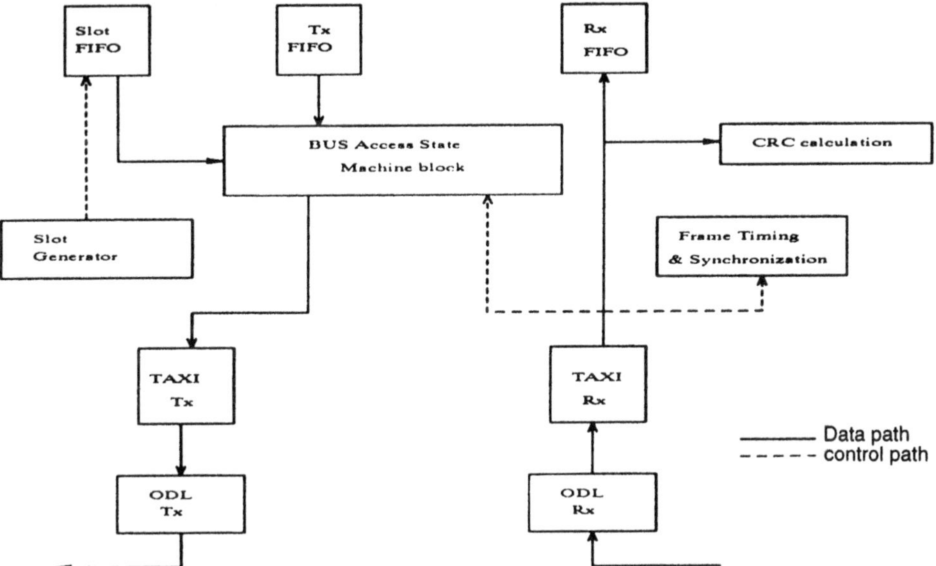

Figure 4: Physical Layer Protocol Block

The DQDB standard recommends mapping of the DQDB physical layer to the DS3 carrier. The DS3 signal consists of 699 octets per 125 μsec time period. One bit in 85 bits is used for the DS3 overhead, which leaves 690.78 octets for the information payload. We have closely alligned our physical layer to the DS3 signal, though we do not implement the DS3 overhead bits. We use a 125 μs frame containing 12 slots at the rate of 43.776 Mb/s. Each slot contains 57 octets. The information payload is 57 * 12 = 684 octets in 125 μsec, hence each slot time is 10.41667 μsec. The first two octets, A1 and A2 are slot delineating octets which are assigned hexadecimal values of F6 and 28. The next two octets are the path overhead octets. The first four octets are used by the physical layer management entity. We use the term *cell* to denote the 52 octets that follow the four management octets in a slot and the Access Control Field (ACF) whose fields are as shown in Fig. 2. We have chosen to implement only one of the four queue priorities provided by the four bit request field.

The data path in the physical layer block is traced in Fig. 4. Reception and transmission of signals to/from the optical fiber is done using AMD's **T**ransparent **A**synchronous **X**mitter Receiver **I**nterface (TAXI) [TAXI Datasheet] and AT&T's Optical Data Link (ODL) [ODL Datasheet] chips. For transmission, the transmit TAXI serializes the 8-bit parallel data at the input and outputs 10-bit ECL signal units to the Tx-ODL. For reception, the Rx-ODL converts the optical signal into ECL level electrical signal which is fed to the Rx-TAXI serial input. The output of the Rx-TAXI is the 8-bit parallel data units at TTL signal level.

The transmit TAXI does a 4/5 encoding of the data presented to its parallel input. The serial output rate of the TAXI is n times the input rate of the clock to the TAXI. The factor n depends on the width of the parallel input data to the TAXI. For 8 bits wide input data the factor n is 10. To achieve the line rate of 43.776 Mbps the input clock to the TAXI should be 43.776 * (5/4) * (1/10) = 5.472 MHz. Hence, on every rising edge of the clock of 5.472 MHz, a byte of data has to be given to the Tx TAXI. Correspondingly the receiver TAXI decodes the 4/5 encoded serial input and gives a 8 bit parallel output every 182.75 nsec.

A card can be configured as Head of Bus (HOB) or an intermediate node. For the bus for which it is the head node, the data path is from the slot FIFO to the Tx TAXI. The HOB node generates and transmits on the bus, twelve, fifty-seven octet slots every 125 μsec. As part of the initialization of the HOB node, the host software writes 12 slots of fifty-seven octets into the slot FIFO. Each of the slots starts with two framing octets and two management octets and fifty-three zero valued octets representing the fifty-three octet cell. The slot generator in the HOB node reads sucessive slots from the slot FIFO and routes the octets to the Tx TAXI for transmission. This is repeated every 125 μs providing a constant stream of slots on the DQDB bus. When the HOB node has data to transmit on the bus, the first five octets are read from the slot FIFO, the fifth octet is routed through the Bus Access State Machine block and the rest of the fifty-two octet cell is read from the Tx FIFO and written to the Tx TAXI.

Synchronization of the receiver to the start of each slot is done by the frame syncronization block. After the initialization of the card, consecutive octets are compared for the value of the A1 and A2 octets. If consecutive octets match with A1 and A2, the receiver is assumed to be synchoronized with the incoming slot. To maintain synchronization, the A1 and A2 patterns are matched at the start of every 57 octet slot.

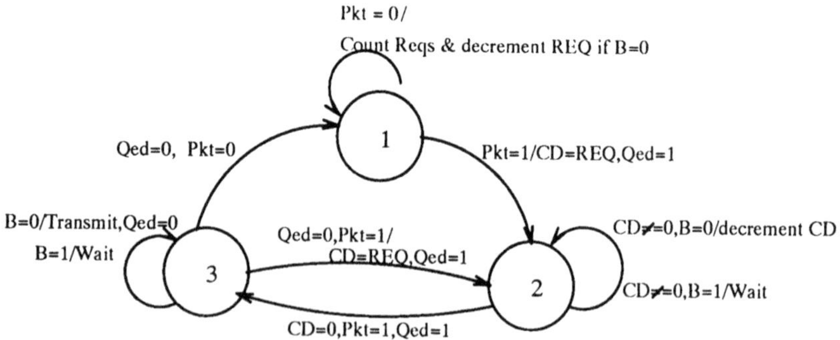

Figure 5: DQDB Bus Access State Machine

The DQDB MAC protocol is implemented as a state machine shown in Fig. 5 We describe the functioning of the state machine when a node want to transmit on bus A. When the node does not have anything to send, it is in state 1, the

IDLE state. In this state the requests going on bus B are counted and for every empty slot that passes on Bus A, the request counter is decremented.

When the node has a cell to transmit, the variable Pkt is set to one and transition to the next state take place. During this transition, the contents of the Request counter are transferred to the Count Down counter and the request counter is reset to zero. Simultaneously, the variable called Q'ed will be set to indicate that the current cell is queued to be sent and a request will be sent on bus B. If the current slot on bus B has the request bit already set, then this request will be held back and sent when the opportunity arises. Inability to send a request to transmit a cell does not prevent that cell from being queued and transmitted. In this condition the request counter will continue to count the requests on bus B but is not decremented. For every empty slot on bus A, the countdown counter is decremented. When the value of the counter reaches zero, transition to state 3 takes place.

State 3 in the state diagram is the transmit state. While in this state, if an empty slot arrives then the busy bit in the ACF is set to 1 and the cell is transmitted. If the busy bit is already set to 1, the state machine waits till the next empty slot arrives. Once the transmission is complete, the Que'd and Pkt variables are reset to zero. If there is another cell to be transmitted, then the state machine moves to state 2, otherwise it moves to state 1, i.e. Idle state. All state changes are triggered by the arrival on the bus of the fifth octet in a slot, i.e. the ACF.

3.2 Transmit Block Design

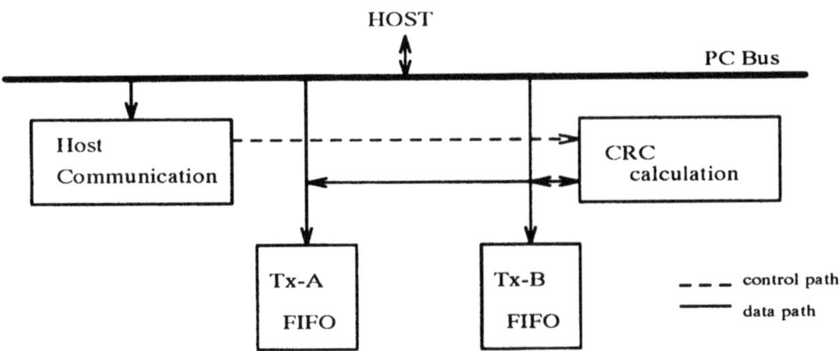

Figure 6: Transmit Block Diagram

This section describes the design of the Transmit machine of the DQDB network. Fig. 6 is the block diagram of the transmit block. As mentioned in Section 2, the DQDB layer that does the segmentation of the IMPDU is implemented in software on the host. After the DMPDU is created from the IMPDU payload, a routing decision is made as to which bus the cell is to be transmitted on. The host checks the transmit counter which holds the value of the number of cells queued to transmit for a particular bus. The 1Kx8 bit transmit FIFO (Tx-FIFO) can queue nineteen cells to be transmitted. If the transmit counter value equals nineteen, then the host has to wait for the transmit counter value to decrement, otherwise the host copies the DMPDU into the Tx-FIFO. The last ten bits of the DMPDU are the payload CRC. The host writes a 16 bit word to the I/O port on which the Tx-FIFO is mapped. The 16 bit word is split into two octets and written into the FIFO. In parallel the octets are fed to an 8 bit shift/storage register that shifts the octets out serially and feeds the bits to a CRC block for the CRC calculation. At the end of the payload, the 10 bit CRC is appended along with a 6 bit payload length into the Tx-FIFO. The transmit counter value is incremented by one to indicate the number of cells queued for transmission. We have provided the option of copying the cell octets from the host memory into the transmit FIFO using the byte mode DMA or through a programmed loop involving the CPU.

3.3 Receive Block Design

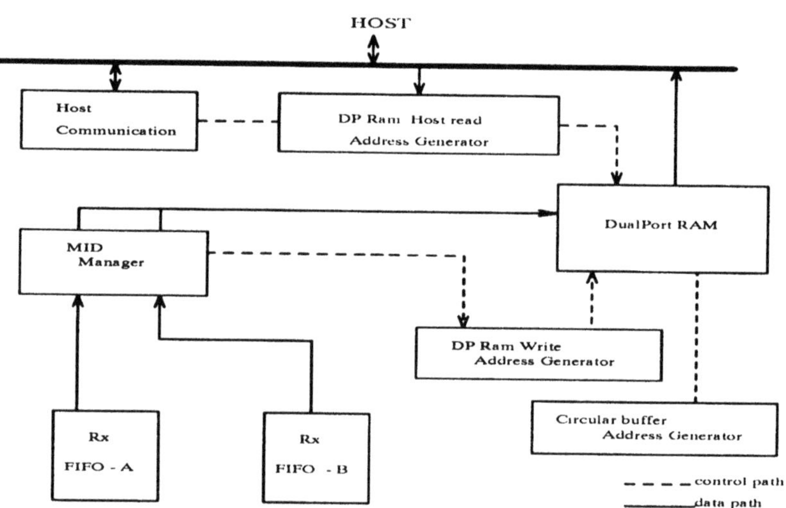

Figure 7: Receive Block Diagram.

All cells arriving on the bus with the Busy bit set to 1 are copied into the receive FIFO (Rx-FIFO) of the bus it has arrived on. The receive block's per cell processing time of 5.21 μs (half the cell time) is required to handle the maximum cell arrival rate on both the buses. After copying a complete cell from the bus into the Rx-FIFO, the physical layer block increments a counter which indicates the number of cells queued in it. The receive block reads octets out of the FIFO and processes them. The MID management block manages the MIDs that are being reassembled. Our design supports concurrent reassembly of five MIDs. The MID values being reassembled are stored in latches which are compared for an MID match with the sixth octet of a cell being read out of a Rx-FIFO. When the DMPDU segment type is BOM and if one of the MID latches is free, the MID value associated with the incoming BOM cell is copied into the available MID latch. The destination address, which is the eighteenth octet of the incoming cell, is compared for a match with the address of the node. If the address does not match, the reassembly is stopped and the MID latch into which the incoming MID value was copied is reset. Memory management of the Dual Port RAM is done by statically assigning six KBytes of dualport RAM to each MID value that is being reassembled. Hence the LLC sublayer can submit to the DQDB layer a six KByte MSDU for transmission.

The address generation block generates the address into which the received octets corresponding to a particular MID are written into the RAM. When an incoming MID value matches, it enables the address generation counter associated with that particular MID latch. Hence cells with the same MID value are copied contiguously in the RAM. After the EOM cell for a particular MID is copied into the RAM, the MID value is reset. The host is communicated by means of an interrupt when the EOM cell associated with a particular MID is copied and the complete IMPDU is now available in the RAM. On being interrupted the host reads the status register to know where in the RAM the reassembled IMPDU can be copied from.

Single Segment Message (SSM) are cells that fit a whole IMPDU within a DMPDU. On receipt of an SSM, the receiver checks the destination address (DA) of the received cell for a match. If the DA address matches and there is memory available in the RAM, the cell is copied into the dualport RAM. The transfer of SSMs from the network to the host is by using 2 KBytes of the dualport RAM as a circular buffer. The write pointer to the circular buffer is incremented when an SSM for which the DA address matches is copied into the RAM. The read address of the circular buffer moves as the hosts reads off SSMs from the dualport RAM. The *ssm-count* counter keeps count of the number of SSMs present in the circular buffer. This counter is readable by the host through a I/O port. The host is interrupted only when the maximum number of SSM cells that can be accomodated in the circular buffer is reached. The host is not interrupted for every received SSM, but is expected to find out the number of SSM cells in the buffer by polling the *ssm-count* register.

An incoming cell, whose MID value matches with one being reassembled but whose CRC does not match is discarded and the corresponding reassembly is aborted. Another event which aborts reassembly of an MID and causes an MID latch to be set free is the expiry of the MID timer. There is a timer associated with each MID being reassembled. This timer sets an upper limit on the time that can elapse between the receipt of consecutive cells with the same MID value. The timer expiry value is set to 0.7 seconds as recommended by the DQDB standard.

The host communication block facilitates communication between the host and the card. The dualport RAM memory on the card is mapped on to an I/O ports of the host. There is a read address counter on the card that the host loads by writing to a I/O port. Read operations on the I/O port serve as a clock to the read address generator, and increment the address the RAM is read from. The read address also enables the dualport RAM; hence data in the DP Ram corresponding to the address from the read address generator is available on the PC bus.

4 FUTURE WORK - DQDB AS AN ATM MULTIPLEXER

To be able to provide distributed access to a high speed ATM link a multipoint to single point access function is required. DQDB fits in very well to serve this need due to the close alignment of the DQDB and ATM cell structure. Reference [Takase *at al.*] describes an experimental B-ISDN system in which DQDB is used to multiplex cells to an ATM link. Reference [Byrne *et al.*] provides a good overview of the evolution of MAN to B-ISDN. Reference [IEEE Network] is a collection of articles on the deployment of ATM. Our effort at development

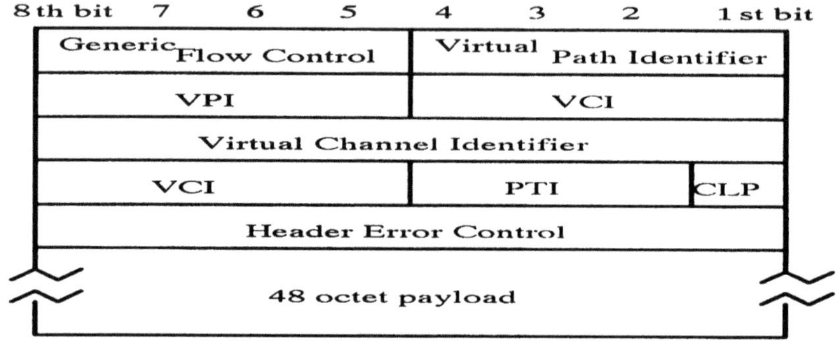

Figure 8: ATM cell structure.

of the DQDB card is proposed to be integrated into a B-ISDN test-bed, as a distributed access multiplexer. Our future work will be aimed at using DQDB to provide multipoint access to an ATM link. The ATM cell structure at the User Network Interface (UNI) is as shown in Fig. 8. A DQDB network can be connected to the T_B interface of the B-ISDN connection architecture as shown in Fig. 9. The other method would be to connect a DQDB network to an ATM switch through a terminal adapter (TA).

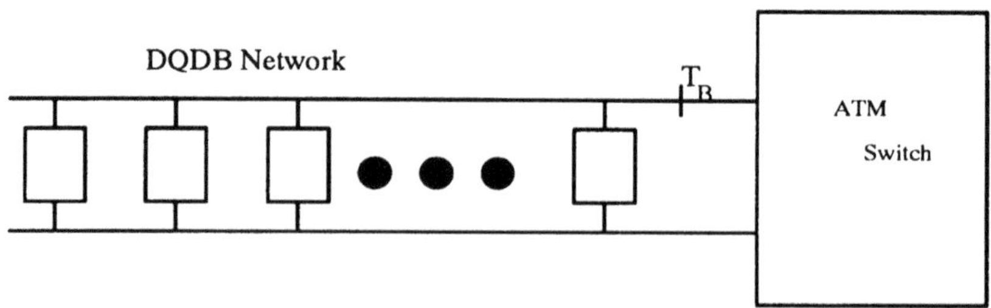

Figure 9: DQDB as an ATM link multiplexer.

To implement the first option, the DQDB and ATM headers have to match. The GFC field in the ATM header at the UNI is for providing multipoint access to a single physical link. While ATM provides for 4 bits at the UNI for the single to multipoint access protocol, DQDB requires the 8 ACF bits for the implementation of the DQDB MAC protocol. If the VPI field at the UNI is not used and a 16 bit VCI is used instead of a 20 bit VCI as defined in the DQDB header, then the direct connection of the DQDB network to a ATM switch is possible. For common VCI values to be used across the DQDB network and the ATM switch, a common signalling protocol for VCI value allocation will have to be used amongst the DQDB nodes and the ATM switch. The connection methodology shown in Fig. 9 is suitable only for data traffic. To transport isochronous traffic between a DQDB network and an ATM switch, the DQDB network will have to be connected to an ATM switch through a terminal adapter (TA). We are studying the implications and feasibility of the above two options to use DQDB as a distributed multiplexer to a ATM link.

5 CONCLUSIONS

We have described the implementation of a DQDB network interface card. Though our card is designed for the PC-AT bus, it can be easily adapted to an EISA platform that provides a higher I/O bus transfer rate. Physical space limitation of the card has restricted the functionality that the card provides. We feel that it would be better to have two physical cards with the physical layer functionality on one card and the transmit and receive block functions on the second card. The card has been debugged and its functionalities shown working. Presently device driver and transport layer software to integrate the card to the host is being developed.

ACKNOWLEDGEMENTS

We would like to thank Dr. Bharat Doshi, AT&T Bell Labs, Dr. T. R. Vishwanathan, AT&T Bell Labs Dr. S. Keshav, AT&T Bell Labs, and Dr. Inder Gopal, IBM Watson Research Center for their useful suggestions during their visit to the ERNET Project, IISc.

REFERENCES

[DQBD Draft] Project 802 - Local and Metropolitan Area Network "Proposed Standard: Distributed Queue Dual Bus (DQDB) - Draft D12 1990"

[Newman *et al.*,1988] R M Newman, Z L Budrikis and J L Hullet, "The QPSX Man", *IEEE Comm. Magazine*, April 1988.

[Newman *et al.*,1986] R. M. Newman and J. L. Hullet, "Distributed Queueing: A Fast and Efficient Packet Access Protocol for QPSX", *Proceedings of the Eighth International COnference on Computer Communication, Munich*, pp 294-299, Elsevier Science Publishers, Sept. 1986.

[Pillai *et al.*] R. R. Pillai, H. Adiseshu, U. Mukherji and V. Sharma, "Distributed Queue Dual Bus Metropolitan Area Network: A Review of Performance Analysis of the Medium Access Control Protocol," *Jl. of IETE - Special Issue on Telematics*, Vol. 39, No. 2, pp. 111-124, Mar-Apr. 1993.

[Byrne *et al.*] W. R. Byrne, G. W. R. Luderer, G. Clapp, B. L. Nelson, H. J. Kafka, "Evolution of Metropolitan Area Network to Broadband ISDN", *IEEE Comm. Magazine*, Vol. 29, No. 1, pp. 69-82, January 1991.

[IEEE Network] *IEEE Network* , Vol. 7, No. 2, March 1993.

[Takase *at al.*] A Takase, J. Yaagi, Y. Sakurai and Y. Miyamori, "An Experimental B-ISDN System for MAN Application Study", *GLOBECOM 91*, pp. 2100-2104, 1991.

[Altera Databook] Altera EPLD databook, 1993

[IEEE JASC] *IEEE Journal on Selected Areas in Communication*, 93.

[Davie] B S Davie, "The Architecture and Implementation of a High-Speed Host Interface", *IEEE Journal on Selected Areas in Communication*, Vol. 11, No. 2, pp 228-239, February 93.

[Traw *et al.*] C B Traw and J M Smith "Hardware/Software Organization of a High-Performance ATM Host Interface" *IEEE Journal on Selected Areas in Communication*, Vol. 11, No. 2, pp 240-253, Feb. 93.

[Solari] Edward Solari, "*AT BUS Design* " Annabooks, 1990.

[TAXI Datasheet] "*Am7968/Am7969 TAXI chip*" AMD Data Sheet, 1987.

[ODL Datasheet] "*ODL 200 Lightwave Data Link with Flag*" - AT&T Data Sheet, 1988.

BIOGRAPHIES

Lakshman Narayanswamy has a B.E. from the University of Bombay and a M.S. from the University of Virginia, both in Electrical Engineering. He has been with the ERNET Project since 1989, where he is involved in Network management, networking software and hardware development.

Jasvinder Singh has a B.E. from Delhi College of Engineering, University of Delhi and M.Tech. from the University of Delhi. He worked from Feburary 1991 to December 1992 as Scientist/Engineer Grade 'SC' with the Department of Electronics. He has been with the ERNET Project, Indian Institute of Science since Dec 1992 where he is involved in the ERNET's network admistration, hardware and software development.

Umesh Kulkarni is a M.E. student in the Electrical Communication Engg department, Indian Institute of Science. He got his B.E. from Walchand College of Engg, Sangli.

Keerthi Mitra got his B.E. from Malnad College of Engineering, Hassan. He worked at the ERNET Project from 1990 to August 1994. He is currently working with Deutsche Software Indian Ltd.

Anurag Kumar has a B.Tech. in E.E. from the Indian Institute of Technology at Kanpur, and a PhD from Cornell University. He was a Member of Technical Staff at AT&T Bell Labs, Holmdel for over 6 years. Since 1988 he has been with the Indian Institute of Science (IISc), Bangalore, in the Dept. of Electrical Communication Engineering, where he is now Associate Professor. He is also the Coordinator at IISc of the Education and Research Network Project, which has set up a country-wide computer network for academic and research institutions, and conducts R&D in the area of communication networks. His own research interests are in the area of modelling, analysis, control and optimisation problems arising in communication networks and distributed systems.

Utpal Mukherji received his B.Tech. degree in electrical engineering from the Indian Institute of Technology, Bombay, in 1980 and the S.M., E.E., and Sc.D. degrees in electrical and computer science from the Massachusetts Institutes of Technology in 1982, 1984, and 1986, respectively. He was a Member of Technical Staff at AT&T Bell Laboratories, Murray hill, from 1986 to 1989. Since 1989, he has been as Assistant professor in the Department of Electrical Communication Engineering at the Indian Institute of Science, Bangalore. His research interests are in communication networks.

PART TWO

Multimedia Systems and Applications

MULTIMEDIA SYSTEMS AND APPLICATIONS

Multimedia systems are fast becoming an integral part of any computing environment. Managing multiple media such as video, voice, data, image and graphics raises interesting research issues relating to storage, transmission, synchronization, multi-party interaction, multicasting, traffic prediction, traffic shaping, flow control, congestion control and so on. In this session, there are three papers addressing some of these issues.

The first paper, "Vartalaap: A Network based multimedia presentation system" by Vijay Talati and Mehndiratta describes a network based multimedia presentation system (called Vartalaap) which is a hierarchical distributed system that closely approximates a classroom environment. The second paper "Communication channels for data multicasting in multi-service networks" by Ravindran deals with a comprehensive design and implementation of a network architecture for multicasting. The last paper titled "Single and multiple frame video traffic prediction using neural networks models" by Radu Drossu, Lakshman, Zoran Obradovic and Raghavendra presents a novel approach to video traffic prediction using neural networks. Experimental results using video teleconferencing data, are also presented. Such predictive models can be useful in dealing with resource allocation problems in multimedia networks and also in shaping the traffic.

7

Vartalaap: A Network Based Multimedia Presentation System

Vijay Talati and S.L.Mehndiratta
ERNET Project
Department of Computer Science and Engineering,
IIT, Powai, Bombay, India.
Tel : +91-022-578 3177, Email : vijay, slm@iitb.ernet.in

Abstract

Vartalaap is a hierarchical distributed system for multimedia communication over a LAN, implemented to provide an environment that closely approximates a classroom, where interaction, in the form of text, graphic images and voice is two way. This paper presents some details of Vartalaap with a special reference to its support for multimedia based 2-way interaction between a "Presenter" and a "group of participants". The Presenter and the participants may be situated anywhere, as long as they are connected through a LAN. Users access the system via a workstation which has "good" graphics capabilities and built-in support for audio. Issues covered in this paper include an overview of a presentation scenario which Vartalaap models, User's view of Vartalaap, System's view of Vartalaap, and Vartalaap protocol. A status report on the current implementation of Vartalaap is also given, along with its comparison with some other systems. We conclude with a discussion on the limitations of the current implementation and directions for future work.

1 INTRODUCTION

Recent advances in networks and workstation technologies have opened up avenues for range of applications which use multimedia communication. Workstations are available which have good graphics capabilities, built-in support for audio and in some cases built-in support for video. There are many organizations which have clusters of such machines that are connected by a network, whose quality (in terms of reliability and bandwidth) is improving day by day. All these have lead to development of various systems like distance-learning systems, group decision support systems; all of which come under "groupware" or "computer supported cooperative work (CSCW) "[Engelbrat, 1988; Kling, 1991; Grudin, 1991; Bui *et al.*, 1986]. Applications like teleconferencing, co-authoring systems, telemarketing, teleshopping are typical examples [Watabe *et al.*, 1991; Robinson *et al.*, 1991].

Distance learning systems [Elmore, 1992] provide a boost to community education programs by opening up avenues to a vast cross-section of society. In large countries like India, with shortage of experts in various fields, this technology can help to provide the services of few experts to wide range of people for training and education in thrust areas such as Medicine, Software Engineering, etc. Computer based distance learning systems differ from the conventional video-based and TV-broadcast programs in that the former allows a higher level of user interaction. To be effective such systems need a good support from networks to carry multimedia traffic and from workstations to present multimedia data in real-time. In this paper, we present one such system, which is basically a network-based multimedia presentation system called Vartalaap. This is a hierarchical distributed system for multimedia communication over a LAN, implemented to provide an environment that closely approximates a class room, where interaction, in the form of text, graphic images, and voice data is two-way between a "Presenter" and a "group of Participants". The presenter and participants may be situated anywhere, as long as they are connected through a LAN. Users access the Vartalaap system via a workstation which has good graphics capabilities and built-in support for audio. It can be easily seen that this system can be used to

1. improve quality and availability of trained manpower.
2. conduct user-friendly and cost-effective group discussions on products/topics of interests. and
3. undertake computer supported cooperative work.

We assume that the required bandwidth for such applications would be available to the system.

Vartalaap relies on some of the ideas/concepts reported in [Lashkari *et al.*, 1993; Robinson *et al.*, 1991; Wallace, 1991; Little *et al.* , 1990; Elmore, 1992] and has various features such as the following :

- The system runs as an application under the well known client-server paradigm.

•It ensures the avoidance of broadcasting by using multicasting at software level . This is acheived based on ideas reported in [Lashkari *et al.*, 1993].

•The system allows multimedia based interaction between presenter and participants in a WYSIWIS (What You See Is What I See) mode.

•It takes care of synchronization constraints across different media types, using event based synchronization approach.

•It compresses both image data and audio data before transmitting over the network.

•It supports both monochrome and color (8-bit or 4-bit) displays and also allows them to talk to each other.

The rest of this paper is organized as follows. Section 2 provides an overview of a presentation scenario which Vartalaap models. Section 3 is devoted to the proposed system along with how this system provides all major functionalities required for the presentation environment. It also includes a brief overview of important packet-formats. Section 4 gives a status report on the current implementation of Vartalaap, along with its comparison with some other systems. Conclusions and future directions are given in the last section.

2 AN OVERVIEW OF A PRESENTATION SCENARIO

Let us first examine what all is involved in a real-life class-room based presentation to a group of students. A presentation of this type would consist of the following objects :

•**Presenter** Presenter is the person who is holding the floor and making the presentation to a class. The whole session would be mostly dominated by the Presenter. He/She also has control over various other objects such as projector, slides etc.

•**Other Participants** All other persons in the class participating in the presentation, are mostly passive. They listen to the Presenter and do not have much control over most of the objects involved. They can however interrupt the presentation to make queries or seek clarifications, and at times can also be given control of certain objects with permission from the Presenter.

•**Projector and Screen** The Projector and Screen together are used by the Presenter to project slides in aid of the topic in discussion. These objects are usually controlled by the Presenter.

•**Slides** The image information is stored in the form of transparencies(referred to as slides, henceforth in the paper). The Presenter owns a set of slides which he/she projects on the screen using the overhead projector. The Presenter can replace or overlap slides.

He/She can mark over slides, point out on the slides or can scribble notes over it using marker pens.

- **Marker Pens** As mentioned above, these objects are used by the Presenter for online editing of slides. These are totally controlled by the Presenter.
- **Pointing device** This object is used to draw the attention of the participants to particular part of information on the projected slide or whiteboard. This is also controlled by the Presenter.
- **Whiteboard and pens** In any conventional presentation environment, there is a common whiteboard over which usually the Presenter can write/draw using pens/chalks. This writing/drawing is visible to all participants in an online fashion.(i.e. while it is done).
- **Microphone and Speaker(s)** The audio media support would be provided through these objects. These objects are not absolutely essential though, as in a small place, one might do without it. However, for the sake of generality, we have included these objects. The microphone is totally under the Presenter's control, but can be passed onto a participant on request; the speakers may or may not be under the control of participants.
- **Video player** Some presentations also rely on video players using which the Presenter can show video clips (e.g. pre-recorded ones). This object is again controlled by the Presenter and the clips shown are visible to all participants.

Many other objects (for example, a video camera) are also used in different Presentation environments. However, in most cases, only objects mentioned above are used.

The various media used during a real-life presentation would be :

- Text : Using whiteboard or slides, the Presenter can communicate through text medium.
- Images/Graphics : Images could be shown using prepared slides and projecting them during presentation.
- Audio : Audio medium is the most natural medium in which the Presenter would communicate. Participants would also use this medium to interrupt and seek clarification.
- Video : Presenter might use this medium through video player to aid the topic of discussions.

3 VARTALAAP: A PROPOSED SYSTEM

The presentation scenario described in Section 2 has all the participants and the Presenter located at one place. Consider a variant where they are situated at different locations. Vartalaap attempts to model this new variant as closely as possible to enable users to have a "presentation" over LANs through their workstations, using text, images/graphics and audio media. A workstation is assumed to be equipped with "good" graphics capabilities and built-in support for audio.

3.1 User's View of Vartalaap:

Here we describe Vartalaap from the user's point of view and indicate how the various objects as listed in Section 2 are realized in our system. Presenter is a user with special privileges. In Vartalaap, whosoever starts the presentation using his/her workstation, becomes the Presenter. Presenter provides information on the content of presentation and its duration. Many such "Presentations" can be in progress at the same time. A user, if desires, can participate in all the "Presentations" in progress. He/She can become a participant by "joining" any of them at any time.

Each Vartalaap user gets an X-Windows [Scheifler *et al.*, 1991] based WYSIWIS(What-you-see-is-what-i-see) interface which is easy to learn and use. The interface (see Figure 1) is logically divided into following sections :

- *The Information Section:*
  This section has a window which displays the participant's photograph and some personal information such as his/her name. It also shows if the participant is currently having control of pen or microphone, using icons.
- *The Text Subwindows:*
  There is a subwindow for each participant in which he/she is allowed to key in text. This appears on every user's monitor.
- *The Shared Whiteboard:*
  This window provides a common area for all participants. This area can be visualized as a whiteboard as well as a screen mentioned in Section 2. There is a common pen controlled by the Presenter. Any participant can get the pen with permission from the Presenter. A participant holding a pen can use this common area to draw graphics, write text, load pre-stored images (slides) or cut and paste from his/her screen.
- *The Control Panels:*
  The interface has two major sets of control panels. A generic control panel for each Presentation, and a panel specific to the shared whiteboard. The generic control panel (see Figure 2) has a set of buttons which allow a participant to "CatchUp" if he has joined late, freeze the presentation for some specified time (on permission from the Presenter) using "Hold", "Leave" the presentation, and activate the audio control panel. Audio control panel allows the participant to control the record gain, play volume, and output device (speakers or jack) of the local audio device.

  The whiteboard control panel (see Figure 3) has a set of drawing tools. They allow participants to generate graphics, text and do annotations on the whiteboard. In most

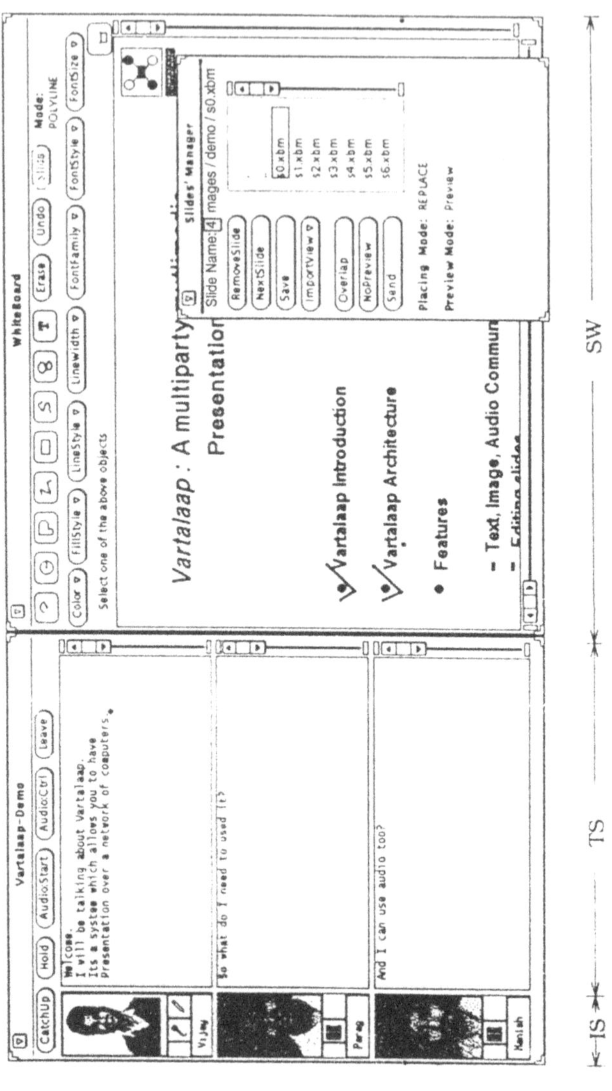

IS : INFORMATION SECTION
TS : TEXT SUBWINDOW
SW : SHARED WHITEBOARD

Figure1: The users view of Vartalaap

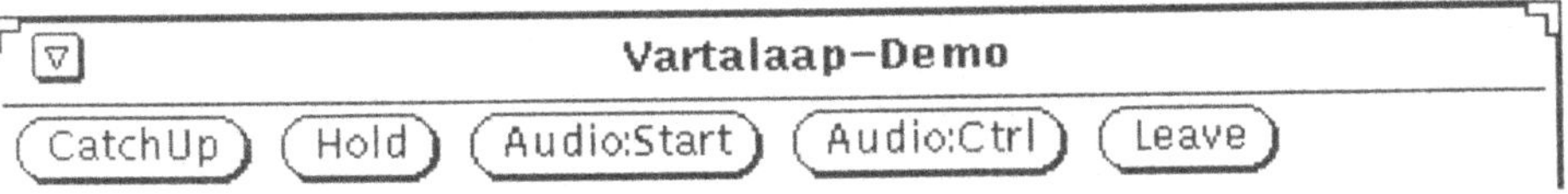

Figure 2: Generic Control Panel

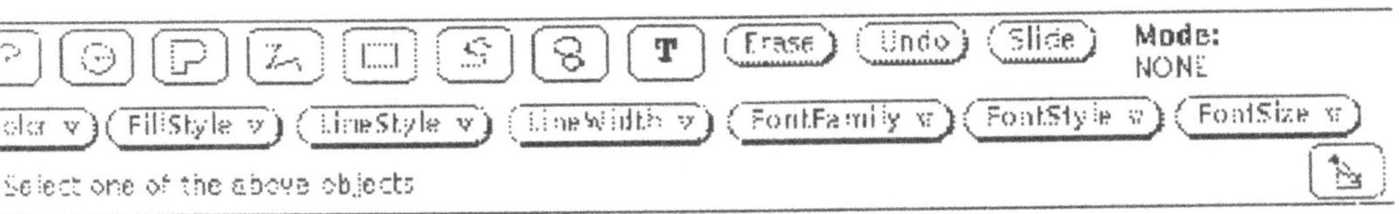

Figure 3: WhiteBoard Control Panel

of the buttons we have avoided using text-strings but have used pictorial representation in order to make the interface easily internationalizable. As the text-strings are used minimally, the language barriers need not deter the user from using the system.

A Slides' Manager panel can be activated using the "Slide" button. Slides' Manager allows to load, overlap, remove or save slides from the whiteboard. The Presenter is assumed to have a repository of slides which would aid him/her during presentation. Slides are represented as images stored as standard graphics-format files. A Slides' Manager Window on the display monitor of the Presenter shows the stack of slides to be used during the presentation scheduled. A marker indicates the slides already displayed and the slide which is next in the queue.

In Vartalaap, each workstation is assumed to have microphone/speakers. However, microphones or Audio-input devices are made totally controlled by the Presenter because of obvious reasons. On the other hand, each participant has been allowed to have control over his/her own speaker (audio-output device). Presenter can speak using his/her workstation's microphone, which is transmitted and played on the speakers of all participants.

3.2 System's View of Vartalaap:

Before giving the design of the system, we give the list of some important goals which motivated such a design. We aimed at having a system such that

•Broadcasting of messages over the network should be avoided.

•System should be run in user mode.

- It should provide support for continuous media like audio and video.
- It should be independent of the type of communication network.
- Users should be able to run the system on easily available popular kind of workstations.
- It should be robust and extensible.
- Users should be able to start using the system after a short demonstration.

In the following sections we will give the internals of Vartalaap system and discuss various design issues.

3.2.1 The System Architecture

Vartalaap uses the well known client-server paradigm. It runs over the Transmission Control Protocol/Internet Protocol (TCP/IP) stack in a LAN environment[Comer, 1991; Stevens, 1992]. Figure 4 shows the Vartalaap's architecture. The system consists of a presentation manager (server) which manages the whole system, and the user-end program - Vartalaap (client) connects to this manager. Both presenter and participants are clients to the presentation manager. To make the server more robust, we have a backup (or mirror) server which mirrors the active server. In case of server failure, the backup server takes over. This happens without the user's intervention.

3.2.2 The Vartalaap Protocol :

Vartalaap follows a simple byte-stream protocol between client and server. The protocol consists of request, reply, error and data packets. (See Figure 5 for few sample packet formats). The first field of each packet identifies the type of the packet. The remaining fields are specific to the packet type. The request and reply packet contains the requests from the clients and the replies from the server respectively. Each data-type is first packetized before sending. The data is put in data packets. Figure 6 gives the basic data transfer algorithm which involves preparing text, image and audio packets. The error packets are generated to indicate invalid requests or abnormal conditions.

3.2.3 Server internals

The presentation manager acts as a server, continuously listening at a well known port. The various functionalities of the server are as under

- Maintain the information of current presentations.
- Maintain the list of participants in each presentation.
- Route the data to appropriate nodes.
- Maintain archives of the information transferred and provide "history" on request. User may generate a "CatchUp" request when they join in late. In response to this request, the server sends the history data.

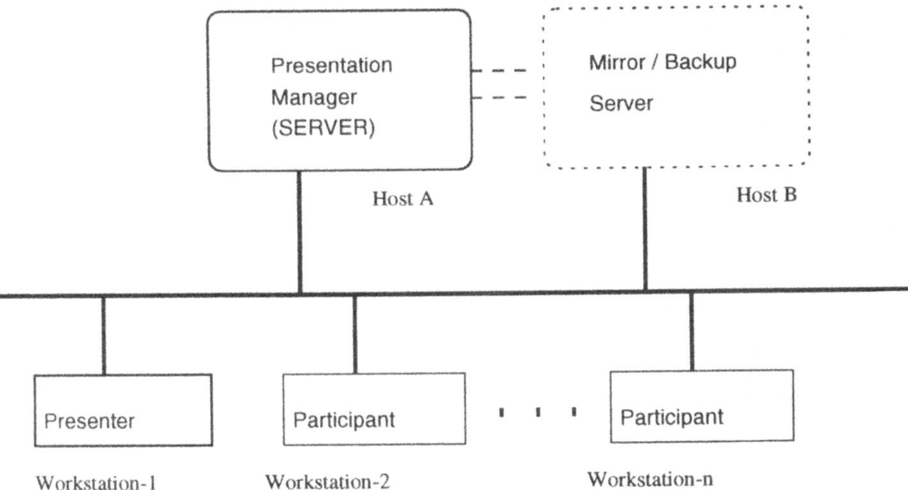

Figure 4: System Architecture

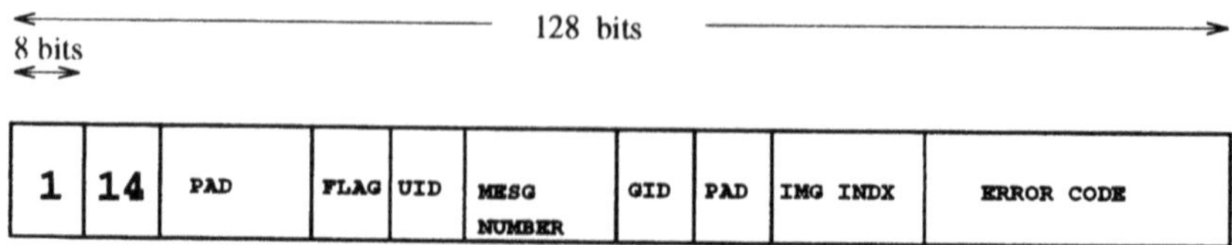

A. Make-Group Reply Packet

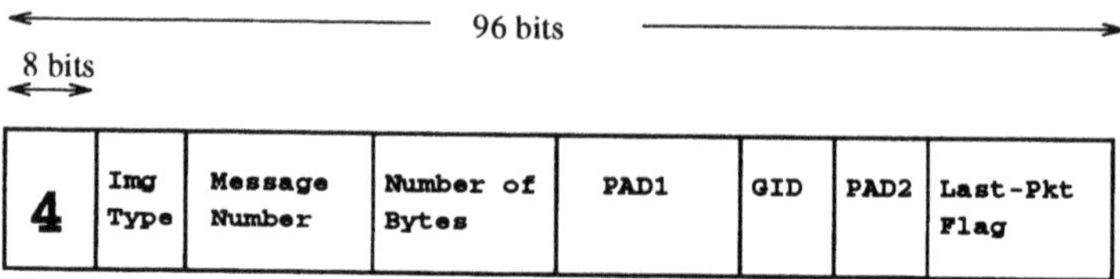

B. Send-Data Request Packet

6	GID	OPCODE	LENGTH	FROM	REQNUMBER

C. Error Packet

7	GID	SENDER	DATA

D. Data Packet

Figure 5: Sample Packet Formats

Text packet Transfer

Repeat

Read in the text data

Form the text packet (as shown in fig. 5) with group identifer and message number fileds set.

Write the packet to the text channel.

Until end-of-text-data.

Image packet Transfer

Repeat

Read the image file contents in a buffer.

Determine the image type and compress using appropriate compressor.

Divide the compressed data into fixed size packets (4K bytes currently).

Form the image packets. Set the image header, with image type , group identifier, last-packet flag values set.

Write the packet to the image channel.

Until end-of-image-file

Audio packet Transfer

Repeat

Read audio data from the audio device and store into buffer.

Do a silence detection. If silence then update the silence gap cntr and go to start of the loop.

If non-silence data, then form the audio packet.

Put a time-stamp. Set other header fileds like group identfiers, previous silence gap.

Set the silence gap count to zero.

Write the packet to audio channel.

Until end-of-audio-data.

Figure 6: Pseudocode for Data Transfer Algorithms

- Serve various other client requests. Clients generate various other miscellaneous requests like "Hold" or "PenRequest" which are to be appropriately served.

It may perhaps be noted that a lot of book-keeping is done by the server and hence the machine running as server should be a powerful one.

3.2.4 Client internals

The presentation client (Vartalaap) is the one which is visible to the user. It can be viewed in terms of two separate layers : The lower layer which communicates with the server and the upper interface layer which provides an intuitive interface for users. The lower layer is written using the set of primitives provided in the Vartalaap library similar to one in [Lashkari *et al.*, 1993]. The client functionalities can be grouped as under

- Allow users to join or start a presentation.
- Allow participants/presenter to leave a presentation.
- Allow participants/presenter to transfer and receive text/image/audio/video data.
- Provide convenient way of generating requests for server. This has been discussed in the Section 3.1 , which gives the User's View of Vartalaap.

3.2.5 Images and Compression

The slides presented by the Presenter are assumed to be in some standard graphics format. The XBM, GIF, PPM, JPEG [Wallace, 1991] are the formats available. Image files (slides) tend to be large. Hence to save on bandwidth, it is desirable to compress the image data before transmitting over the LAN.

Vartalaap compresses the image files depending on their formats (all those mentioned above are supported). Monochrome XBM files are compressed using standard unix compress, which compresses up to 10-30% depending on the content of the image. The color images (GIFs/PPMs) are compressed using the JPEG compression algorithms to achieve compression up to 10% (PPM files) to 30% (GIF files). Images are packetized into smaller chunks and sent over the network. These packets are reassembled at the recipient end.

Vartalaap supports both monochrome and color (8-bit or 4-bit) displays and also allows them to talk to each other. The necessary conversion of depth of the image is done at the recipient end. Color images are ditherized and shown on monochrome displays. On 8-bit color displays, all available colors are preallocated using 6:6:6 palette. Of course this would not reproduce the closest colors, but since we allow overlapping of slides, we could not keep dynamic color allocation[Scheifler *et al.*, 1991].

3.2.6 Audio transmission

Audio data is known to be very bulky. On SunSparcs with 8-bit u-law encoding , we get around 500Kbytes of data generated every minute. Transmitting such a large amount of data would flood the LAN . In order to control this situation in Vartalaap, we have encoded silence into non-silence audio data and transmitted only this encoded form. Audio data is also packetized and sent with time-stamps which help in synchronization.

3.2.7 Synchronization between various media:

Synchronization is an essential requirement in systems supporting a mix of time dependent and time independent data which are subjected to random delays as they travel over the network. In Vartalaap, we use a simple mechanism for this, which works without making any assumptions about underlying network or operating system. The time sequence between audio, image and text data is maintained using this mechanism. All the data is assumed to be generated from live sources, with no prior knowledge about their inter-relationship in temporal domain.

Each client does a coarse synchronization of clock with the server's clock at the start up time. During communication, each packet is tagged with a timestamp generated from local clock (clock of the data source). The recipients buffer the incoming data packets and plays these buffered packets sequencing them as per their timestamps. Thus the temporal relationship between the different streams is maintained.

The critical issue is the size of buffer required to achieve synchronization. We have proposed to determine the buffer size using simulation runs for a given traffic pattern over a live network. This approach is not only different from that in [Little *et al.* , 1990; Little *et al.* , 1991] but also found to be practical! Once the buffer size is determined, it is pre-allocated at each site and we do not require any special algorithm to determine it at runtime.

4 IMPLEMENTATION ISSUES AND RELATED WORK

Vartalaap's current version has been implemented on SunSparc Workstation running SunOS 4.1.1. The workstations are connected by an Ethernet LAN. The current version can be easily ported onto any platform which has good networking support, adequate GUIs, and devices like microphone and speakers.

Unlike some other systems like IRC [Oikarinen *et al.*, 1993], which have a set of distributed servers, Vartalaap has a single central server backed up with a mirror server. IRC, as it has distributed servers, is more fault tolerant. But because of multiple servers the messages may travel through a longer path and hence there is a possibility of unpredictable delays. IRC is text-based system so these delays are tolerable. But in multimedia based systems like Vartalaap , with the multimedia traffic , these delays can not be tolerated. So we opted for a single central server based architecture with a backup mirror server, for Vartalaap. However, in the current version of Vartalaap the mirror server has not yet been implemented.

Vartalaap does multicasting at application level using the ideas reported in [Lashkari *et al.*, 1993]. Extensions are being proposed to support multicasting at IP itself [Deering, 1989] and many systems [Jain *et al.* , 1993] propose to use IP multicasting as a base. However, this requires that all nodes on the network support the particular extension which is not being widely supported yet. Due to this we believe that the multicasting at application level as is done in Vartalaap would be beneficial at this stage.

The current version of Vartalaap supports audio communication via half duplex single audio channel. Such a support is found to be adequate in the class-room presentation scenario that Vartalaap models. But a detailed study is required to be conducted to compare its performance with those systems where multiple audio channels are used [Jain *et al.* , 1993].

Vartalaap provides a single pen for annotating, drawing and editing the slides in the WhiteBoard. Only the holder of the pen can use the drawing tools and make annotations or edit displayed slides. It is required to be explored to use two sets of annotation tools; one for the Presenter and other for the participants [Jain *et al.* , 1993]. In [Robinson *et al.*, 1991] a tray of annotations tools (called pens) are provided. But allowing more than one user drawing at the same time makes it necessary to provide some color code to each user. Vartalaap is designed so that it can work even on lower end monochrome workstations . Hence we have provided a single pen which is controlled by the Presenter. Participants can use it on permission from the Presenter, and this may avoid unnecessary confusion and bring more discipline in the presentation.

In Vartalaap, event based synchronization between different media is achieved by buffering the data at the recipient end and playing as per the timestamps (Section 3.2.7). The buffer sizes required are determined apriori using simulation runs as mentioned earlier (Section 3.2.7). In [?; Little *et al.* , 1991], petri-nets have been proposed to model the temporal relationship between various data types. However, this approach requires real-time support from the underlying operating system, bandwidth reservation capabilities from the network and run time algorithms to determine buffer requirements. The simple approach adopted in Vartalaap has produced satisfactory results on LAN without making any assumptions about the above mentioned requirements.

Because of the fixed size data structures at the server in Vartalaap and the display monitor size, its current version limits the number of participants to be 5, and the number of presentations to be 3. However these numbers can be changed and Vartalaap can be recompiled for use, but there would be no more than five information sections on the monitor at any given time.

5 CONCLUSIONS AND FUTURE DIRECTIONS

The paper has presented a network based multimedia Presentation system. This system has been modeled to handle a class-room type presentation scenario, where more interaction takes place between the presenter and participants. It can be easily seen that Vartalaap has all the features required for video-conferencing and distance-learning on LAN.

The current version of Vartalaap ensures the avoidance of broadcasting, allows multimedia based interaction between presenter and participants in a WYSIWIS mode, takes care of synchronization constraints across different media types, compresses image and audio data and supports monochrome and color displays. Furthermore, this differs from the systems based on ideas contained in [Robinson *et al.*, 1991; Oikarinen *et al.*, 1993; Little *et al.* , 1990; Little *et al.* , 1991; Jain *et al.* , 1993]. An earlier version of the system was placed in the public domain, and the feedback obtained so far has been very encouraging. Some of the suggestions which came out of this feedback, have been incorporated in the version reported in this paper.

Attempts are under way to incorporate the mirror server, multiple audio channels, and the background required for full-motion video. The latter one may require the new version of Vartalaap to run on high speed LANs like FDDI. Furthermore, an exploration is under way towards making Vartalaap suitable for distance education on WAN, using group communication standards. All this will be reported in subsequent papers.

References

[Lashkari *et al.*, 1993]Lashkari, Y., Ramachandran, V., Malpani, S., Mehndiratta, S.L. Vartalaap: a Distributed Multicast Communication System. *Software Practice and Experience Vol 23 No 27 (July 1993)* pp 799-811.

[Engelbrat, 1988]Engelbrat,D., Lehtman, H. Working Together. *Byte Vol 13 No 13 (Dec. 1988)* pp 242-252.

[Kling, 1991]Kling, R. Cooperation, Coordination and Control in Computer-Supported Work *Communications of the ACM Vol 34 No 12 (Dec. 1991)* pp 83-88.

[Grudin, 1991]Grudin, J. CSCW, Introduction *Communications of the ACM Vol 34 No 12 (Dec. 1991)* pp. 30-34.

[Bui *et al.*, 1986]Bui, T.X., Jarke, M. Communications design for co-oP: A group decision support system *ACM Trans. Off. Inform. Syst. Vol 4 No 2 (April 1986)* pp. 81-103.

[Watabe *et al.*, 1991]Watabe, K., Sakata, S., Maeno K., et. al. Distributed Desktop Conferencing System with Multiuser Multimedia Interface *IEEE Journal on Selected Areas in Communications Vol 9 No 4 (May 1991)* pp 531-539 .

[Robinson *et al.*, 1991]Robinson J., Rubinov E., Toulson C., Prasada B., et. al. A Multimedia Interactive Conferencing Application for Personal Workstations *IEEE Transactions on Communications Vol 39 No 11 (November 1991)* pp 1698-1708.

[Wallace, 1991]Wallace, G.K. The JPEG Still Picture Compression Standard *Communications of the ACM Vol 34 No 4 (April 1991)* pp 30-44.

[Comer, 1991]Comer, Douglas E. Internetworking with TCP/IP Vol 1: Principles, Protocols, and Architecture. Second Edition, Prentice-Hall International, Inc. 1991.

[Scheifler *et al.*, 1991]Scheifler, Robert W., Gettys, James X Window System , The Complete Reference to Xlib, X Protocol (X11R4) Digital Press. 1991.

[Stevens, 1992]Stevens, W. Richard Unix Network Programming Prentice-Hall of India Pvt. Ltd. 1992.

[Oikarinen *et al.*, 1993]Oikarinen, J., Reed, D. Internet Relay Chat, Internet RFC 1459 Network Information Center, (May 1993).

[Little *et al.* , 1990]Little T.D.C, Ghafoor Arif Synchronization and Storage Models for Multimedia Objects *IEEE Journal on Selected Areas in Comm. Vol 8 No 3 (April 1990)* pp 413-427

[Little *et al.* , 1991]Little T.D.C, Ghafoor Arif MultiMedia Synchronization Protocols for Broadband Integrated Services *IEEE Journal on Selected Areas in Comm. Vol 9 No 9 (December 1991)* pp 1368-1381.

[Jain *et al.* , 1993]Jain B.N., Anand S.S., Rao S. Tele-Training System using Multi-Media communcation over Computer Networks: A Pilot Implementation. *Proceedings of International Conference on Globalization of Computer and Communications: Perspectives for Developing Economies (October 1993)*

[Deering, 1989]Deering S. E. Host Extensions for IP Multicasting Internet RFC 1112 Network Information Center, (August 1989).

[Elmore, 1992]Elmore G.C. Integrated Technologies: An Approach to Establishing Multimedia Applications for Learning *EDUCOM Review Vol 27 No 1 (January/February 1992).*

Vijay Talati is a Research Engineer at IIT, Bombay. He did his B.E. in Computer Engineering from Gujarat University. He has been working with the ERNET since last 3 years and is also pursuing his MTech studies in Computer Science and Engineering at IIT, Bombay.

Dr. S.L.Mehndiratta is a Professor, Department of Computer Science & Engineering and Head, Computer Centre at IIT Bombay. He has a number of research publications to his credit and has guided a number of students towards their Ph.D. and M.Tech. degrees. His research interests are in the areas of computer networks, protocol verification, theoretical computer science and distributed systems.

Communication Channels for Data Multicasting in Multi-service Networks*

K. Ravindran,
Department of Computing & Information Sciences,
Kansas State University Manhattan, KS 66506 (USA).

Abstract

The paper generalizes the abstraction of tree structured communication channels used in wide-area data multicasting by introducing the notions of: i) *acyclic graph* structured channel to provide connectivity among user entities in an application through network nodes and links, and ii) user specifiable attributes of data flow through a channel, namely, *data directionality* and *data transfer rates.* The abstraction is useful for multi-service networks where applications have diverse transport requirements such as multi-source broadcasting to a common set of destinations (e.g., video conferencing) and bidirectional/unidirectional data flow (e.g., broadcast audio). The directionality is used by the network to create unidirectional or bidirectional edges in the graph over the links between nodes through which data may flow in only one or both directions, as the case may be. The data rate may be mapped to resource demands on the network, viz., link bandwidth and node buffers, to transport data over the edges. The data directionality and data rate information are part of user level *functional binding* to graph structured channels in the network, and allow the network to effectively 'micro-manage' the resources. This becomes desirable, particularly with a large number of users and diverse transport requirements of applications, as in multi-service networks. Using the multicast graph abstraction, the end-systems in the network can directly support many-to-many communication structures for applications such as video conferencing and digital TV.

Key words

Multicast routing, flow specification, backbone network topology, resource reservation protocol, network-wide resource control.

---

*Network modeling and system development & implementation components of this work are supported by **US AirForce Rome Laboratory** under contract: F30602-94-C-0241.

1 INTRODUCTION

Recent research on data multicasting in wide-area networks has proposed *tree structured channels* whereby the switching system in the network replicates data from the source at root of a tree and forwards the data over pre-established multipoint paths to the destinations at leaves of the tree [1, 2]. See Figure 1. Given a tree in the network rooted at a source, multicasting requires setting up a path segment from a destination d to the tree through a set of intervening switches and links so that d can receive the data sent over the tree. The multicasting function in a switch then has two components:

- Selecting a set of links through which a data received on a link is to be sent out on its way to destinations (*routing*);
- Allocation of switch buffers and link bandwidths to receive data over the incoming link and send data over the outgoing link(s) attached to the switch (*resource allocation*).

Accordingly, a tree structured channel may be viewed as indicating: i) logical connectivity between a source and its destinations through a network-wide data path consisting of switches and links, and ii) network-wide resource allocations at the switches and links in this path, to support data flow from the source to destinations.

Tree structured channels have been successfully used in the network to support multicasting for many computer applications such as mail distribution and decentralized 'object location' in a network [1, 3]. In this paper, we study the abstract properties desired of these network channels to meet the diverse transport requirements of an application in multi-service networks such as multi-source broadcasting of data to a common set of destinations (e.g., client access to a replicated database), bidirectional and unidirectional data transfers among entities (e.g., video conferencing, audio distribution), and widely varying transfer rates of data from sources (e.g., audio, video and graphics).

The canonical communication structure of a multi-service application at the multicast data transport level may be viewed as a flat grouping of source and destination entities, with the network establishing logical connectivity among these entities and allocating resources in the underlying path, to enable flow of data between them. With use of a tree structured channel as the basic network level object to embody the data multicasting functions, the network needs to create multiple trees to support the application, where each tree is rooted at a distinct source and a destination connects to each of these trees separately to receive data from the various sources ('tree 1' and 'tree 2' rooted at sources s_1 and s_2 respectively in Figure 1). The network however cannot relate these trees as supporting a single application, without maintaining additional state information about the transport level grouping of source and destination entities connected by different trees. Accordingly, the data multicasting overhead in the network (viz., link cost incurred to support multiple path segments, channel identifier space consumed and extent of routing control actions required) for the application increases linearly with the number of sources even when these trees have overlapping path segments. Secondly, the application level communication structure needs to compose multiple network channels for exercising the required data transport from various sources to a common set of destinations across the user-network interface (UNI), thereby requiring separate local bindings to each of the channels across the UNI (such as 'call reference numbers' in Broadband ISDN Signalling Protocols). This is inflexible in comparison to the case where a single instance of *functional binding* to channels, i.e., binding based on transport attributes supported on channels, across the UNI suffices for exercising the data transport and allows uniformity in the communication structure. For example, a dynamic conferencing group that allows varying number of sources/destinations can be uniformly structured with a functional interface to the network level multicast transport. Thus it is desirable to provide a flexible multicast service model that allows transmission of the information on transport attributes of an application from the users to the network to parameterize the channel characteristics and reduce the multicasting overhead.

Our model is based on a network level abstraction of *acyclic multicast graph* that embodies logical connectivity across different sources and destinations and resource allocations in the underlying paths. A graph has vertices in the switches and edges over the links, with each user residing in a vertex and data flowing from sources to destinations in a given application through various edges. The direction of data flow projects a graph into multiple trees with root at various sources and leaves at the destinations. In other words, a graph is an unrooted tree. A user is allowed to specify *data directionality*, i.e., whether a user can only send or only receive or both send and receive data, through a graph. The directionality is used by the network to create unidirectional or bidirectional edges in the graph, as may be necessary. A user may also specify *data rate*, i.e., the bit rate at which the user can send and/or receive data, through the graph. The data rate may be mapped to resource demands on the network, viz., link bandwidth and switch buffers, to transport data from and/or to the user. Effectively, the transport attributes of data flow (viz., data directionality and data rate) across all user entities in the application allow the network to 'micro-manage' the resources. Thus a graph structured channel is a network level object that can be accessed by user entities through a functional interface.

With the multicast graph abstraction, there is potential for an underlying routing algorithm to reduce the network overhead for data transport by combining the data from various sources to flow over common links towards destinations thereby amortizing any fixed link costs and routing control actions across these sources. This becomes desirable, particularly with a large number of users and diverse transport requirements of applications, as in multi-service networks. Using multicast graphs, the end systems in the network can build a variety of high level many-to-many communication structures for

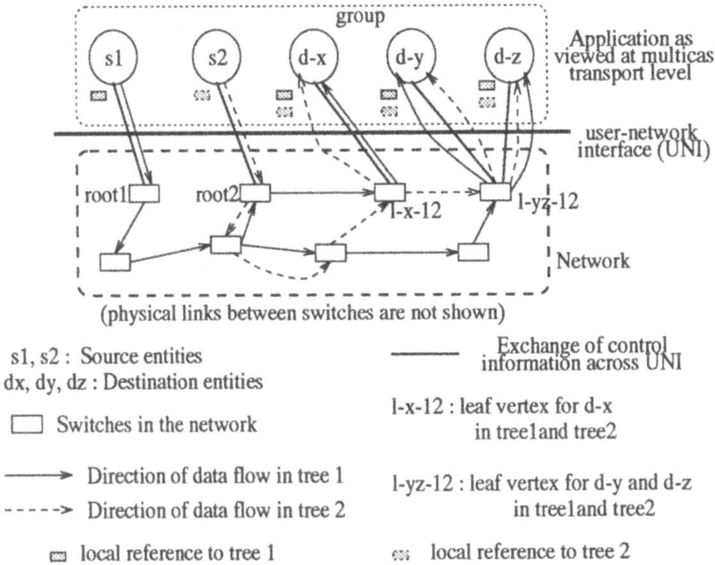

Figure 1: Tree structured communication channels.

applications such as conferencing, video distribution and distributed computing. One-to-one communication (or unicasting) for applications such as video-telephony is elegantly subsumed into this approach. The paper also deals with resource allocation control in the network to support our graph abstraction.

The paper is organized as follows: Section 2 discusses related works. Section 3 describes our model of multicasting using acyclic connectivity graphs. Section 4 gives a view of multicast graphs oriented towards resource allocations in the network. Section 5 describes the functional elements needed in the network to support the model. Section 6 shows a realization of sample applications using the model. Section 7 concludes the paper.

2 RELATED WORKS

Consider the multicast models for low-to-moderate speed wide-area networks such as ARPANET and interconnected LAN's [1]. These models support primarily applications with less diverse transport requirements such as distributed databases and mail distribution, and hence are inadequate for the evolving applications such as video conferencing, digital TV and multi-user graphic visualization. For instance, these models do not allow user level flow specification of the data to the network. The recent work on multicasting of real-time audio over wide-area internets [4] is primarily along the lines of providing additional support elements in the network (e.g., buffer reservation at various switches in the multicast tree) rather than providing a generic higher level service model that can systematically augment the network with the support elements.

The Internet Stream Protocol ST-II [5] and the resource reservation protocol RSVP [6] incorporate new features expressly designed to support the evolving applications, namely, allowing a flow specification (primarily, data rates) to be systematically transmitted from users to the network switches. The ST-II allows the flow specification from a source to be propagated to agents located at the vertices of a tree rooted at the source so that these agents may make resource allocations in the switches to support the required flow. But the protocol by itself does not provide a systematic mechanism by which a given switch can handle the flow specifications from multiple sources that send data through the switch (though an implementation of the protocol can install local procedures to handle the multi-source flows). The RSVP allows multiple sources in an application to specify their flows to the network; however, it does not deal with how the multi-source flow specification can be incorporated into a network abstraction that allows a systematic use of this flow information by the switches. In this sense, our work provides elements of a canonical network model that can be instantiated into appropriate flow specification protocols (such as ST-II and RSVP) executing at the UNI.

3 OUR MODEL OF MULTICASTING

The network consists of a set of nodes (or switches) $\mathcal{V}$, interconnected with one another through high speed communication links. Users, viz., service providers and subscriber terminals, may be attached to various network nodes. See Figure 3. A user is capable of exchanging data with one or more users, such as in TV broadcasting, video conferencing and voice phone. The data switching functions of network nodes are provided by the backbone network such as ATM switches in MANs, routers in interconnected LANs and packet switches in high speed WANs.

The user entities implementing an application are the end-points of communication, with nodes $\mathcal{U} \subseteq \mathcal{V}$ containing the user entities forming a *multicast set*. A user may have both send and receive capability (UPDOWN), only send capability (UP) or only receive capability (DOWN). These capabilities depict end-to-end directionality of data flow through the network. Users in a multicast set indicate transport requirements on the network in the form of a *data generation rate* by UP or UPDOWN users (i.e., sources) and a *data consumption rate* by DOWN or UPDOWN users (i.e., destinations). In an example of conference, the source is the initiator of a conversation and the destinations are the other participants in the conference browsing the conversation. The data rates may be part of a user-specified *quality of service* (QOS) required of the network. The transport

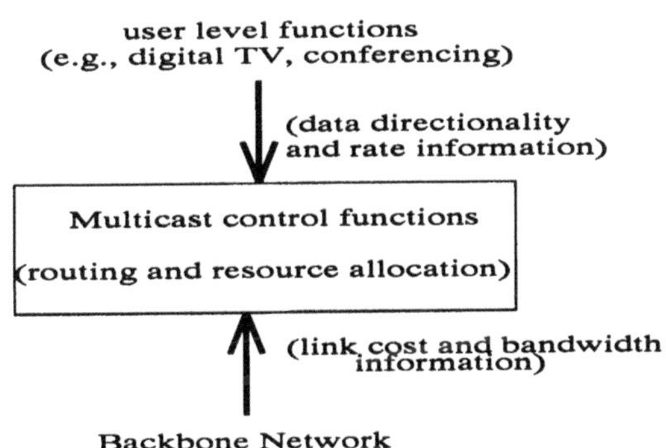

Figure 2: Layers of functions in network architecture.

attributes, viz., data rates and data directionality, are used in defining an appropriate network control structure to support multicasting.

The underlying communication architecture consists of: i) user-level functions to generate and consume data, ii) backbone network that manages physical connectivity among switches $\mathcal{V}$, and iii) ***multicast control functions*** that allow routing of data from sources to destinations over a path consisting of a set of switches and links in the backbone network. See Figure 2. A multicast data transport incurs a ***cost*** on the backbone network in the form of link bandwidth allocation and link management overhead. The multicast control functions define an abstraction, viz., ***graph structured channel***, for data transport. The abstraction allows the routing functions to systematically utilize the data rate/directionality information from the users and the link cost/topology information from the backbone network so that a network-wide performance index can be optimized (e.g., finding a minimal cost path from the sources to destinations).

We discuss below the multicast graph abstraction (hereafter, the term 'network' includes the multicast control functions, unless stated otherwise).

3.1 Multicast connections

A *connection* supports multipoint communication among users through a set of network nodes $\widehat{\mathcal{U}}$, where $\mathcal{U} \subseteq \widehat{\mathcal{U}} \subseteq \mathcal{V}$. The network views the connection as an acyclic graph $\mathcal{G}(\widehat{\mathcal{U}}, \widehat{\mathcal{E}})$ providing logical connectivity among users, with vertices residing in $\widehat{\mathcal{U}}$ and edges $\widehat{\mathcal{E}}$ mapping to the intervening links. Refer to Figure 3. The flow of data from a node x to a neighboring node y over the link between these nodes, where $x, y \in \widehat{\mathcal{U}}$, manifests as traversal of the edge $e(x,y) \in \widehat{\mathcal{E}}$ connecting the adjacent vertices in x and y. A route between vertices, say, $\widehat{u}, \widehat{u'} \in \widehat{\mathcal{U}}$ consists of a set of adjacent edges $[e(\widehat{u},t), e(t,\_), \cdots, e(\_,q), e(q,\widehat{u'})]$. Figure 3 shows a graph that connects user entities U_a, U_b and U_d through a backbone net-

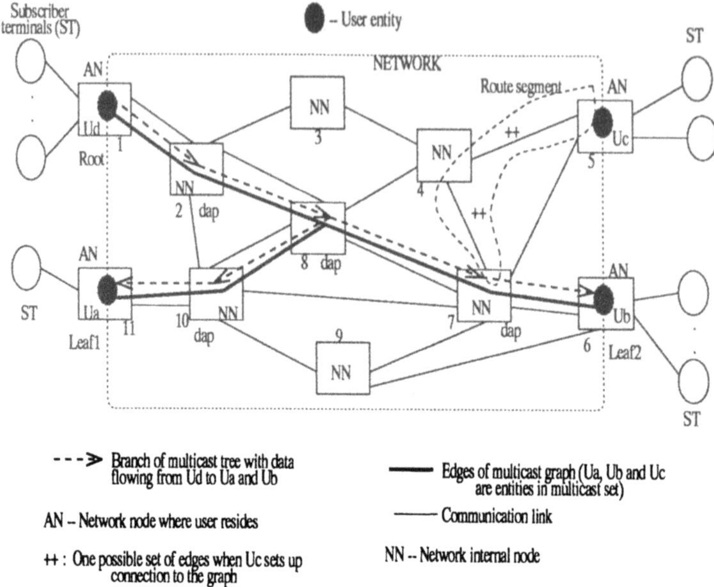

Figure 3: Graph structured communication channels.

work with $\mathcal{V} = \{1, 2, \cdots, 10, 11\}$. For this graph, $\mathcal{U} = \{1, 6, 11\}$, $\widehat{\mathcal{U}} = \{1, 2, 8, 7, 6, 10, 11\}$ and $\widehat{\mathcal{E}} = \{e(1,2), e(2,8), e(8,10), e(10,11), e(8,7), e(7,6)\}$.

The network may support one or more graphs, with a switch capable of supporting a vertex belonging to each of these graphs. Typically, each graph may provide connectivity among users in a single application. For example, all participants in a video conference may be connected by a single graph. For the purpose of this paper, we assume a single graph in the network without loss of generality.

3.2 Multicast routes

The vertices and edges in $\mathcal{G}$ participating in a data flow may be projected as a multicast tree $\mathcal{T}_s(\mathcal{G})$ with the root at node $s \in \mathcal{U}$ where the source resides and the leaves at nodes $\{d\} \subseteq (\mathcal{U} - s)$ where the destinations reside. In this tree, the flow of data from s to d can be viewed as a traversal of the edges $e(s,t), e(t,\_), \cdots, e(\_,q), e(q,d)$ through the vertices in $s, t, \cdots, q, d \in \widehat{\mathcal{U}}$ in that sequence, with $e(s,t)$ originating from (i.e., directed from) s and $e(q,d)$ incident on (i.e., directed towards) d. The multicast route in $\mathcal{T}_s$ is a merge of the paths $[e(s,t), e(t,\_), \cdots, e(\_,q), e(q,d)]_{\forall d}$, i.e., the union of all these paths such that the common edges in the paths appear in $\mathcal{T}_s$ exactly once. $\mathcal{T}_s$ is thus a spanning tree for the nodes $\{s, t, \cdots, q, d\}_{\forall d}$, with no loops in the data flow along various paths. A vertex $x \in \mathcal{T}_s$

constitutes a *branch point* if there is more than one edge directed from x; the data arriving at the branch point x gets replicated along the edges $\{e(x,\_)\}$. Figure 3 illustrates a tree projected from a graph, with root at the node 1, leaves at the nodes 6 and 11, branch point at node 8 and the multicast route as $[e(1,2), e(2,8), [e(8,10), e(10,11)], [e(8,7), e(7,6)]]$.

There can be more than one $\mathcal{T}_s$ projected from $\mathcal{G}$ with root at various nodes $\{s\} \subseteq \mathcal{U}$ such that $\bigcup_{\forall s}\ s + \{d\} = \mathcal{U}$; the data from various sources flow along the edges of $\{\mathcal{T}_s\}$ to reach the destinations. A $\mathcal{T}_s$ may have one or more of its edges overlapping with that of other $\mathcal{T}_s$'s, implying that the multicast routes in the various trees may share one or more links. A vertex $y \in \mathcal{G}$ is a *junction point* if there is more than one edge directed towards y; the various data arriving at the junction point y along the edges $\{e(\_,y)\}$ get combined with one another for forwarding. In Figure 3, the node 8 is the junction point for the data arriving along the edges $e(2,8)$ and $e(10,8)$. The multicast route in $\mathcal{G}$ is then a merge of the multicast routes in each of $\{\mathcal{T}_s\}_{\forall s}$. Thus, $\mathcal{G}$ is an unrooted spanning tree for $\hat{\mathcal{U}}$, representing[1] an end-to-end view of data flow between user entities through the network[2].

3.3 Data access points

Each node $u \in \hat{\mathcal{U}}$ supports a certain level of data rate by allocating link resources (e.g., buffers) to handle the data arriving at and/or sent by u. This allows data to be 'injected' into and/or 'tapped' from $\mathcal{G}$ through u by user entities, so u can be considered as a *data access point* (DAP). When a user U requests connection to $\mathcal{G}$, a routing algorithm in the network locates a node u that can serve as DAP for U and sets up a path from U to u. The path setup involves:

- Creating a *route segment* $[u_1, u_2, \cdots, u_l]_{l \geq 1}$ that consists of:

  Vertices in nodes $u_1, u_2, \cdots, u_l$ adjacent to one another, with u_1 containing U and $u_2, \cdots, u_l$ providing physical connectivity between u_1 and u;

  Edges $e(u_1, u_2), e(u_2, \_), \cdots, e(\_, u_l)$ over the links intervening various nodes;

- Interconnecting the route segment $[u_1, u_2, \cdots, u_l]$ to u with an edge $e(u_l, u)$.

The new graph now is $\mathcal{G}((\hat{\mathcal{U}} + \{u_1, u_2, \cdots, u_l\}), (\hat{\mathcal{E}} + \{e(u_1, u_2), e(u_2, \_), \cdots, e(\_, u_l), e(u_l, u)\}))$.

[1] $\mathcal{G}$ need not span all the nodes $\mathcal{V}$ in the network but only sufficient number of nodes $\hat{\mathcal{U}}$ to connect the users attached to nodes $\mathcal{U}$ (i.e., $\mathcal{U} \subseteq \hat{\mathcal{U}} \subseteq \mathcal{V}$). In this sense, our notion of multicast graph is unlike the existing notions of implementing multicasting on top of a 'network-wide spanning tree' [7, 8].

[2] A connection refers to a control association between various entities to bind them to a logical unit for data transport, viz., multicast set. It does not refer to any data level association between entities that may be needed to provide a desired degree of reliability in data transport such as 'virtual circuit' and 'acknowledged data transfer'.

Thus the network has the perspective of adding new vertices and edges to $\mathcal{G}$, while the data sources and destinations involved in the connection have the perspective of embedding new trees on the physical topology.

3.4 Network-wide directionality of data

The multicast graph model allows the various data sources $\{s_i\}_{i=1,2,\cdots,N}$ and the destinations to reside in vertices of a single connected acyclic graph $\mathcal{G}$. From a destination point of view, the data from various sources get combined over a common set of links and switches in their path towards the destination. From a source point of view, its data flowing through a switch gets replicated over one or more links attached to the switch in its path towards various destinations. The graph model unifies these two views within a single abstraction, with the various switches where data are combined/replicated and the data flow over the intervening links represented as vertices and edges respectively of a graph. Thus the graph model can be used to provide network level support for multi-source broadcasting of data to destinations.

The various trees $\{T_{s_i}(\mathcal{G})\}_{i=1,2,\cdots,N}$ may have one or more edges of $\mathcal{G}$ common between them. The data from two different sources that share a common edge $e(x,y)$ may flow either in the same direction, say from x to y, or in opposite directions between x and y. The edge $e(x,y)$ is *unidirectional*, directed from x towards y, in the former case and is *bidirectional*, directed from x towards y and vice-versa, in the latter case. Since the combining of data from sources and the replication of data towards destinations may occur at various vertices of $\mathcal{G}$, some edges of $\mathcal{G}$ may be unidirectional with the remaining edges being bidirectional. Consider the graph illustrated in Figure 3 as example. Suppose U_a is an UPDOWN entity, U_b is a DOWN entity and U_d is an UP entity. Along the tree T_{U_d}, the data from the source U_d attached to node 1 flows along the edges $[e(1,2), e(2,8)]$, gets replicated at the vertex in node 8, and then flows along the edges $[e(8,10), e(10,11)]$ and $[e(8,7), e(7,6)]$ to reach the destinations U_a and U_b attached to the nodes 11 and 6 respectively. Along the tree T_{U_a}, the data from U_a flows along the edges $[e(11,10), e(10,8), e(8,7), e(7,6)]$. So in the graph, the edges $e(10,11)$ and $e(8,10)$ are bidirectional, while the edges $e(1,2)$, $e(2,8)$, $e(8,7)$ and $e(7,6)$ are unidirectional; the data from U_a and U_d get combined at the vertex in node 8 to flow over the edges $[e(8,7), e(7,6)]$. Thus for a given set of data rates from s_i's, the logical topology of $\mathcal{G}$, viz., the nodes $\widehat{\mathcal{U}}$ that form vertices of $\mathcal{G}$ and how the various edges $\widehat{\mathcal{E}}$ spanning these nodes are directed, in general reflects the source-destination configuration in the physical topology (i.e., the placement of sources and destinations relative to one another in the physical topology) and the link cost constraints enforced by the multicast routing algorithm.

We now describe how a multicast graph can characterize resource allocations in the network.

4 NETWORK-WIDE RESOURCE ALLOCATIONS

A switch S that contains a vertex $u \in \mathcal{G}$ allocates resources for one or more links of S through which edges of $\mathcal{G}$ are incident on u and/or originate from u. The transport resource allocations for a link include reservation of buffers at a *port* in S through which the link is attached to S and[3] assignment of bandwidth by the backbone network for sending and/or receiving data over the link.

4.1 Data rates of multi-source data

A general scenario of many-to-many communication consists of one or more sources $\{s_i\}_{i=1,2,\cdots,N}$, with each s_i generating data at the rate of q_i, and one or more destinations consuming each the combined data at the rate given by $q_1 + q_2 + \cdots + q_N$, i.e., the transfer rate of a combined data is the sum of the rates specified in each q_i (this holds for peak and average data rates). Consider, for example, a multimedia conferencing with 5 participating window-based workstations. Each workstation may generate video and audio with data rate requirements of 2 *mb/sec* and 64 *kb/sec* respectively. The various workstations may form a multicast set with 10.32 *mb/sec* as the combined data rate requirement. A destination can connect to the s_i's only if it can consume data at the combined rate[4].

The data traffic sent from various s_i's over a graph get mixed at one or more vertices. See Figure 4. Consider a vertex in switch B with edges through ports p_a, p_b and p_c. The data arriving in p_a and p_b are sent out through p_b, p_c and p_c respectively, thereby appearing as a mixed data traffic over the link attached to p_c and the downstream paths from p_c. If q_1 and q_2 indicate the data rates for the data arriving in p_a and p_b respectively, the transfer rate of the mixed data sent out through p_c is $(q_1 + q_2)$.

The network may allocate resources in various vertices of a graph to meet the combined data rate requirement. So after mixing the data traffics from various s_i's, the network need not distinguish between the data from individual s_i's for the purpose of resource allocations[5].

4.2 Relationship between data rates and network resources

We assume the availability of a functional relation to provide the mapping between a data rate and the underlying resource needs at a switch to support data transfer at this rate,

[3] Typically, a port maps to the network interface to a link in a switch.

[4] The data belonging to different sources are separable from the mixed traffic by a higher level filter protocol at a destination.

[5] The distinction may however need to be maintained by the network to handle other transport attributes such as delay jitter and loss sensitivity of various data for determining packet scheduling priorities.

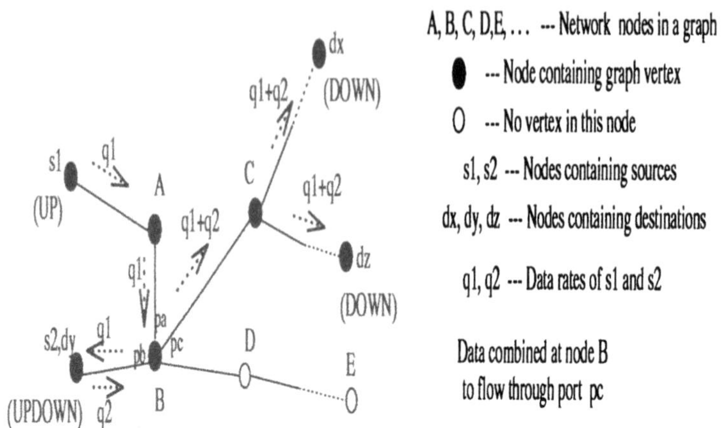

Figure 4: Multi-source data flowing over a graph.

given as below:

$$\mathcal{F} \, : \, \mathrm{q} \rightarrow \mathcal{R},$$

where $\mathcal{F}$ provides one-to-one mapping of a data rate to resource needs $\mathcal{R}$, viz., packet buffers and link bandwidth [9]. Consider the earlier example of multimedia conferencing. At the point of delivery of video and audio data to a workstation window, the backbone network should support a link bandwidth of at least 10.32 mb/sec. In general, q is specified as a pair (a, m) indicating the average data rate a and the peak data rate m.

The resource allocations by a switch to meet the requirements of data from various sources $\{s_i\}_{i=1,2,\cdots,N}$ satisfy the *linearity* property, indicated as: $\mathcal{F}(q_1 + q_2 + \cdots + q_N) = \mathcal{F}(q_1) + \mathcal{F}(q_2) + \cdots + \mathcal{F}(q_N)$. The linearity property allows incremental resource allocations and de-allocations when sources connect and disconnect respectively from a graph.

The cost of data transport over a link consists of two components:

(I): Network resources allocated for the data, namely, the link bandwidth assigned and buffers reserved at the port to transport the data;

(II): Overhead in the backbone network for maintaining a connection over the link.

The cost component **(I)** depends on the data rate q over the link, indicated as $\mathcal{F}(q)$, while the component **(II)** depicts a fixed cost that is solely based on the internals of the backbone network (e.g., a lease rental for the 'connect time' over the link). Thus we have the net cost as

$$h.\mathcal{F}(q) \; + \; \text{overhead cost}, \tag{1}$$

where h is a normalization constant that maps the resource needs to a cost[6]. Using equation (1) at various vertices of a graph, the network-wide cost of data transport across various sources and destinations connected by the graph can be computed.

4.3 Resource allocation control

Since only UP/UPDOWN entities can generate data, the vertices in the path from an UP/UPDOWN entity need to support the data rate of only this entity up to the vertex at which data traffic from another source is combined. Similarly, the vertices in the path towards a UPDOWN/DOWN entity need to support a data rate for the combined traffic from all sources. The creation and deletion of edges $\{e\}$ when users connect to and disconnect from $\mathcal{G}$ manifest in allocation and de-allocation of resources respectively in various vertices. Thus $\mathcal{G}$ may be viewed as an indicator of resource allocations at various vertices $\hat{\mathcal{U}}$, with each edge $e \in \hat{\mathcal{E}}$ providing an appropriate transport capacity. A reservation protocol handles the resource allocation and release along the data distribution path in $\mathcal{G}$ to ensure that the end-to-end QOS requirements are met.

When a user is to be connected to $\mathcal{G}$, choosing a DAP in $\mathcal{G}$ and a path to the DAP is a resource control policy decision. For instance, the routing algorithm for a user with UP directionality may choose a DAP such that the sum of the 'distance' from the DAP to all destinations and the 'distance' from the switch containing the user to the DAP meets a cost constraint. Typically, the routing algorithm may evaluate alternate paths in the physical topology and then set up a route segment over a path of minimal 'distance'. Thus a dynamic change in the graph size may be viewed as a network-wide resource control activity. For a given source-destination configuration in the physical topology, specific aspects of resource control such as how much resources are allocated and how the resource allocations are distributed across various vertices primarily depend on the data rates of traffic flowing network-wide from various sources to destinations.

We now discuss various elements of the multicast graph model from a resource allocation point of view.

5 FUNCTIONAL ELEMENTS OF MULTICAST GRAPH MODEL

A user specifies a pair $(u\_dir, \{s\_rate, r\_rate\})$ in its request for connection setup to a graph $\mathcal{G}$, where s_rate and r_rate indicate the sending and receiving data rates respectively and u_dir indicates the data directionality. The specification of these transport attributes is part of the functional interface to the network, and constitutes the user's reference to $\mathcal{G}$ across the UNI. We describe in this section how the directionality of data flow interacts with resource allocations in the network.

[6] The costs may be payable by users as a 'tariff' to the network provider.

5.1 Switch level resource needs

Consider a switch S containing a vertex u of $\mathcal{G}$ with edges through ports $\{p\} \subseteq \mathbf{Set\_of\_Ports}(S)$, where $\mathbf{Set\_of\_Ports}(S)$ refer to the set of ports in S. For a port p linked to a neighboring switch S^p, u may allocate two sets of resources: $\mathcal{R}_s(p,u)$ to send data to and $\mathcal{R}_r(p,u)$ to receive data from the adjacent vertex u' in S^p. An edge through p is bidirectional carrying data both towards and away from p if $\mathcal{R}_s(p,u) > 0 \wedge \mathcal{R}_r(p,u) > 0$, unidirectional carrying data towards p if $\mathcal{R}_s(p,u) = 0 \wedge \mathcal{R}_r(p,u) > 0$, and unidirectional carrying data away from p if $\mathcal{R}_s(p,u) > 0 \wedge \mathcal{R}_r(p,u) = 0$. The following conditions should then hold for a stable data transport over the edge $e(u,u')$:

$$\begin{aligned} \mathcal{R}_r(p',u') &\geq \mathcal{F}\Big(\sum_{\forall x \in (in\_edge(u)-p)} in\_rate_{x,u}\Big) \\ \mathcal{R}_s(p,u) &\geq \mathcal{F}\Big(\sum_{\forall x \in (in\_edge(u)-p)} in\_rate_{x,u}\Big) \\ \mathcal{R}_r(p,u) &\geq \mathcal{F}(in\_rate_{p,u}), \end{aligned}$$

where $in\_rate_{x,u}$ is the data rate for the traffic flowing into u through port x in S, p' is the port in S^p through which S is linked, and $in\_edge(u)$ is the set of ports in S through which an edge is directed towards u. These conditions should hold in the presence of addition/deletion of edges to u and/or changes in the resources allocated for various edges (this is possible with the linearity property of resource allocations). See Figure 5 for an illustration.

5.2 Resource reservation protocols

The aggregation of resource allocations at various switches in the data distribution path, as described above, needs to be embodied into a reservation protocol. Such a protocol may be activated by user invocations for network multicast service across the UNI. The execution model of the protocol may consist of *agent entities* installed in the switches that interact with one another over a control channel (e.g., B-ISDN signaling channels) to coordinate the resource allocations in a decentralized manner. Accordingly, the protocol is specifiable in terms of:

- *Control messages* exchangeable across the agent entities to carry the QOS information pertaining to the flow of data from various sources and the resource availability information at switches;
- *Tables* maintained by the agent entities to describe the mapping function $\mathcal{F}$ in the form of QOS values (viz., average and peak data rates) and the corresponding resource needs.

See Figure 6. Examples of control messages are the 'connect' and 'accept'/'refuse' messages used in ST-II [5]. The explicit representation of $\mathcal{F}$ in the form of tables is needed

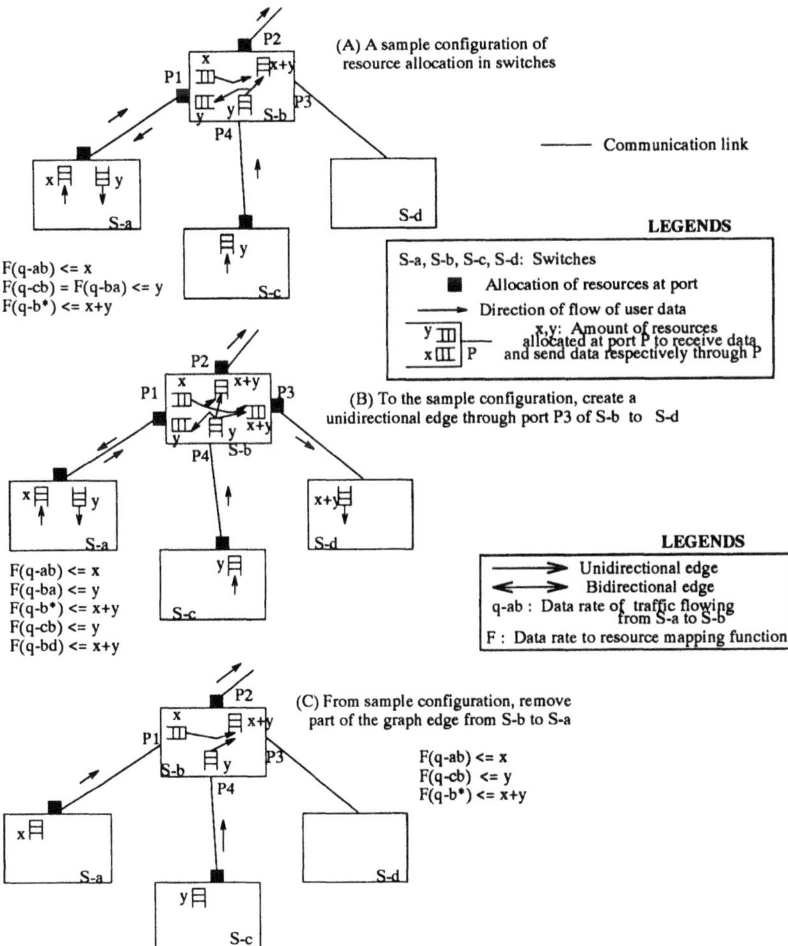

Figure 5: Resource allocation in switches and its representation as directional edges in the network.

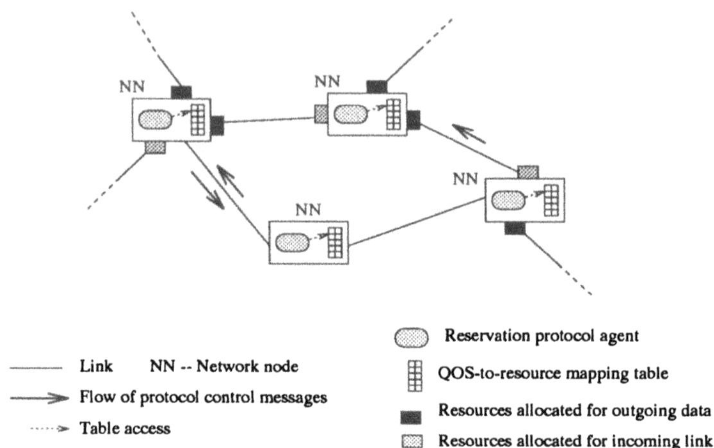

Figure 6: Execution model of resource reservation protocol.

in our graph model to allow dynamic changes in the traffic levels flowing through various switches due to addition/removal of users in the multicast configuration.

Suppose (a, m) is the QOS value currently supported by the reservation agent in a switch, with resources allocated in the amount given by $\mathcal{F}((a, m))$. When the level of traffic flowing into the switch has to be increased by $(\delta a, \delta m)$, the agent computes the resource needs $\mathcal{F}((a+\delta a, m+\delta m))$ from the table and determines the admissibility of the additional traffic by checking if enough resources are available both locally and at other downstream agents to satisfy the increased needs. If the check is successful, the additional resources are committed at appropriate switches in the path. When the traffic level has to be decreased by $(\delta a, \delta m)$, the agent computes the resource needs $\mathcal{F}((a-\delta a, m-\delta m))$ from the table and deallocates the excess resources both locally and at other downstream agents. When resource allocations satisfy the linearity property, the incremental amount of resources may be simply estimated as $\mathcal{F}((\delta a, \delta m))$.

For resource allocation purposes, each agent maintains a local state variable *c_rate* indicating the traffic level flowing through its switch. The inter-agent coordination to allocate/de-allocate resources and to update *c_rate* is achieved by exchanging control messages.

5.3 Routing control information

From our foregoing discussion, we see that a routing algorithm requires, to varying degrees, a global knowledge of the following resource control factors:

- Source-destination configuration in the physical topology;

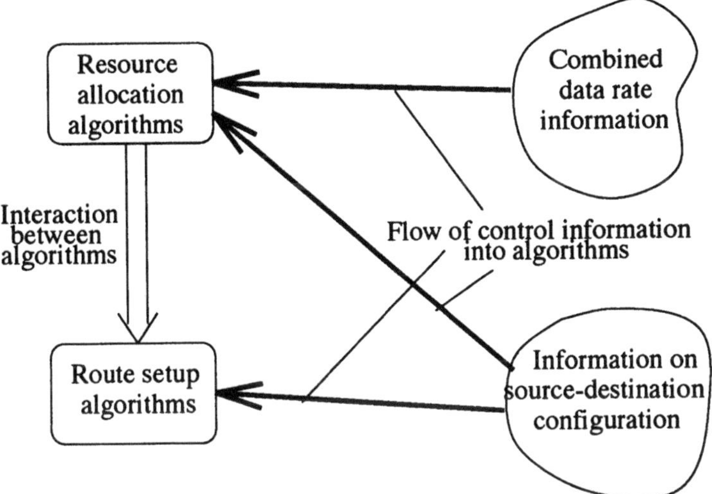

Figure 7: Network level view of resource control algorithms.

- Data traffic flowing network-wide from various sources to destinations.

The information on these factors may be obtained from the data directionality and data rates of user entities and the physical topology of backbone network. So the incorporation of data directionality and data rates as part of the multicast semantics in our graph model allows the required information to be made available to the routing algorithm (often acquired and processed in a distributed manner)[7]. See Figure 7.

A systematic availability of the information on data directionality, data rates and physical topology to the resource control algorithms allows the latter to optimize the cost of multicast data transport. Computation methods studied elsewhere [2, 10, 11] to enforce cost constraints on multicast data transport through tree structured channels may be appropriately incorporated into a chosen routing algorithm executed by the network in an implementation of our multicast model.

It is beyond the scope of this paper to deal with the design of multicast routing algorithms as such (see [1, 2] for routing algorithms studied elsewhere in different network environments). Rather, the paper focuses on providing a network model of multicasting

[7]It should be noted in this context that the network-wide configuration and data flow related information are not directly available to the resource control algorithms when the network employs tree structured connectivity.

that allows systematic acquisition and use of resource control information by a chosen routing algorithm (say, for a network-wide path search [12]).

We now give some sample scenarios to illustrate the dynamic formation of graphs in our model.

6 RESOURCE ALLOCATION TRACE FOR SAMPLE APPLICATIONS

For ease of description, we denote: i) the creation of an out-going edge from a vertex u in a switch through a port p as invoking a network level function **Assign_Edge**(p) by u, and ii) the removal of an out-going edge through p as invoking a function **Deassign_Edge**(p) by u. Note that the creation of the edge requires allocation of resources $\mathcal{R}_s(p,u)$ at p satisfying the data rate requirements and the removal of the edge causes de-allocation of these resources. We also assume a function **to**(S^p) provided by the backbone network that is invocable by a switch S to obtain the port p through which S is linked to the neighbor S^p (the function may also return a link cost value that may be used by a routing algorithm).

Let $\{S_0, S_1, \cdots, S_n\}_{n \geq 0}$ be the switches, in that order, in the path to a DAP, with the user U requesting connection attached to S_0 and the DAP residing in S_n. When $n = 0$, U can simply start sending and/or receiving packets through S_0, as the case may be.

6.1 Digital TV

A TV station initially creates a single-vertex graph with $c\_rate = 0$. A TV broadcaster is an UP entity connecting to a vertex u through unidirectional edges from the TV broadcaster vertex to u, created by invoking **Assign_Edge(to(**S_{l+1}**))** in each switch S_l and setting c_rate to the traffic level indicated by u, where $l = 0, 1, \cdots, n-1$. A TV receiver is a DOWN entity connecting to a vertex u' through unidirectional edges from u' to the TV receiver vertex by invoking **Assign_Edge(to(**S_l**))** in each switch S_{l+1} and setting $c\_rate = in\_rate_{\mathbf{to}(S_{l+1}),-}$ at S_l. See Figure 8.

6.2 Conference

The conference manager initially creates a single-vertex graph with $c\_rate = 0$. An active participant who can send and receive data is an UPDOWN entity connecting to a vertex u through bidirectional edges created between u and the participant vertex: i) by invoking **Assign_Edge(to(**S_{n-1}**))** in S_n and setting its $c\_rate = c\_rate + in\_rate_{\mathbf{to}(S_{n-1}),-}$, ii) by invoking **Assign_Edge(to(**S_1**))** in S_0 and setting its $c\_rate = s\_rate + in\_rate_{\mathbf{to}(S_1),-}$, and iii) when $n \geq 2$, by invoking **Assign_Edge(to(**S_{l+1}**))** and **Assign_Edge(to(**S_{l-1}**))** and setting its $c\_rate = in\_rate_{\mathbf{to}(S_{l-1}),-} + in\_rate_{\mathbf{to}(S_{l+1}),-}$ in each S_l, where $l = 1, 2, \cdots, n-1$.

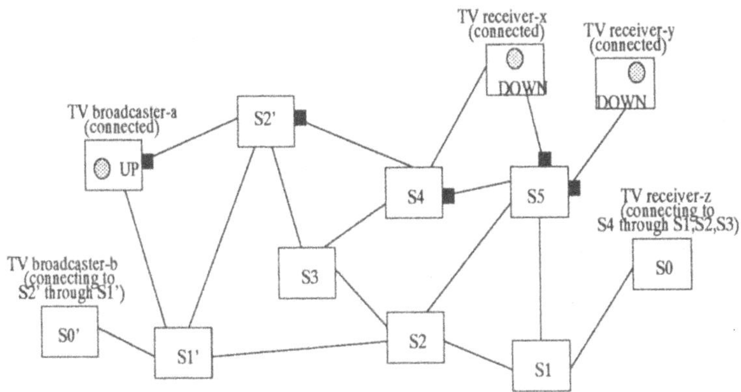

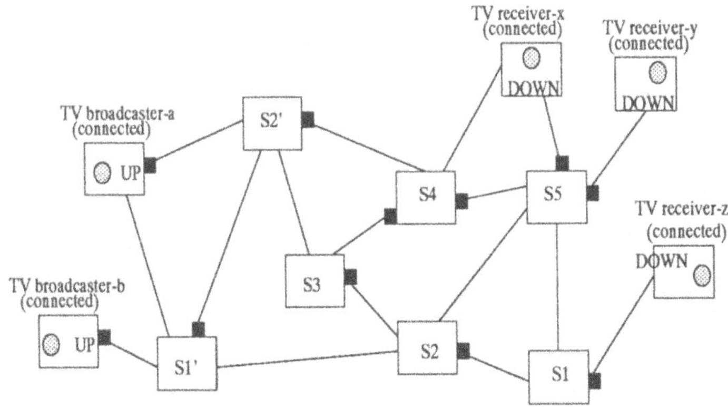

■ **Assign_Edge** on port --- corresponds to switch internal resource allocations
(amounts are not shown)

Figure 8: Scenarios before and after multicast connections in TV.

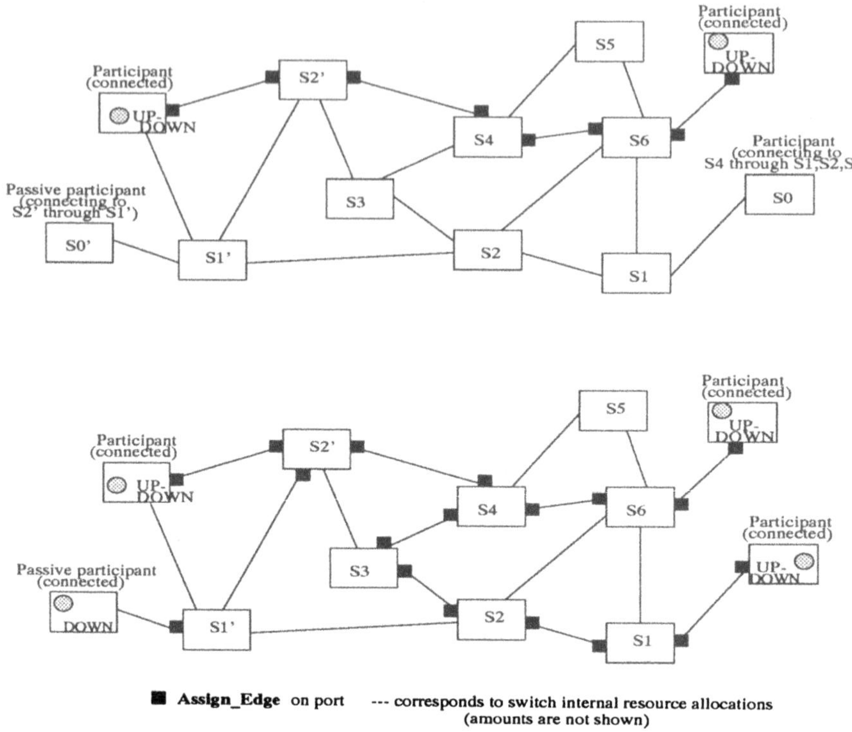

Figure 9: Scenarios before and after multicast connections in conference.

A passive participant who only receive data (e.g., observer) executes similar to the TV receiver discussed earlier. See Figure 9.

The creation of edges connecting the broadcaster in the TV example and the active participant in the conference example to u is preceded by an incrementing of c_rate in all vertices of the graph, including u, by an amount s_rate.

As can be seen, the resource demands on the network largely depend on the physical topology of the network, the location of the connected user entities and the directionality of the user entities.

7 CONCLUSIONS

The paper described a canonical model for a network to offer multicast services to applications. The model incorporates a functional form of the UNI to flexibly allow the network meet the diverse transport requirements imposed by the evolving multi-service applications. These requirements include support for widely varying data rates in communication channels, delivery of data to a large number of destinations, unidirectional and bidirectional data transport, and multi-source broadcasting of data to a common set of destinations.

The paper generalized the abstraction of tree structured channels currently used for wide-area packet multicasting by introducing the notions of: (i) acyclic graph structured channel that connects various users, viz., source and destination entities, in an application through a set of vertices and edges realized over switches and links respectively, and (ii) user-specifiable data directionality and data rates that manifest in allocation of resources, viz., switch buffers and link bandwidths, at various vertices and edges respectively, in the graph. A graph structured channel can be accessed by a user at any of its vertices for sending and/or receiving data over the channel. The user specified directionality projects a graph into multiple trees, where each tree is rooted at a source and has leaves at its destination(s), with appropriate resources allocated in each of the underlying paths to support data flow from the concerned source. Thus a graph structured channel specifies the connectivity among multiple sources and destinations through switches and links, and also the resources allocated for the combined data flow from various sources to destinations.

In contrast, multicast models using tree structured channels [2, 5] often require the network to set up multiple channels for supporting bidirectional data flow and multi-source broadcasting of data in an application. This increases the network overhead for multicasting of data because of the need to manage each channel separately. Multicast models that employ graph structured channels but do not use the directionality information [1] often require the network to support bidirectional edges (i.e., a switch port containing an edge of the graph allocates resources to both send and receive data) by default and require the end-systems to enforce directionality at the point of admitting data traffic into and from the network. In this context, the use of data directionality in the network saves resources in a switch because a switch need not allocate resources in both directions of a link for unidirectional flow of data over this link. Accordingly, our graph abstraction is quite useful for multi-service networks since many applications have unidirectional data transfer requirement (e.g., broadcast video, passive participants in a conference).

Our model of a multicast graph incorporating the directionality/rates of data flow through the graph is usable as an elegant building block for the many-to-many communication among a set of users required in many applications. The model allows the network to map the transport requirements of applications to resource demands at various switches and links. It also optimizes the cost of multicast data transport across the entire application by reducing the data multicasting overhead in the network due to sharing of links by multiple paths.

We believe the model presented in the paper will lend insight into system architectures that need to support data multicasting, at the network level, in an environment of diverse multi-service applications.

8 CURRENT STATUS OF WORK

The functional model described in the paper is basically a beginning of a comprehensive design and implementation of network architecture for data multicasting. There are many facets to the future activity: formulation of user level primitives at the network interface, development of network internal routing algorithms, embedding of network architecture on operational backbone networks (e.g., ATM-based testbeds), and demonstration of prototype applications (e.g., video conferencing) using the network architecture. This further activity is supported and funded by the **US Airforce Rome Laboratory** through a **R & D** contract.

9 REFERENCES

[1] S. E. Deering and D. R. Cheriton. **Multicast Routing in Datagram Internetworks and Extended LANs.** In *ACM Transactions on Computer Systems*, Vol.8, No.2, pp.85-110, May 1990.

[2] B. M. Waxman. **Routing of Multipoint Connections**, In *IEEE Journal on Selected Areas in Communications*, Vol.SAC-6, No.9, pp.1617-1622, Dec. 1988.

[3] D. R. Cheriton and S. E. Deering. **Host Groups: A Multicast Extension for Datagram Internetworks.** In *9th Data Communications Symposium*, ACM SIGCOMM, pp.172-179, Sept. 1985.

[4] S. Casner and S. E. Deering. **First IETF Internet Audiocast.** In *Computer Communications Review*, Vol.22, No.3, pp.92-97, July 1992.

[5] C. Topolcic. **Experimental Internet Stream Protocol, version 2 (ST-II).** In *RFC 1190*, CIP Working Group, Oct. 1990.

[6] L. Zhang, S. E. Deering, D. Estrin. S. Shenker and D. Zappala. **RSVP: A New Resource ReSerVation Protocol** In *Network*, IEEE Communications Society, pp.8-18, Sept. 1993.

[7] M. D. Schroeder and et al. **Autonet: A High Speed Self-configuring Local Area Network Using Point-to-point Links.** In *Tech. Report*, Digital Equipment Corporation (System Research Center), Palo Alto (CA), April 1990.

[8] D. Bertsekas and R. Gallager. **Routing in Data Networks**. Chapter 5 in *Data Networks*, Prentice Hall Publ. Co., 1992.

[9] A. A. Lazar, A. Temple, and R. Gidron. **An Architecture for Integrated Networks that Guarantees Quality of Service**. *Tech. Report*, Center for Telecomm. Research, Columbia University, 1990.

[10] D. C. Verma and P. M. Gopal. **Routing Reserved Bandwidth Multi-point Connections**. In Proc. *Communication Architectures, Protocols and Applications*, ACM SIGCOMM, pp.96-105, Sept. 1993.

[11] V. P. Kompella, J. C. Pasquale and G. C.Polyzos **Multicasting for Multimedia Applications**. In proc. *Computer Communications*, INFOCOM'92, Italy, 1992.

[12] K. Ravindran and M. Sankhla. **Multicast Models and Routing Algorithsm for High Speed Multi-service Networks**. In proc. 12th *Intl. Conf. on Distributed Computing Systems*, Yokohama (Japan), June 1992.

10 BIOGRAPHY

Dr. K. Ravindran is currently a faculty member at the dept of Computing & Information Sciences, Kansas State University. He received his Ph.D from university of British Columbia, Canada and M.E./B.E. degrees from IISc Bangalore. He has held positions at ISRO (Bangalore), Bell-Northern Research (Canada) and IBM European Networking Center(Germany). His research interests are in Distributed systems modeling and design, high speed network architectures and protocols.

9

Single and Multiple Frame Video Traffic Prediction Using Neural Network Models

R.Drossu
School of Electrical Engineering and Computer Science
Washington State University, Pullman WA 99164-2752

T.V. Lakshman
Bellcore,Red Bank,NJ 07701

Z.Obradovic *
School of Electrical Engineering and Computer Science
Washington State University, Pullman WA 99164-2752

C.Raghavendra
School of Electrical Engineering and Computer Science
Washington State University, Pullman WA 99164-2752

Abstract

The objectives of this paper are to investigate applicability of neural network techniques for single and multiple frame video traffic prediction. In the *single* and *multiple* frame traffic prediction problems, the information of previous frame sizes is used to predict either the following or several following frame sizes respectively. Accurate traffic prediction can be used to optimally smooth delay sensitive traffic [Ott *et al.*, 1992] and increase multiplexing gain in asynchronous transfer mode (ATM) networks. Neural network models for both single and multiple frame traffic prediction problems are proposed. Two important types of video sequences are considered - video teleconferencing and entertainment video. An off-line learning method is suggested for simple traffic and an on-line learning method for complex one. Simulation studies of cell losses in an ATM multiplexer using recorded variable-bit-rate coded video teleconference data indicate reasonably good predictions for buffer delays between 0.5 and 5 ms.

*Research sponsored in part by the NSF research grant NSF-IRI-9308523

1 INTRODUCTION

In the last few years a substantial amount of research has been performed in the area of neural networks. The experimental work indicates their potential for practical applications where traditional computation structures have performed poorly (e.g. ambiguous data [Le Cun et al.,1989] or large contextual influence [Eberhart *et al.*,1990]).

In addition to the tradition analytical approximation techniques, recently neural networks have also been proposed for some ATM control problems. For example in [Hiramatsu 1990;Takahasni *et al.*,1990] a neural networks learning method is suggested for service admission control in the ATM networks. The objective of this application is to keep the requested service quality parameters by rejecting some of the call set-up requests while connecting as many calls as possible. In [Nordstrom1993], a neural network is used to refine a traditional analytical approximation technique for admission control for improved bandwidth efficiency. Another promising application for neural networks in ATM is call control [Neves eta al.,1993]. Simulation results show that neural networks can lead to a compromise between the maximum number of calls accepted and the satisfaction of the quality of services negotiated for established calls.

In this paper we explore whether neural networks prediction techniques can be used for video traffic prediction. Accurate source traffic models and traffic predictions are needed for effective utilization and control of ATM networks. Video traffic consists of a periodic arrival of frames (25 frames per sec for our traffic sequence) with a variable number of ATM cells per frame. The number of cells per frame varies significantly. Predictions of the number of cells per frame can be used for improving ATM network efficiency by incorporating the predictions in schemes for multiplexing, routing, smoothing and bandwidth allocation. As an example, statistical models of video sources developed in [Heyman *et al.*,1992] have been used to optimally smooth video traffic [Ott *et al.*,1992] and to increase multiplexing gain. We use neural networks to predict the number of cells per frame for two important classes of video applications - video teleconferencing and entertainment video.

The computational model used in the paper is a neural network with ***sigmoidal*** computing units. Each unit computes the function $1/(1 + e^{-\beta x})$, where x is the units weighted input sum. The computation units are organised in layers with connections only among units in adjacent layers. In the simplest form of such a ***layered network***, there is just one ***input layer*** of source units that projects onto an ***output layer*** of computation units. In ***multilayer networks*** there is one or more ***hidden layers***, whose computation units are accordingly called ***hidden units***. The network is said to be ***fully connected*** if each node in each layer of the network is connected to every other node in the adjacent forward layer. By adding one or more hidden layers, the network is enabled to extract higher order statistics from the data (to approximate more complicated functions). A network is called *feedforward*, if all the directed links between different layers of units start in a layer that is closer to the input layer and end in a layer that is closer to the output layer.

The proposed neural network model and learning algorithms for single frame prediction are explained in Section 2, followed by the multiple frame prediction in Section 3. In Section 4 the experimental results are presented.

2 SINGLE FRAME PREDICTION

Let $u(k)$ be the frame size at time k and $y(k+1)$be the predicted frame size at time $k+1$. We assume that $y(k+1)$ is a function of $u(k), u(k-1), \ldots u(k-n+1)$. The model we use for predicting this function is a fully connected i feedforward multilayer neural network (FNN) with sigmoidal computation units. In our model, the FNN has n input units, one layer of hidden units and one output unit. The architecture is shown in Fig. 1, where the blocks denoted by z^{-1} represent one step delay elements. Learning is performed by the back propagation algorithm (BP), which is a gradient-descent iterative method for minimization of the total squared error on a set of training examples[Rumelhart *et al.*,1986]. Here, each example has the form $< u(t-n), ..., u(t-1), u(t) >$, where $< u(t-n), .., u(t-1) >$ is used as an input to the neural network and $u(t)$ is used for comparison to the predicted frame size $y(t)$.

A viable alternative to FNN with BP is the radial basis function network (RBFN) [Moody *et al.*,1988] which is a local approximation technique. Time series prediction research [Narendra 92] indicates that this approach might give very good results and fast response if the information needed for prediction is contained in a small number of previous frames. Otherwise, the prediction would require too large radial basis networks.

Both networks have advantages and disadvantages: FNN with BP learns slowly because of the gradient descent technique used for training, but is usually the choice when dealing with complex structures having a large number of inputs, whereas RBFN is very fast, but the number of radial basis units tends to increase exponentially with the dimension of the input space, so that this approach becomes practically infeasible when the dimensionality of the input space is high.

2.1 Off-line learning for simple video signal

For video teleconference traffic, the recorded traffic sequence does not have any abrupt fluctuations because there are no significant scene changes. So we first test if simple *off-line learning* is appropriate for the video teleconference sequence prediction problem. A sequence of T examples $< u(t-n), \ldots, u(t-1), u(t) >$, where $n+1 \leq t \leq T$ is used for neural network training. After a training phase of N epochs the weights are frozen and the system is used for prediction on the rest of the available sequence (for $t > T$ on input $< u(t-n), \ldots, u(t-1) >$, the predicted output is $y(t)$).

The advantage of this approach is that it is extremely fast (after the initial training phase) which makes it applicable to real-time forecasting. The drawback is that the method assumes a fixed distribution signal, which is not true for complex video sequences such as entertainment video.

2.2 On-line learning for complex video signal

For video traffic with a number of sudden scene changes, frame cuts, zooms, rapid movement of objects, the ***continuous on-line learning*** seems more appropriate. The learning is performed for M epochs on S consecutive examples $< u(t-n), \ldots, u(t-1), u(t) >$, where $n+1 \leq t \leq S$ and the network is used to predict $y(S+1)$ which is the size of the next frame. Whenever a new frame becomes available, the training set is shifted to include this new frame and the oldest frame is discarded ($t \leftarrow t+1$). A new learning session of M epochs is performed on this new training set and the next frame size prediction is obtained. The process is then repeated. The weights are adjusted continuously during the whole process. After the training set is shifted, the new training session of M epochs starts from the existing weight values rather then from random values. Therefore it is reasonable to expect convergence in a significantly smaller number of epochs as compared to the off-line learning ($M \ll N$). Since the learning has to be performed in real time, we also reduce the number of training examples as compared to the off-line learning ($S < T$). The on-line learning algorithm can also have an additional initial learning phase, similar to the learning phase of the off-line algorithm, so that the on-line learning starts with properly initialized weights.

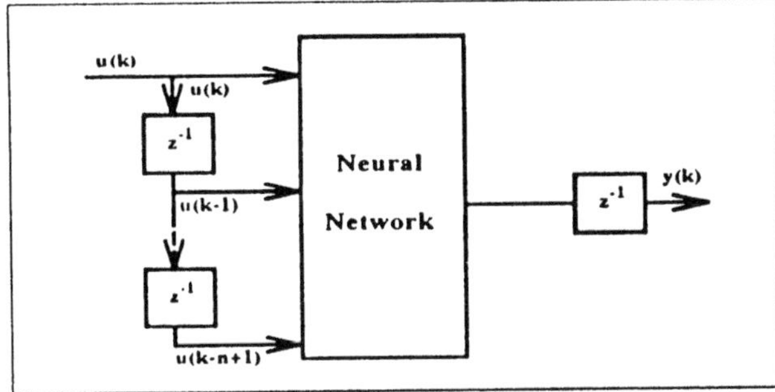

Figure 1: Neural Network Structure Used in the Simulation

The advantage of the on-line approach is the continuous learning on new examples that should cope with changes in the distribution for complex sequences. The problem of this approach is that it might be computationally too expensive for real-time response if experiments show that for accuracy reasons M or S have to be large. In such a case, an alternative to standard back propagation learning is more efficient parallel learning that we have proposed earlier and applied successfully to other problems of large scale [Mehr *et al* .,1994;Obradovic *et al.*,1994;Venkateswaran *et al.*,1994]

Another possible learning approach for complex signals is the recently proposed growing cell structures network [Fritzkel1992]. This local approximation algorithm is very fast and in contrast to regular RBFN, it is still appropriate for larger dimension problems. For very complex signals, an appropriate approach is to combine domain specific prior knowledge and constructive learning from examples into a hybrid system, as proposed in [Fletcher *et al.*,1993a;Fletcher *et al* .,1993b].

3 MULTIPLE FRAME PREDICTION

To optimally smooth delay sensitive traffic we examine to what extent neural networks can do a multiple frame size prediction. More precisely, in an s- frame prediction problem given the acutal frame sizes $u(k), u(k), \ldots, u(k-n+1)$, the network has to predict not only $y(k+1)$ as earlier, but also $y(k+2), \ldots, y(k+s)$, where $s > 1$. Obviously, this is a more challenging problem compared to the previous 1-frame prediction, and so the accuracy might be compromised if s is too large. It is an experimental question of how many previous frames are needed for s-frame prediction and how many frames ahead (maximum value for s) neural networks can predict without a significant decay in accuracy.

The first approach for multiple frame prediction (*incremental approach*) has a training phase as earlier (using an off-line) or on-line technique depending on the signal complexity). In the prediction phase $y(t+1)$ is computed using $< u(t-n), \ldots, u(t)) >$ as the network input. Prediction for the frame size at time $t+2$ is computed using $< u(t-n+1), \ldots, u(t), y(t+1) >$ (oldest actual frame size is discarded and the previous prediction -$y(t+1)$ is used instead of the corresponding actual value). Similarly, $y(t+3)$ is predicted using $< u(t-n+2), \ldots, u(t), y(t+1), y(t+2) >$, and the process continues until $y(t+s)$ is computed using $< u(t-n+s-1), \ldots, u(t), y(t+1), \ldots, y(t+s-1) >$.

Another approach for multiple frame prediction (*direct approach*)is to learn a mapping $f : N^n \rightarrow N^s$ from the previous n to the next s frames. In this model the network has s output units rather than one. The training phase uses examples of the form $< u(t-n), \ldots, u(t+s) >$ where the first n components are used as the network input and the rest to compare the computed response versus the actual frame sizes. The advantage of this direct approach is that the prediction error is not accumulating as in the incremental approach. On the other side this approach has more free parameters in the system. So

Table 1: Simulation using small buffer and predicted teleconference sequence

buffer	0.5	1	2	3	4	5
predicted	4.104e-05	3.830e-05	3.447e-05	3.109e-05	2.802e-05	2.495e-05
actual	7.444e-05	6.988e-05	6.184e-05	5.592e-05	5.183e-05	4.808e-05

the learning process is computationally more demanding and also a larger training set is needed for a good generalization.

4 EXPERIMENTAL RESULTS

A number of experiments were performed by varying the training set size, the network input layer size, the number of hidden units, the training time (number of epochs) and the learning rate. This section contains a short summary of the obtained results. Sections 4.1 and 4.2 present the results for the single frame prediction problem, while Section 4.3 presents the multiple frame prediction.

4.1 Video Teleconference Data

The experiments on video teleconference sequence forecasting, were done by using recorded and compressed real video conference data consisting of 40,000 frames previously used in [Heyman *et al.*,1992]. The first 1000 frames are shown in Fig.2.

Reasonably good prediction results are obtained using the off-line learning on a neural network with 5 inputs, 5 units in one hidden layer and single output unit. Learning is performed on the first 666 frames for 10000 epochs and testing is done on the next 329 frames. Fig. 3 shows the quantile-quantile plot of predicted versus acutal data. The generalization stays approximately the same for smaller number of training epochs and longer test sequences (up to all available 40,000 frames). Fig. 4 shows the results obtained using an identical network and the same training examples as previously, but with 2000 training epochs and a prediction of 9000 frames. The autocorrelation function for the actual and the predicted data corresponding to Fig. 4 is shown in Fig. 5. The autocorrelation functions are exponentially decreasing and the fit between actual and predicted data is good.

The predicted sequence is used in simulations in which 25 multiplexed video sources feed a finite buffer queue which models the queue at the output port of an ATM switch. The output rate from the queue is 56Mbps. The buffer size at the queue is specified in terms of maximum delay that can be tolerated at the queue. Maximal delay is varied from 0.5 ms to 35 ms. Incoming traffic (in the form of ATM cells) is lost if it arrives to a full buffer. The results of a number of simulations is shown in Table 1 and 2. In these

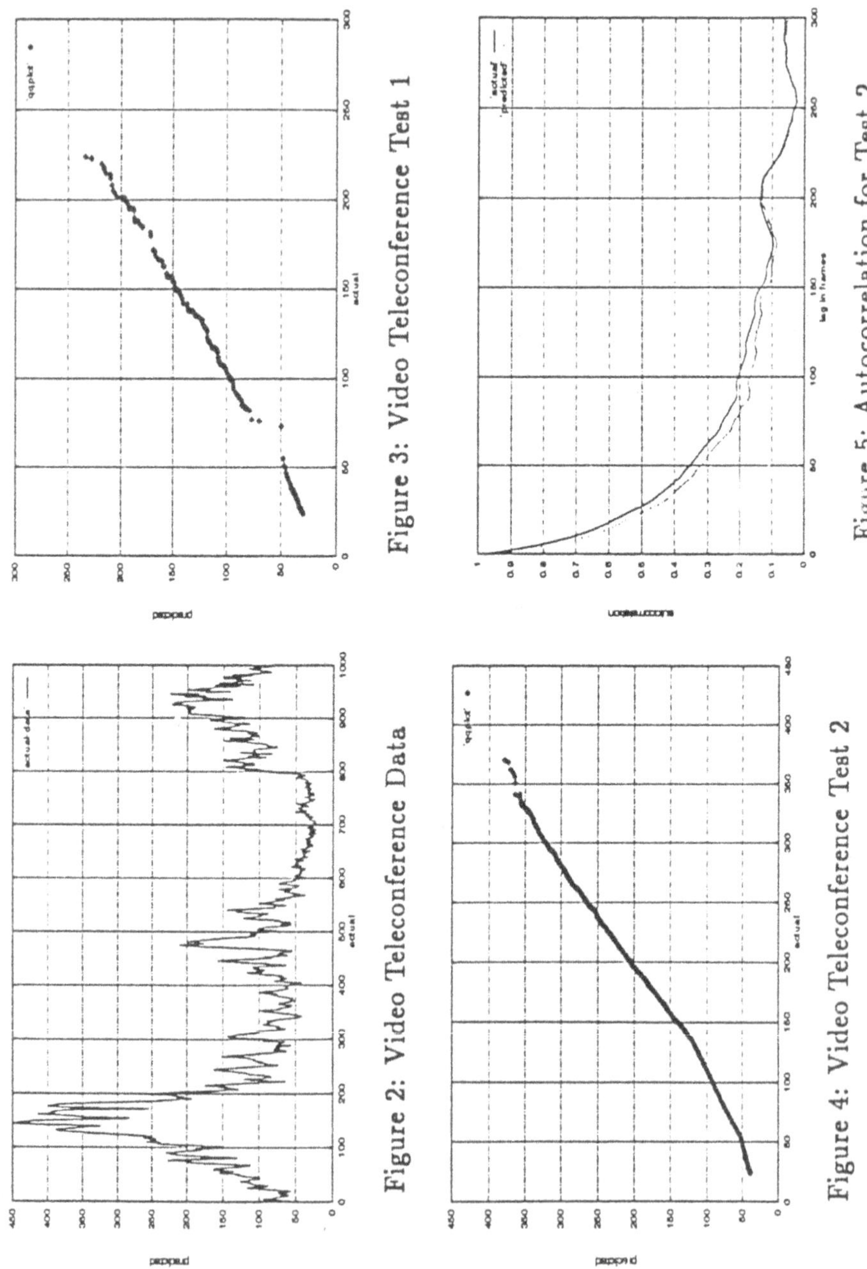

Figure 2: Video Teleconference Data

Figure 3: Video Teleconference Test 1

Figure 4: Video Teleconference Test 2

Figure 5: Autocorrelation for Test 2

Table 2: Simulation using large buffer and predicted teleconference sequence

buffer	11	12	13	14	15
predicted	1.156e-05	1.002e-05	8.486e-06	6.937e-06	5.768e-06
actual	3.647e-05	3.497e-05	3.347e-05	3.196e-05	3.046e-05

tables the buffer size is specified in terms of maximal delay(i.e.0.5ms,1 ms,2 ms etc) and the cell loss probabilities corresponding to these buffer sizes are specified for the predicted and the actual sequences. The fit is reasonably good for traffic purposes for small buffers (accurate to within a factor of 2 or 3) but for large buffers the actual and predicted losses diverge.

Slightly better results were obtained by using on line learning. Fig.6 shows the quantile-quantile plot of predicted versus actual data for predicting 1000 values. The neural network used has 5 input units, one hidden layer with 5 units and one output unit and learning is performed using a window of S=100 frames and M=200 epochs.

4.2 Entertainment Video Data

Entertainment video has frequent scene changes which makes the prediction problem very difficult. Supervised neural networks learning seems quite appropriate for such noisy prediction problems. The experiments on entertainment video sequence forecasting, have been done by using recorded and compressed real data consisting of 12,000 frames. The global distribution for this data set is unknown, although certain small subsequences of data can be identified to have a gamma distribution. The first 4000 frames are shown in Fig.7.

Fig.8 Shows the quantile-quantile plot of predicted versus actual data obtained using the off-line learning on a neural network with 5 inputs, 5 units in one hidden layer and a single output unit. Learning is performed on the first 666 frames for 2000 epochs and testing is done on the next 3000 frames. From Fig.8 it is clear that the losses are caused manily by the scene changes(peaks). The autocorrelation function for the acutal and the predicted data used in the entertainment video test corresponding to Fig.8 is shown in Fig. 9. It can be observed that the fit between actual and predicted data is still good, although the exponential character of the curves is lost.

The on-line experiments were similar to those performed for the video teleconference data, but the results were fairly poor. Although the predictions had a tendency of sensing the scene changes, the the errors were excessive. We believe that the reason for this behaviour is the slow back propagation learning.

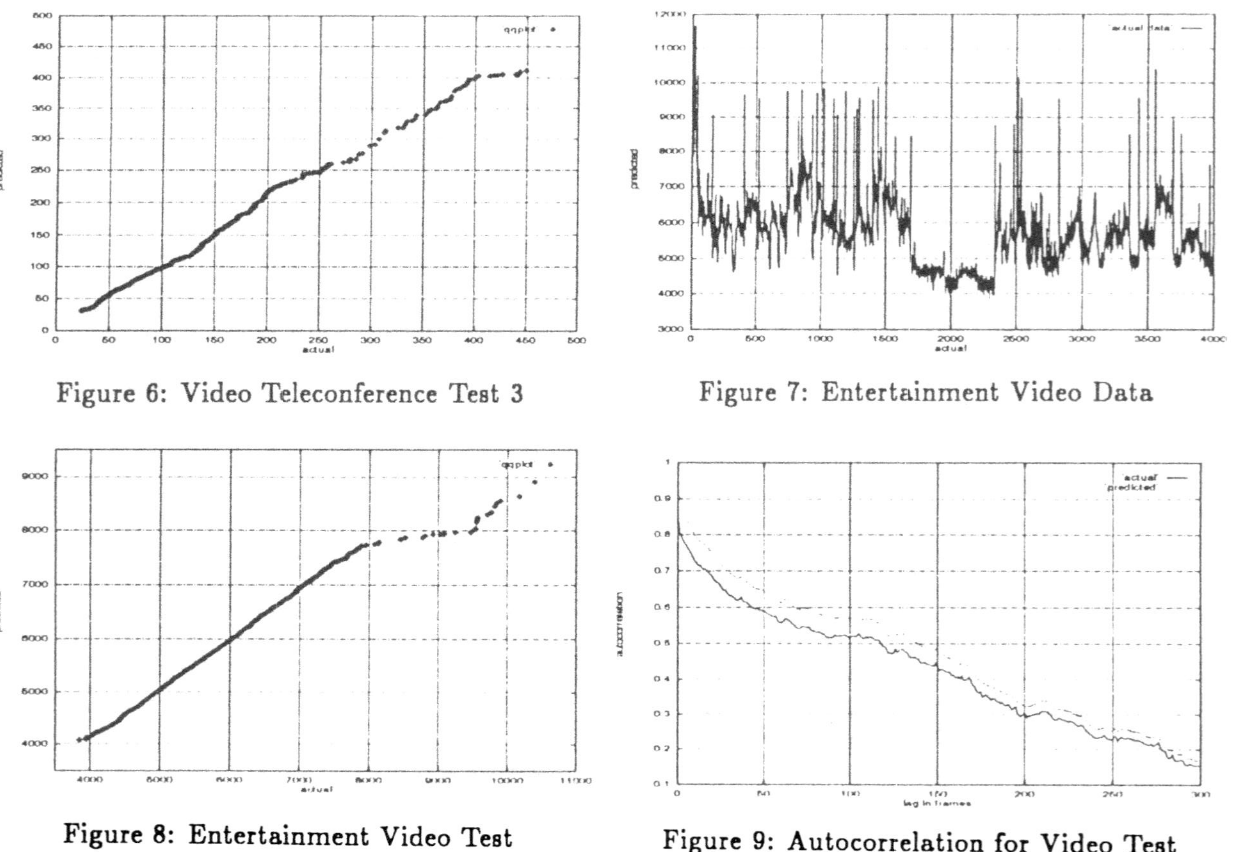

Figure 6: Video Teleconference Test 3

Figure 7: Entertainment Video Data

Figure 8: Entertainment Video Test

Figure 9: Autocorrelation for Video Test

4.3 Multiple Frame Prediction for Video Teleconference Data

The previously explained incremental approach is used for 2-frame prediction experiments on video teleconference data. The neural network and the data set used were exactly the same as those used in Test 2 (corresponding to Fig 4). Figures 10 and 11 present the quantile-quantile plots of predicted versus actual data for the 2-frame and the 3-frame problems respectively. In the 2- frame problem, the accuracy of the prediction is measured by comparing the second predicted frame size versus the corresponding actual value. Similarly, in the 3-frame problem the accuracy is measured by comparing the third predicted frame size versus the corresponding actual value. By comparing Figures 4,9 and 10 it can be observed that in the s-frame prediction problem the accuracy of the prediction is decreasing as s increases.

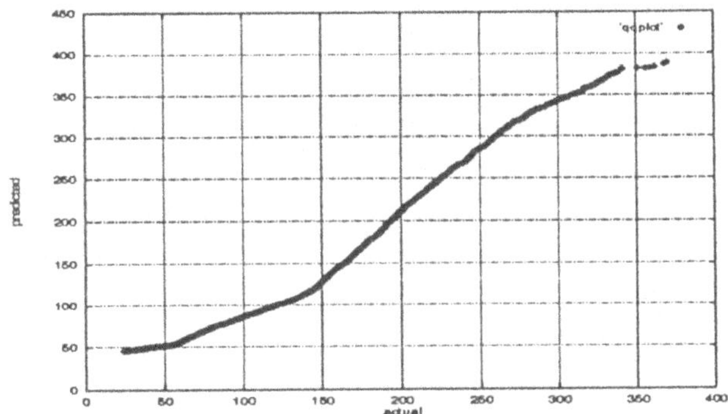

Figure 10: 2-nd Frame for Test 2

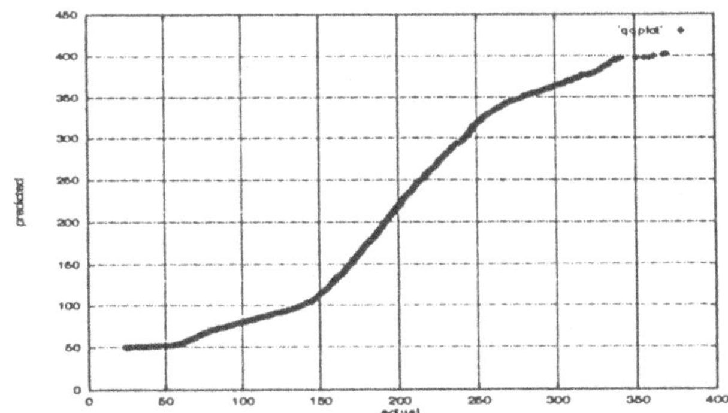

Figure 11: 3-rd Frame for Test 2

These preliminary results on the multiple frame prediction problem suggest that the direct approach might give better results since the error accumulated in the incremental approach is too large even for predicting the third following frame.

5 CONCLUSION

The experimental results indicate that the single frame problem for the video teleconference data can be predicted with reasonably good accuracy using both on-line learning methods. The entertainment video data is more difficult to predict but the off-line learning method still gives promising results. The on-line learning method in its actual form can not be used for predicting bursty data in real-time.

Preliminary results on the multiple frame problem using the incremental learning approach indicate the method is inappropriate for prediction using only information contained in a small number of previous frames. The direct learning approach for this problem is still under investigation. Also, the experimentation with radial basis function networks for on-line learning of the video entertainment data is in progress.

Traffic prediction has application in smoothing video traffic in an ATM network, which is useful in network traffic management. Probabilistic models used for video traffic prediction have been successfully applied to traffic smoothing with some constraints [Ott *et al.*,1994]. The application of the neural networks based traffic prediction model to traffic smoothing and to other traffic management problems are currently under consideration.

6 REFERENCES

[Le Cun et al.,1989] Y.Le Cun *et al.*, Handwritten Digit Recognition: Applications of Neural Network Chips and Automatic Learning, *IEEE Communications*, pages 41-46,1989.

[Eberhart *et al.*,1990] R.C. Eberhart, R.W.Dobbins. Case Study I: Detection of Electroencephalogram Spikes, *In Neural Networks PC Tools*, ed. R.C.Eberhart, R.W.Dobbins, San Diego, CA, Academic Press, 1990.

[Fletcher et al.,1993a] J.Fletcher, Z.Obradovic. Combining Prior Symbolic Knowledge and Constructive Neural Networks, *Connection Science Journal*, vol 5, Nos 3 and 4, pages 365-375, 1993.

[Fletcher *et al.*,1993b] J.Fletcher, Z. Obradovic. Parallel Constructive Neural Network Learning, *Proc. IEEE 2nd Int. Symp. on High-Performance Distributed Computing*, Spokane, WA, pages 174-178,1993.

[Fritzkel1992] B.Fritzke. Growing Cell Structures - A Self-Organizing Network in K Dimensions, *In Artificial Neural Networks 2*, eds. I. Aleksander and J. Taylor, North-Holland, pages 1051-1056, 1992.

[Heyman et al.,1992] D.P.Heyman, A.Tabatabai, T.V.Lakshman. Statistical Analysis and Simulation Study of Video teleconference Traffic in ATM Networks, *IEEE transactions on Circuits and Systems for Video Technology*, Vol.2, No.1, pages 49-59, March 1992.

[Hiramatsu1990] A.Hiramatsu. ATM Communications Network Control by Neural Networks, *IEEE transactions on Neural Networks*, Vol.1, No.1, pages 122-130, March 1990.

[Mehr et al.,1994] I.Mehr, Z.Obradovic. Parallel Neural Network Learning Through Repetitive Bounded Depth Trajectory Branching, *Proc 8th IEEE Int. Parallel Processing Symposium*, Cancun, Mexico, pages 784-791, 1994.

[Moody et al.,1988] J.Moody, C.Darken. Learning with Localized Receptive Fields, *Connectionist Models Summer School*, Morgan Kaufmann, pages 133-143, 1988.

[Narendra92] K.S.Narendra. Adaptive Control of Dynamical Systems Using Neural Networks, *In Handbook of Intelligent Control*, ed. D.A. White and D.A.Sofge, Van Nostrand Reinhold, pages 141-183, 1992.

[Neves *et al.*,1993] J.E.Neves, L.B.de Almeida, M.J.Leitao. ATM Call Control by Neural Networks, *Proc. Int. Workshop on Applications of Neural Networks to Telecommunications*, Princeton, N.J., 1993.

[Nordstrom 1993] E.Nordstrom. A Hybrid Admission Control Scheme for Broadband ATM Traffic, *Proc. Int. Workshop on Applications of Neural Networks to Telecommunications*, Princeton, N.J., 1993.

[Obradovic et al.,1994] Z. Obradovic, R. Srikumar. Evolutionary Design of Application Tailored Neural Networks, *Proc. IEEE Int.Symp. on Evolutionary Computation*, Orlando, FL, 1994.

[Ott *et al.*,1994] T.J. Ott, T.V.Lakshman, A.Tabatabai. Smoothing Traffic with Real-Time Constraints Using Arrival Forecasts, *Manuscript submitted to IEEE Transactions on Communications.*

[Ott *et al.*,1992] T.J. Ott, T.V. Lakshman, A. Tabatabai. A Scheme for Smoothing Delay-

Sensitive Traffic Offered to ATM Networks, *Proc of INFOCOM*, pages 776- 785, 1992.

[Rumelhart *et al.*,1986] D.E.Rumelhart, G.E.Hilton and R.J.williams. Learning Internal Representations by Error Propagation, ***Parallel and Distributed Processing***, Eds. D.E.Rumelhart and J.L.McClelland, Cambridge, MA, MIT Press, 1986.

[Takahasni et al.,1990] T. Takahasni, A. Hiramatsu. Integrated ATM Traffic Control by Distributed Neural Networks, *Proc. 13th Int. Switching Symp.*, Vol. 3, pages 42-56.

[Venkateswaran *et al.*,1994] R. Venkateswaran, Z. Obradovic. Efficient Learning through Cooperation, *Proc. 1994 World Congress on Neural Networks*, San Diego, CA, 1994.

PART THREE

Protocol Engineering

PROTOCOL ENGINEERING

The advent of OSI standards and the consequent interest generated in the industry to produce "Standards-conformant" products lead to the growth of Protocol Engineering with special emphasis on conformance testing. There are two papers in this session one dealing with test sequence generation and the other comparing all the well known testing methods.

The paper titled "Protocol test sequence generation using MUIOS based on TSP problem" by Prem Kumar and Venkataraman proposes a test sequence generation method based on MUIOS, and the proposal is shown to be efficient compared to the other existing methods. The second paper titled " Protocol conformance testing -A survey" by Neelakantan and Raghavan focuses on the optimization test sequences and stresses the need for maximum overlap between test sequences. The paper highlights multilevel method along with illustrations.

10

Protocol Test Sequence Generation Using MUIOS Based on TSP Problem

G. Prem Kumar and P. Venkataram
Department of Electrical Communication Engineering
Indian Institute of Science; Bangalore - 560 012, INDIA
(Tel : (+91) (080) 3340855; Fax : (+91) (080) 3347991
e-mail : {prem, pallapa}@ece.iisc.ernet.in)

Abstract

In this paper, a method for optimal test sequence generation using Multiple Unique Input/Output Sequences (MUIOS) is presented. The test sequence generation problem using MUIOS is viewed as an asymmetric travelling sales person problem and is solved using *simulated annealing* methodology. This technique is applied to various test cases and the results are compared with the results obtained by using rural Chinese postman algorithm.

Keywords

Protocol conformance testing, formal methods for test sequence generation, multiple input/output sequences, travelling sales person problem, simulated annealing, rural Chinese postman algorithm.

1 INTRODUCTION

Conformance testing for communication protocols is to test whether a protocol under test, conforms to the specifications laid down during the design (Aho, 1991). The conformance testing of protocols is being done in two ways (Sarikaya, 1984) (Sarikaya, 1987) (Linn, 1989/1990). First approach is to build a test system, which simulates the protocol. These test systems designed by users and suppliers to their requirements, often have following disadvantages : (a) it is a costly effort to build individual testers, and (b) standardization of protocol products is difficult. Second approach deals with generation of efficient test cases for testing the implemented protocol product. The desirable features are short test sequences and less human intervention. Following the second way of conformance testing, this paper gives an efficient method of test sequence generation.

The formal methods used to generate conformance test sequences, that are modeled as Finite State Machines (FSMs) (Kohavi, 1978) are Transition Tours (T-method) (Bosik, 1991) (Naito, 1981) Distinguishing Sequence (D-method) (Gonenc, 1970) (Ural, 1992) (Chow, 1978), Characterizing Sequences (W-method) (Bosik, 1991) and Unique

Input/Output Sequences (U-method) (Aho, 1991) (Bosik, 1991) (Sidhu, 1989) (Ural, 1992) (Sabnani, 1988) (Miller, 1993).

We formulate the problem of test sequence generation (TSG) based on MUIOS as an asymmetric travelling sales person problem (we call this as TSP) where sub-sequences (defined later) corresponding to an edge of the FSM are considered as a node and bridge length between the nodes as elements of the TSP-matrix. We solve the TSP problem thus formulated by using Simulated Annealing (SA) methodology.

In Section 2, the notation used in this paper and an overview of some of the protocol test sequence generation methods are introduced. In Section 3, the asymmetric travelling sales person problem and its solution using Simulated Annealing methodology are described. Section 4 explains the method of test sequence generation. Section 5 illustrates the proposed method with an example. Section 6 presents the results and comparison of the proposed method with the one solved using Rural Chinese Postman (RCP) problem. Finally, some concluding remarks are given in Section 7.

2 NOTATION AND OVERVIEW OF FORMAL METHODS FOR TEST SEQUENCE GENERATION

Before giving an overview of formal methods, we introduce the notation related to finite state machines, graph theory and protocol conformance testing, that is followed in this paper.

2.1 Definitions

Definition 1: A protocol represented as a FSM consists of a set of states and a set of transitions. It can be modelled as a directed graph $G = (V, E)$, where V is a set of vertices corresponding to the states of the FSM and E is a set of edges corresponding to the transitions of the FSM. G is called as *FSM graph.*

Definition 2: A FSM M, is *deterministic* if there is at most one transition defined for each input at each state of M.

Definition 3: States s_i and s_j of M are *equivalent* if they have identical edges.

Definition 4: M is *minimal* if none of its states are equivalent.

Definition 5: M is *strongly connected* if there exists a path between every pair of states.

Definition 6: M is *fully specified,* if every state can accept all the possible inputs. Otherwise, M is *partially specified.*

Definition 7: A state s_i has a *self loop* if it has a transition that starts and ends in s_i.

Definition 8: A *tour* in a graph $G(V, E)$ is a sequence of consecutive edges which starts and ends at the same vertex in V.

Definition 9: A graph $G(V, E)$ is *symmetric* if for every vertex $v_i \in V$, the indegree of v_i equals its outdegree.

Definition 10: An *Euler tour* in G is a tour that contains every edge of G exactly once.

Definition 11: *Chinese Postman Problem* is defined as finding the minimum cost tour of a strongly connected directed graph G, that contains every edge of G at least once.

Definition 12: *Rural Chinese Postman (RCP) problem* is to find a tour in a strongly connected directed graph G such that it traverses some subset of edges in G at least once with minimum cost.

Definition 13: *Controllability* of a protocol is the ability to bring the implementation into a desired state.

Definition 14: *Observability* of a protocol is the ability to observe the current state of the protocol (this is generally done by looking at the output(s) generated).

In a protocol specification using FSM, testing an edge $(s_i, a_p/x_q, s_j)$ is done in three steps. In the first step, place the FSM in state s_i, second step is to apply the input a_p and observe the output x_q and the third step is to verify that the FSM entered into state s_j.

Definition 15: *Reset edges* are those edges that bring M from any state of the FSM to the initial state.

We discuss, briefly, four formal methods for test sequence generation and their implementation as automated software tools.

2.2 T-method

Transition Tour (Bosik, 1991) (Naito, 1981) is an input sequence of FSM M, called TT, that starts at some state s_i, visits all the edges of M at least once and comes back to s_i. One method of generating TT is to make M (i.e., its graph G) symmetric (call that obtained graph as $G*$) such that it replicates minimum number of edges of G, which is solved as a linear programming problem, and finding an Euler tour of $G*$.

2.3 D-method

In Distinguishing Sequence method (Gonenc, 1970) (Chow, 1978), there is a sequence of inputs I, called DS, when applied to all the states of M, results in different outputs. One method of generating DS is to find the minimal covering of @(i,DS-diagram) (Kohavi, 1978) (where $i \in I$ and the symbol '@' denotes the concatenation operation on strings). One of the necessary conditions for a FSM to have DS is that, the FSM be fully specified.

2.4 W-method

Characterizing Sequence method (Bosik, 1991) is used only when M does not have a DS. The idea here is to concatenate some more input transitions to form a *partial DS*, called CS, that distinguishes a state from a subset of those states which are responding with same output. The set of such input sequence of a FSM is called the characterizing set W of the FSM. Once W is obtained, the test sequence can be built using the method used in DS method. Like DS method, in this case also, the FSM needs to be fully specified.

2.5 U-method

In this method, a Unique Input/Output (UIO) sequence is obtained for each state of the FSM. For example, if UIO_j is $(a/x, b/y)$ then no state other than j will generate the output sequence x,y for the input sequence a, b. To each edge e_{ij}, UIO_j is concatenated (together is called a *sub-sequence*, defined in detail later) and all such sub-sequences are joined in one of the several ways (Aho, 1991) (Bosik, 1991) (Miller, 1993) by using ***bridge sequences*** to build a test sequence. In (Sabnani, 1988), it is shown that U-method can be used to generate test sequences if combined with the rural Chinese postman algorithm, an NP-complete problem, for the graphs with two sufficient but not necessary conditions : FSM has reset feature and each state of the FSM has a self loop.

In a case, when some states have more than one UIO sequences (of minimal length), they are termed Multiple Unique Input/Output (MUIO) sequences. Then, for each state, choose one of the MUIO sequences, that gives the shortest test sequence. Many refinements of MUIOS are reported all of which attempt to minimize the length of the test sequence and yet have the properties of controllability and observability (Aho, 1991) (Lombardi, 1992) (Shen, 1992).

3 THE TSP PROBLEM

Travelling Sales Person problem is an example of a computationally hard problem (Rosenkrantz, 1977). In this problem, a sales person starts at an initial city and visits all n cities exactly once so that the effective length of the tour is optimal (minimal). In asymmetric travelling sales person problem, where $d_{ij} \neq d_{ji}$, there are (n-1)! distinct tours with different lengths. Here, d_{ij} represents the distance between the cities i and j. Many solutions have been proposed to solve the TSP problem. Some of them are Nearest Neighbour approach, Branch and Bound algorithm, Linear Search, and Simulated Annealing. Since Simulated Annealing is used in this work, we briefly present about the first three methods and describe Simulated Annealing methodology in detail. In Nearest Neighbour approach, starting from a given city, the unvisited nearest neighboring city is visited next till the tour is complete (Rosenkrantz, 1977). In Branch and Bound algorithm (Gillet, 1979), an attempt is made to find a solution using Assignment Problem (AP). If AP fails to find a solution and results in sub-tours then Branch and Bound strategy is again applied to solve all the sub-problems using AP. The process is repeated till a tour with smallest bound is obtained. In Linear Search method (Kanellakis, 1980), the TSP problem is formulated as no-wait flowshop scheduling problem.

3.1 Simulated Annealing

Simulated Annealing concept is derived from annealing of metals in metallurgy. In annealing, metals are heated to high temperatures and are cooled down slowly and gradually. At each decrement in the value of temperature, the metal is allowed to reach the stable state, so that when it is cooled completely, it reaches thermal equilibrium. This concept is used to solve many optimization problems (Aarts, 1990) (Kirkpatrik, 1983). We use this technique to solve the test sequence generation problem by formulating it as a TSP problem. We discuss, here, the Simulated Annealing procedure (TSP-SA), to solve TSP problem.

3.1.1 Functioning of *TSP-SA*

The procedure TSP-SA (see Figure 1) uses two basic functions : *Obj-fn()* and *Perturb()*. *Obj-fn(R)* returns length of the given tour R of a graph with n nodes. Initially tour R is either selected randomly with a fixed initial node or taken from the user. p is the number of current outer loop iterations. $Perturb(R^0, p)$ perturbs the route vector R^0 at pth iteration. Here two nodes, i, j ($1 \leq$ i,j $\leq$ n-1), of the route vector R^0 are randomly selected and swapped. (i,j $\neq$ 0 since the initial city is fixed.) s is the cooling rate of the annealing process. $random(0,1)$ generates a value between 0 and 1. The iterative values, T_i and T_f are analogous to the initial and final temperatures used in annealing of metals. T_i and T_f have to be selected properly based on the value, $(V_0 - V_1)$ (Aarts, 1990). We select T_i and T_f values as 1000 and 10 respectively for $(V_0 - V_1)$ of order 10. s should be less than and very close to 1. We set s = 0.975.

```
Procedure TSP-SA(input: Ti,Tf, Max-ite; output: R(1..n))
begin
    var p,k : int
        V0, V1 : real
    Initialize R to R0,
    V0 = Obj-fn(R0), p = 0
    while (Ti > Tf)
    begin
        k = 0
        while (k < Max-ite)
        begin
            R1 = Perturb(R0 , p),
            V1 = Obj-fn(R1), /* TSP(n,n) is used here */
            If((V1 < V0) Or (exp(-(V1 -V0)/Ti) > random(0,1)))
            begin
                R0 = R1, V0 = V1,
                if(Obj-fn(R0) < Obj-fn(R))
                begin
                    R = R0
                end
            end
            k = k + 1
        end
        p = p + 1, Ti = s * Ti
    end
    return R
end
```

Figure 1 Procedure Simulated Annealing.

4 TEST SEQUENCE GENERATION USING MUIOS, THE PROPOSED METHOD

In this section, for a protocol given in the form of a deterministic, strongly connected, minimal FSM, we formulate the conformance test sequence generation problem as an asymmetric travelling sales person problem and find a solution to this problem by using Simulated Annealing methodology as described in the previous section.

First, we introduce some terminology related to the proposed method and llustrate them using FSM *P1* (see Figure 2).

Definition 16 : *Sub-Sequence* (SS_{mk}) is a concatenation of an edge $e_m \in E$ of the FSM graph, G(V,E), and kth UIOS of its tail state. For example, the sub-sequences for the edge $e_2(s_1$,b/y,$s_3)$ of FSM $P1$ are,

$$SS_{21}\text{: } (s_1\text{,b/y,}s_3\text{;}s_3\text{,a/x,}s_2\text{;}s_2\text{,a/x,}s_1),$$
$$SS_{22}\text{: } (s_1\text{,b/y,}s_3\text{;}s_3\text{,b/y,}s_1\text{;}s_1\text{,a/y,}s_2)$$

since there are two UIOS corresponding to the tail state s_3, of e_2. Where as, edge $e_1(s_1$,a/y,$s_2)$ has only one sub-sequence,

$$SS_{11}\text{: } (s_1\text{,a/y,}s_2\text{;}s_2\text{,a/x,}s_1\text{;}s_1\text{,a/y,}s_2).$$

Definition 17 : *Bridge length* (t_{ij}) is the distance between the tail state of sub-sequence SS_{ik} and the head state of sub-sequence SS_{jl}. The t_{ij} is an integer value which measures in terms of number of transitions *essential* to reach from tail state of SS_{ik} to head state of SS_{jl}. The sequence joining them is called a *bridge-sequence*, b_{ij}. For example, bridge length t_{21} between SS_{22} and SS_{11} is 1 and the corresponding bridge sequence b_{21} is 'a/x'. Similarly, the bridge length t_{21} between the sub-sequence SS_{21} and sub-sequence SS_{11} is zero and the corresponding bridge-sequence b_{21} is NULL.

Definition 18 : *TSG-graph* is a directed graph $TSG(N, B)$, where N is a set of nodes, $n_i \in N$, n_i = { SS_{ik}| 1 $\leq$ k $\leq$ U, U is the number of UIOS of the tail state of edge e_i and SS_{ik} are the sub-sequences corresponding to the edge e_i of the FSM } and B is a set of bridge-sequences, $b_{ij} \in B$ (the minimum bridge-sequence between the nodes n_i and n_j is represented as an edge). For example, the bridge-sequence b_{21} entered as the edge of the TSG-graph is NULL and not 'a/x', since former is the minimum possible bridge-sequence between the nodes n_2 and n_1 of the TSG-graph.

Definition 19 : A *tour in TSG-graph* always starts at the initial node, $INODE$ (the sub-sequences in the INODE must have the *initial state* of the FSM as the *head state*) and traverses the minimal path across all the nodes. For example, in FSM $P1$ of Figure 2, the INODE is n_1 or n_2, as both the nodes meet the requirement of having state s_1 as its head state.

Definition 20 : *TSG-matrix* is a 2-D matrix, where each element, $TSG(i, j)$ contains two fields, first field representing the sub-sequence, SS_{ik}, chosen at node n_i to traverse to n_j with minimum bridge-sequence b_{ij} and the second field representing the bridge-sequence b_{ij}. For example, the element, $TSG(2, 1)$ contains SS_{21}| NULL.

Definition 21 : *TSP-matrix* is a 2-D matirx, where each element, $TSP(i, j)$ represents the bridge length, t_{ij}, between the nodes ni and nj. For example, $TSP(2, 1)$ represents the bridge length t_{21}, which is zero.

Definition 22: *Overlap edges* of two sub-sequences are those edges at the end of first sub-sequence, which overlap with same number of edges at the beginning of second sub-sequence.

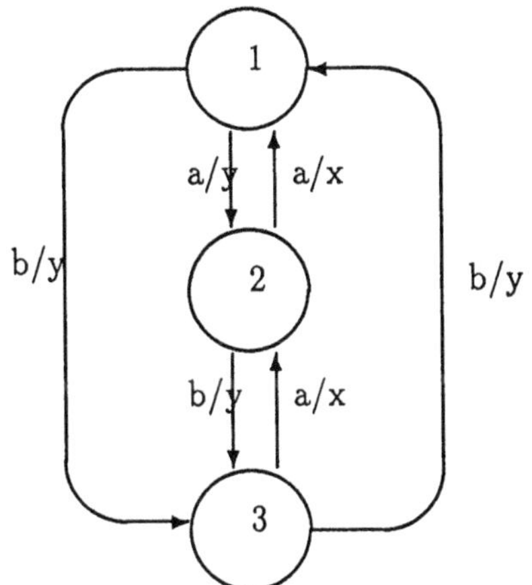

Figure 2 FSM P1.

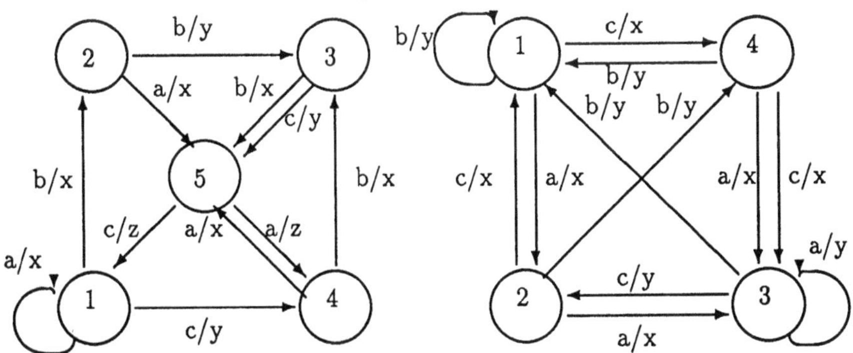

Figure 3 FSM P2.

Figure 4 FSM P3.

Table 1 MUIOS for the FSMs shown in Figures 2-4

FSM P1		FSM P2		FSM P3	
state	UIOS	state	UIOS	state	UIOS
1	a/y	1	a/x,a/x	1	* a/x,c/x
			* b/x,a/x		a/x,a/x
2	a/x,a/y		a/x,b/x	2	* b/y,a/x,c/y
3	* a/x,a/x		c/y,b/x		c/x,a/x,a/y
	b/y,a/y		a/x,b/y		b/y,a/x,a/y
		2	b/y		
		3	* b/x,c/z	3	* c/y
			b/x,a/z		a/y
		4	b/x,c/y	4	* c/x,a/y
		5	* c/z		c/x,a/y
			a/z		

* UIOS selected in case of SUIOS.

Now, the solution to test sequence generation problem is to make a tour of the TSG-graph. Algorithm to generate protocol test sequences using the proposed method is given in Figure 5.

Algorithm TSG-TSP

```
begin
    var i,j,k : int
    1. Generate the MUIOS as given in (Sabnani, 1988).
    2. Construct TSG-graph TSG(N,B).
    3. Select an INODE in TSG(N,B).
    4. Construct TSG(n,n) and TSP(n,n) matrices, where n is
       number of nodes in the TSG-graph.
    5. Call TSP - SA with TSP(n,n) and obtain the route vector R.
    6. Calculate the test sequence, TS, using the route vector, R as follows :
            k = 0, TS = NULL
            repeat
            begin
                  i = R(k),
                  j = R(k+1),
                  TS = @( TS, @(TSG(i,j).sub-sequence, TSG(i,j).bridge-sequence)),
                  k = k+1
            end
            until(k ≥ (N-1))
            TS = @(TS, TSG(N-1,0).sub-sequence)
    7. Output TS.
end
```

Figure 5 Algorithm for test sequence generation using the proposed method.

5 AN EXAMPLE

We now illustrate the method of test sequence generation by the proposed *TSG-TSP* using FSM $P1$ as an input protocol (see Figure 2) for the case of MUIOS without reset and without overlap.

In step 1, we have the MUIOS of the FSM as shown in Table 1.

In step 2, construct the TSG-graph for the edges of the protocol, FSM P1,

$$E = \{s_1,a/y,s_2;s_1,b/y,s_3;s_2,a/x,s_1;s_2,b/y,s_3;s_3,b/y,s_1;\ s_3,a/x,s_2\}.$$

The nodes of the TSG-graph are, $n_i \in N$, where n_i represents a set of sub-sequences corresponding to e_i. They are :

n_1:
SS_{11}: $(s_1,a/y,s_2;s_2,a/x,s_1;s_1,a/y,s_2)$

n_2:
SS_{21}: $(s_1,b/y,s_3;s_3,a/x,s_2;s_2,a/x,s_1)$,
SS_{22}: $(s_1,b/y,s_3;s_3,b/y,s_1;s_1,a/y,s_2)$

n_3:
SS_{31}: (s_2,a/x,s_1;s_1,a/y,s_2)

n_4:
SS_{41}: (s_2,b/y,s_3;s_3,a/x,s_2;s_2,a/x,s_1),
SS_{42}: (s_2,b/y,s_3;s_3,b/y,s_1;s_1,a/y,s_2)

n_5:
SS_{51}: (s_3,b/y,s_1;s_1,a/y,s_2)

n_6:
SS_{61}: (s_3,a/y,s_2;s_2,a/x,s_1;s_1,a/y,s_2).

The edges of the TSG-graph are the minimum bridge-sequences between the nodes (given in TSG-matrix later). The TSG-graph for FSM P1 is shown in Figure 6.

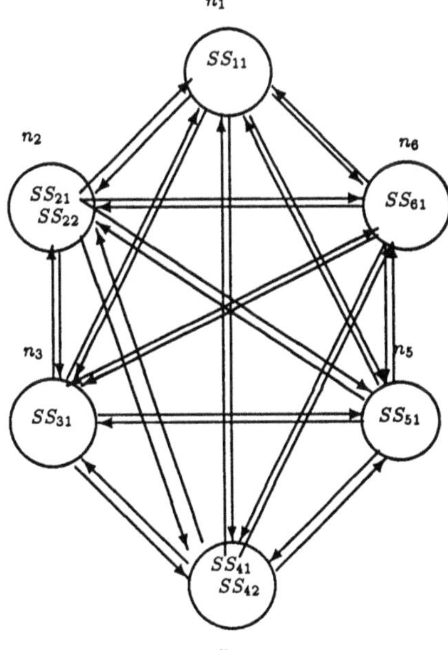

Figure 6 TSG-graph for the FSM P1 shown in Figure 2.

In step 3, select the INODE n_1 from the TSG-graph.

In step 4, the TSG-matrix is :

$$\begin{bmatrix} \text{NULL}|\text{NULL} & SS_{11}|a/x & SS_{11}|\text{NULL} & SS_{11}|\text{NULL} & SS_{11}|b/y & SS_{11}|b/y \\ SS_{21}|\text{NULL} & \text{NULL}|\text{NULL} & SS_{22}|\text{NULL} & SS_{22}|\text{NULL} & SS_{21}|b/y & SS_{21}|b/y \\ SS_{31}|a/x & SS_{31}|a/x & \text{NULL}|\text{NULL} & SS_{31}|\text{NULL} & SS_{31}|b/y & SS_{31}|b/y \\ SS_{41}|\text{NULL} & SS_{41}|\text{NULL} & SS_{42}|\text{NULL} & \text{NULL}|\text{NULL} & SS_{41}|b/y & SS_{41}|b/y \\ SS_{51}|a/x & SS_{51}|a/x & SS_{51}|\text{NULL} & SS_{51}|\text{NULL} & \text{NULL}|\text{NULL} & SS_{51}|b/y \\ SS_{61}|a/x & SS_{61}|a/x & SS_{61}|\text{NULL} & SS_{61}|\text{NULL} & SS_{61}|b/y & \text{NULL}|\text{NULL} \end{bmatrix}$$

and the TSP-matrix is :

$$\begin{bmatrix} 999 & 1 & 0 & 0 & 1 & 1 \\ 0 & 999 & 0 & 0 & 1 & 1 \\ 1 & 1 & 999 & 0 & 1 & 1 \\ 0 & 0 & 0 & 999 & 1 & 1 \\ 1 & 1 & 0 & 0 & 999 & 1 \\ 1 & 1 & 0 & 0 & 1 & 999 \end{bmatrix}.$$

In step 5, the procedure *TSP-SA* is called with $TSP(n, n)$, which returns the route vector $R = (1\ 3\ 5\ 4\ 2\ 6)$.

In step 6, the test sequence is :

a/y,a/x,a/y,a/x,a/y,b/y,b/y,a/y,b/y,a/x, a/x,b/y,a/x,a/x,b/y,a/x,a/x,a/y.

The length of the test sequence is 18.

6 RESULTS

We also present the results for other two FSMs shown in Figures 3 and 4. For each test case, we consider SUIOS without reset edges, with and without overlaps; MUIOS without overlaps, with and without reset edges. The results obtained by the proposed method are compared with results obtained using rural Chinese postman algorithm (see Table 2).

Table 2 Comparison of length of the test sequences obtained by TSG-TSP and RCP

Testcase	Test Sequence Length				
	SUIO w/o reset edges		MUIO w/o overlap edges		RCP with reset and w/o overlap using MUIOS
	overlap	w/o overlap	w/o reset	reset	
P1	13	18	18	25	26
P2	24	34	29	43	43
P3	31	40	37	46	54

7 CONCLUSION

In this paper, a technique for test sequence generation using MUIOS is described. The test sequence generation problem is considered as a TSP problem and is solved by using Simulated Annealing methodology. The results presented demonstrate that the method is *at least as efficient* as the method of test sequence generation using rural Chinese postman algorithm.

ACKNOWLEDGMENT

The authors would like to thank Prof. Sidhu and Dr. Motteler, University of Maryland, USA, for their helpful suggestions.

REFERENCES

Aarts E. and Korst J. (1990) Simulated Annealing and Boltzmann machines - A stochasic approach to combinatorial optimization and Neural computing, John Wiley & Sons.

Aho A. V., Dahbura A. T., Lee D. and Uyar U. (1991), An optimization technique for protocol conformance test generation based on UIO subsequences and rural Chinese postman tours, *IEEE trans. on Comm.*, 39(11), 1604–1615.

Bosik B. S. and Uyar M. U. (1991), Finite state machine based formal methods in protocol conformance testing : from theory to implementation, *Comp. Networks and ISDN Systems*, 22, 7–33.

Chow T. S. (1978) Testing software design modeled by finite-state machines, *IEEE trans. on Software Engineering*, SE-4(3), 178–187.

Gillett B. E. (1979) Introduction to Operations Re search, A computer oriented algorithmic approach, Tata Mc Graw Hill Edition, New Delhi.

Gonenc G. (1970) A method for the design of fault detection experiments, *IEEE trans. on Comp.*, 19, 551–558.

Kanellakis P. C. and Papadimitriou C. H. (1980) Linear search for the asymmetric travelling salesman problem, Operations Research, 28(5), 1086–1099.

Kirkpatrick S., Gelatt J. C. D. and Vecchi M. P. (1983) Optimization by Simulated Annealing, *Science*, 13 , 220(4598), 671–680.

Kohavi Z. (1978) Switching and Finite Automata Theory, Mc Graw Hill, New York.

Linn J. R. J. (1989/1990) Conformance testing for OSI protocols, *Computer Networks and ISDN Systems*, 18, 203–219.

Lombardi F. and Shen Y. N. (1992) Evaluation and improvement of fault coverage of conformance testing by UIO sequences, *IEEE trans. on Comm.*, 40(8), 1288–1293.

Miller R. E. and Paul S. (1993) On the Generation of Minimal-Length Conformance Tests for Communication Protocols, *IEEE/ACM trans. on Networking*, 1(1), 116–129.

Naito S. and Tsunoyama M. (1981) Fault detection for sequential machines by transition-tours, in: *Proc. of* 11th *IEEE Fault Tolerant Computing Symp.*, 238–243.

Rosenkrantz D. J., Stearns R. E., and Lewis P. M. (1977) An analysis of several heursistics for the Travelling Salesman Problem, *SIAM Jl. of Computing*, 6, 563–581.

Sabnani K. and Dahbura A. T. (1988) A protocol testing procedure, *Comp. Networks and ISDN Systems*, 15, 285–297.

Sarikaya B. (1987) Conformance testing: Architectures and test sequences, *Comp. Networks and ISDN Systems*, 17, 111–126.

Sarikaya B. and Bochmann G. v. (1984) Synchronization and Specification issues in protocol testing, *IEEE Software Engineering*, Com-32(4), 389–395.

Shen Y. N., Lombardi F. and Dahbura A. T. (1992) Protocol conformance testing using Multiple UIO sequences, *IEEE trans. on Comm.*, 40(8), 1282–1287.

Sidhu D. P. and Leung T. K. (1989) Formal methods for protocol testing : A detailed study, *IEEE trans. of Software Engineering*, 15(4), 413–426.

Ural H. (1992) Formal methods for test sequence generation, *Comp. Comm.*, 15(5), 311–325.

Prem Kumar Gadey received his B.Tech. (Electronics & Communication Engineering) from Sri Venkateswara University in 1990 and M.Tech. (Artificial Intelligence & Robotics) from University of Hyderabad in 1992. Since then he is a Ph.D. student in department of Electrical Communicaiton Engineering, Indian Institute of Science, Bangalore. His major research interests include Communication Networks, Internetworking, Distributed Computing, Expert Systems and Artificial Neural Networks. Currently he is focussing on applying Artificial Intelligence techniques to the area of Network Management. He is a student member of IEEE Communication Society.

Pallapa Venkataram received his Ph. D. degree from The University of Sheffield, England, in 1986. He is currently an Associate Professor in department of Electrical Communication Engineering, Indian Institute of Science, Bangalore, India. He worked in the areas of Distributed Databases, Communiation Protocols and AI applicaitons in Communication Networks and has published many papers in these areas.

11

Protocol Conformance Testing – A Survey

B. Neelakantan and S.V.Raghavan,
Department of Computer Science and Engineering,
Indian Institute of Technology Madras, Madras 600 036, India.
svr@iitm.ernet.in

Abstract

Testing is an integral part of protocol development cycle. In this paper, we will briefly discuss the protocol conformance testing methodologies and framework proposed by the International Standards Organization. Many efficient test sequence generation methods have been proposed to check the conformance of an implementation of a protocol to its standards. We will discuss these methods briefly. Finally, we will compare different test methodologies based on their fault coverage and the length of the test sequence.

Key words

Black Box Testing, Fault Detection Methods, Fault Diagnosis Methods, Finite State Machine, Protocol Conformance Testing, Test Sequence, Test Suite Design.

1 INTRODUCTION

Protocol Software Engineering has been an active area of research in the field of computer networks and distributed computing. From the time International Standards Organization (ISO) started developing standards for conformance testing of network protocols, a lot of research work has been undertaken in the field of conformance testing.

Conformance testing involves testing both the capabilities and the behavior of an implementation and checking what is observed against the conformance requirements in the relevant recommendations and against what the implementor states the implementations capabilities are [Rayner (1987)]. This is done by applying a test suite to the Implementation Under Test (IUT) and giving a verdict by comparing the observed output with the expected output. Conformance testing ***does not*** asses the performance, robustness or reliability of an IUT.

The protocol specification and its implementation are modeled as deterministic Finite State Machines (FSMs). Also, the implementation of a protocols is considered a *black box* for testing purposes. The internal states of the FSM are not visible and they can be identified only by applying the State Identification Sequences (SISs) or signatures for that state. The aim of the testing process is to find out whether all the states that are in the specification are also there in the IUT and if all the transitions are implemented correctly. This is equivalent to *checking experiments* of hardware testing.

In this paper, we present two important aspects of protocol conformance testing. The first part given in section 3, discusses the Open System Interconnection (OSI) conformance testing standards, the abstract test suites, the test architecture, the test language, test realization and conformance assessment review. These standards have approached the International Standards stage. The second part discusses the test sequence generation for the protocols. This part is yet to be standardized and there is a lot of research work going on in this area and are discussed from section 4 to section 7.

The paper is organized as follows. In section 2, we give some definitions that are used in the paper. In section 3, we discuss the OSI conformance testing methodology and framework. In section 4, we highlight the issues that should be considered during the test sequence generation. In section 5, we give different testing methodologies, which are capable of detecting a faulty IUT. In section 6, we discuss fault diagnostic methods. In section 7, different testing methods are compared with different factors like fault coverage and length. Finally, in section 8, we summarize the work reported in this paper.

2 PRELIMINARIES

2.1 Finite state machine and graph theory

A DFSM M is a quintuple $M = (I, O, S, \delta, \lambda)$, where I, O, S are finite, nonempty sets of inputs, outputs and states respectively; $\delta : I \times S \longrightarrow S$ is the state transition function; λ is the output function such that $\lambda : I \times S \longrightarrow O$. The cartesian product $I \times S$ is the set containing all pairs of elements (I_i, S_j). The state transition function δ associates with each pair (I_i, S_j) an element S_k from S, called the *next state*. The output function λ associates with each pair of (I_i, S_j) an element O_k from O [Kohavi (1978)]. Henceforth in this paper we assume that the FSM's are deterministic.

A *completely specified* FSM is one in which there exists a permissible output ($\in O$) for every input element ($\in I$) at every state. Otherwise, the FSM is said to be *partially specified*. If for every pair of states s_i and s_j of the FSM, there exists an input sequence which takes the FSM from s_i to s_j, the FSM is said to be *strongly connected*. An FSM is *minimal* if no two states of the FSM produce identical output for any possible input.

An FSM can be represented by a directed graph $G = (V, E)$, where the set of nodes $V = \{v_1, v_2, \ldots, v_n\}$ represent the set of specified states of the FSM and E, the set of labeled directed edges, defined as follows: There is a directed edge $e_{ij} \in E$, from v_i to v_j ($v_i, v_j \in V$) labeled a_k/o_l, if and only if there is a transition from v_i to v_j with input label a_k and output label o_l. Since the FSM is deterministic, for each node v_i there are no two edges leaving v_i, with same input labels. We assume that the specification FSM have n states and m edges.

A *transfer subsequence* between any two states is the shortest path between those two states. An input sequence is said to be a *homing sequence*, if the state of the machine after application of the input can be uniquely determined from the machines response, regardless of the initial state [Kohavi (1978)]. A *recovery subsequence* takes the FSM from the home state to a state where from the next test subsequence starts.

A machine is k-distinguishable if, for each pair of states, there is at least one input sequence of length k which, when applied to the pair, yields different output sequences [Chow (1978)]. A faulty transition in an IUT is *directly reachable* by a test subsequence if no other fault is encountered till the input corresponding to *edge under test* is applied [Bochmann (1993)]. Some testing methods assume that a *reliable reset* is present in the IUT, which when applied to the IUT takes it to the *start state*, irrespective of the state in which the IUT exists.

2.2 The FSM fault model

The conformance testing methods consider the specification and the IUT as FSM's and centers around the idea of applying sequence of inputs to the FSM's. If the *observed output* of the IUT is the same as the *expected output*, the IUT is said to have passed the test. However, if the *observed output* differs from the *expected output*, a *fault* is said to have occurred. Faults in an IUT can be classified as follows [Bochmann (1991)].

Output fault : A transition has an output fault if, for the corresponding state and received input, the IUT provides an output different from the expected output.

Transfer fault : A transition has a transfer fault if, for the corresponding state and received input, the IUT enters a different state other than the one specified by the next-state function.

Missing transition fault : An implementation has an missing transition, if for a pair of present state and input, the transition is not present in the IUT.

3 CONFORMANCE TESTING FRAMEWORK

In this section, we will discuss the OSI conformance testing methodology and framework, which is a five part standard. The aim of this standard is to have uniformity in testing the OSI protocols. This will lead to comparability and wide acceptance of test results produced by different testers. A survey of developments in conformance evaluation methodology, test architecture, test language, test generation methodology, formal description techniques and other issues related to protocol testing can be found in [Linn (1989), Rayner (1987)].

3.1 Part 1 : General concepts

Part 1 of the standards specifies the applicable requirements that are to be met by any implementation. According to OSI, conformance is concerned only with the conformance of an IUT with respect to its standards. The conformance requirements are given as follows: (a) Mandatory requirements, which are to be met in all cases; (b) Conditional requirements, which are to be observed only when the conditions specified in standards apply; and (c) Options, which can be selected to suit the implementation [ISO (1989a), Rayner (1987)].

The client should provide two documents to the testing center which will help in obtaining the test suite for the IUT. Protocol Implementation Conformance Statement (PICS) is a statement provided by the implementor, stating the capabilities and options which are implemented, and features which are omitted. Protocol Implementation Extra Information for Testing (PIXIT) contains the information needed by the test operator in order to run the appropriate test suites. It should also provide other specific information which cannot be provided in the PICS. PICS and PIXIT should not contradict each other.

ISO distinguishes four types of testing according to the extent to which they provide an indication of conformance. They are the

1. *Basic interconnection test* checks whether there are any gross violations in the IUT.
2. *Capability test* verifies whether the observed capabilities of the IUT are same as mentioned in the PICS.
3. *Behaviour test* aims to provide a comprehensive testing over the full range of dynamic conformance requirements that are specified in the standard.
4. *Conformance resolution test* does in-depth testing of an IUT to particular requirements and provide a definite yes/no answer and diagnostic information to the specific conformance issues.

The *observed test outcome* is the output provided by the IUT for the applied *test case*. The *foreseen test outcome* is derived from the abstract test case along with the standards.

A *test verdict* is a statement of *pass, fail* or *inconclusive* and it is associated with each *test case*. A real system that is implemented based on OSI protocols, comes from a wide variety of configurations and hence a range of testing methods are defined by ISO, to take care of these variations. The IUT is a part of System Under Test (SUT) which is to be tested for conformance. ISO provides guidelines for testing single layer IUT's and multilayer IUT's [ISO (1989a)].

3.2 Part 2 : Abstract test suite specification

This part gives an approach to provide conformance test suites which are independent of the means of executing those test suites (called as Abstract Test Suites (ATS's)). As a result, the ATS can be useful in comparing the results produced by different organizations which run the corresponding executable test suites [ISO (1989b), Rayner (1987)]. An ATS comprises of a number of test cases. A test case shall include different types of tests like capability test, behaviour tests, etc. Each ATS should have a test purpose which tests a single conformance requirement.

Abstract test methods are described in terms of the Point of Control and Observation (PCO) available to both the Upper Tester (UT) and Lower Tester (LT). The location of PCO differs in different testing methods, and all the methods assume that there is at least one PCO for the LT. There are four types of abstract test methods that are recommended by ISO. They are the (i) Local test method, (ii) Distributed test method, (iii) Coordinated test method and (iv) Remote test method. A new test method was proposed by Zeng, *et al.*, called the Ferry clip method [Zeng (1989)]. All these test methods are explained in detail in [ISO (1989b), Rayner (1987), Zeng (1989)].

3.3 Part 3 : The tree and tabular combined notation

TTCN is an informal notation, through which generic and abstract test cases can be expressed. This notation is independent of the test methods, layers and protocols. TTCN test suites are hierarchically structured. A test suite can be viewed as a combination of test groups which in turn is a combination of test cases and this can be subdivided into test steps and test events [ISO (1989c)].

A TTCN suite has four parts: (i) the Test suite overview part, (ii) the Declarations part, (iii) the Dynamic part and (iv) the Constraint part. The test suite overview gives the purpose of test suite and its individual tests. Declarations are made about PCO, global variables, service primitives and PDUs. The dynamic behaviour table contains behaviour description, label, constraint reference, verdict and comment columns. The values for ASPs and PDUs that are used in the dynamic behaviour tables are given in the Constraints table [ISO (1989c)].

3.4 Part 4 : Test realization

Test realization is the process of producing a means of testing IUT's for conformance to standards with reference to an ATS. The means of testing is a combination of equipment and procedures that can perform: (a) the derivation (b) the selection (c) the parameterization and (d) the execution of the test cases, in conformance with the reference ATS and can produce a conformance log.

In the derivation process, abstract test cases are converted to executable test cases. In the selection process, the appropriate test cases for the IUT are selected based on PICS and PIXIT. In the parameterization process, parameters in the test cases are given appropriate values. Executable Test Suites (ETS's) are obtained after the derivation process [ISO (1989d)]. This part provides guidelines on the realization of Lower Tester, Upper Tester, Test Coordination Procedures, Conformance logs and other details.

3.5 Part 5 : Conformance assessment process

In this part, the role of both the client and the test laboratory are addressed. A broad set of guidelines are provided on the requirements that are to be met both by the client and the test laboratory [ISO (1989e)]. The conformance assessment process consists of three phases. They are:

1. *Preparation for testing :* In this phase, the general administrative steps are taken. Also, the completeness of the documents like PIXIT, PICS are verified.

2. *Test operations :* During this phase, the static conformance review is done. The parameterized executables are identified and they will be executed. The tests like basic interconnection tests, capability tests and behaviour tests are conducted.

3. *Test report production :* This phase is the culmination of the above two phases. Based on the results, an assessment is made about the conformance of the IUT with respect to the standard. This is recorded in System and Protocol Conformance Test Report (SCTR and PCTR).

4 TEST SEQUENCE GENERATION

The test sequence should be *efficient*, i.e., it should be relatively short, fast and easily executable. Also it should also be *effective*, i.e., it should have a fairly large fault coverage. These two properties are conflicting in nature.

The specification usually provides only the legal transitions for all inputs. These transitions are called *core edges.* A lot of state-input combinations are left unspecified and different implementations will have different behaviour for those state-input combinations. These unspecified state-input combinations are called *non-core edges* and it is assumed that the IUT either goes to an error state or remains in the same state and generates null output. This is called the *completeness assumption* [Sabnani (1988)]. Two levels of conformance can be defined for partially specified FSMs. *Strong conformance* tests both core and non-core edges for conformance. *Weak conformance* tests only core edges for conformance.

Each state in the specification is identified by means of a SIS. There are three well known SISs: Distinguishing sequences (D), Characterizing set (W) and Unique Input/-Output sequences (U). An input sequence is a distinguishing sequence of the specification, if the output produced in response to the input sequence is different for each starting state in it [Gonenc (1970)]. The characterization set consists of a set of inputs that can distinguish between every pair of states present in the specification [Chow (1978)]. An Unique Input/Output sequence for a state is the input/output behaviour that is not exhibited by any other state of the specification [Sabnani (1988)]. Multiple SIS's may be present for a state. The test sequence methods use any one of the SIS to identify the states.

The aim of testing is to verify whether the IUT is *isomorphic* to the specification. The IUT is assumed to have same number of states as the specification. For each edge, a test subsequence is derived which should verify whether the *head* and *tail* state of the *edge under test* is the same as that of the specification. Also the *input/output behaviour* of the edge should be the same as that of the specification. Thus, the test subsequence has three parts: *a preamble*, *the edge under test* and *a postamble*. The *preamble* is an input sequence which will take the IUT to the *head state* of the *edge under test*. The *postamble* is the SIS (any one of D, W and U subsequences) for the *tail state* of the *edge under test*. Since the IUT is a black box, the *preamble* and *postamble* are usually long.

The specification should be *strongly connected, minimal* and in most of the cases the IUT is assumed to have *reliable reset*. Test sequence generation methods can be classified into *Fault detection methods* and *Fault diagnosis methods*. In fault detection methods, the aim is to simply find out if the implementation is faulty. Test result analysis does not give any indication of where the error is present. On the other hand, fault diagnosis methods will identify and locate the faulty transition.

5 FAULT DETECTION METHODS

5.1 Transition tour method

A *Transition tour* is an input sequence which takes the specification from its initial state, traverses every transition at least once, and returns to its initial state. If the FSM is strongly connected, a transition tour exists [Naito (1981)]. This method can detect output faults. On the other hand it may fail to detect transfer faults, as it does not verify the head and tail states of all the edges.

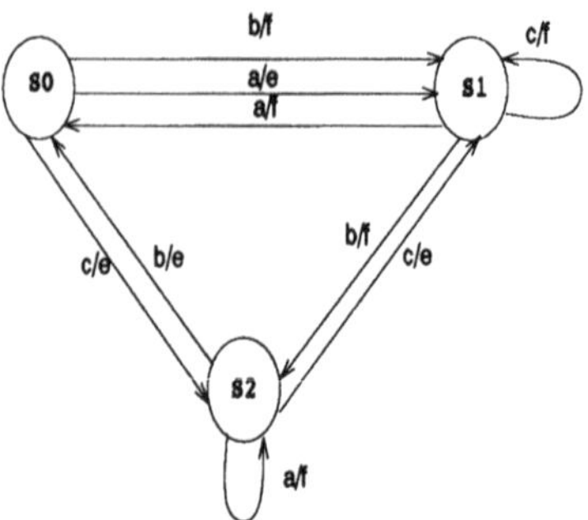

Figure 1: The specification graph.

5.2 Extended transition tour method

The aim of this method (E-method) is to verify if the behavior corresponding to each incoming edge and outgoing edge of every state in the implementation is similar to that of the specification [Kripanandan (1990)]. To achieve this, each incoming edge of the node is followed by every outgoing edge of the node, at least once in the test sequence. This method can detect output faults. It may fail to detect transfer faults if the specification is not 1-distinguishable.

5.3 Test sequence for D W and U methods

In the D, W and U methods, the testing is divided into two phases. Initially, the state verification is done by checking whether all the states that are in the specification are also there in the implementation. Next all the edges are tested by applying their corresponding test subsequences. The test subsequence for each transition e_{ij} is $P_{xi}.e_{ij}.SIS(v_j)$ where P_{xi} is the ***preamble*** to bring the graph to the ***head state*** of the ***edge under test*** from the ***start state*** and $SIS(v_j)$ is the SIS (D, W or U sequences) of the ***tail state***. The complete test sequence is the set of test cases, which includes state verification subsequences and test subsequence for each edge in the specification. If any test subsequence is completely contained in another test subsequence, it can be eliminated.

Discussion and Example : Consider the specification given in Figure 1. Some of the DS's of the specification are {a.a}, {b.b} and {c.c}. The UIO sequences for the state *S0, S1 and S2* are a/e, $a/f.a/e$ and $a/f.a/f$ respectively.

Some of the W-sets for the specification are {a, a.a}, {b, b.b} and {c, c.c}. Let us use UIO sequences for testing. The state verification subsequences are given in Table 1 and the complete test sequences are given in the Table 2.

Table 1: State identification sequence verification

State	Test subsequence	Expected output
S0	r,**a,a**	null,**e,f**
S1	r,a,**a,a**	null,e,**f,e**
S2	r,c,**a,a**	null,e,**f,f**

Table 2: Test subsequence for all the edges

Edge	Test subsequence	Expected output	Edge	Test subsequence	Expected output
S0–a/e–S1	r,**a**,a,a	null,**e**,f,e	S0–b/e–S1	r,**b**,a,a	null,**f**,f,e
S0–c/e–S2	r,**c**,a,a	null,**e**,f,f	S1–a/f–S0	r,b,**a**,a	null,f,**f**,e
S1–b/f–S2	r,b,**b**,a,a	null,f,**f**,f,f	S1–c/f–S1	r,b,**c**,a,a	null,f,**f**,f,e
S2–a/f–S2	r,c,**a**,a,a	null,e,**f**,f,f	S2–b/e–S0	r,c,**b**,a	null,e,**e**,e
S2–c/e–S1	r,c,**c**,a,a	null,e,**e**,f,e			

5.4 Optimization of testing techniques

The complete test sequence for the IUT is obtained from the test subsequences. All the test subsequences generated in section 5.3 start from the start state. As a result, the test sequences have a lot of redundant transitions. Also there is no need to start the test subsequence from the start state. Instead of reset, a transfer subsequence can be used between different test subsequences. In general the problem of identification of an optimal test sequence from the test subsequences is *NP Complete* [Boyd (1991)].

An efficient algorithm exists if the IUT is assumed to have certain properties. By assuming that the IUT has a *reliable reset* or *self loop transitions* on all the states, Aho *et al.*, gave an efficient algorithm based on *UIO sequences* and *rural chinese postman algorithm* to determine the minimum length test sequence from the test subsequences [Aho (1988)].

Multiple SIS (MSIS) exists for some of the states of the IUT which ends at different tail states, and the length of the test sequence produced by using MSIS are in general shorter than the test sequence produced by using single SIS per state [Shen (1989)]. The reduction in the length of the test sequence will be more pronounced only when there are more number of SIS sequences per state which end at different tail states. An improvement to W method called W_p, was proposed in [Fujiwara (1991)]. In this method, instead of using the set W to verify the reached state, a subset of this set is used.

It was pointed out in [Aho (1988)] that, any overlap between different test subsequences can also be used to reduce the length of the test sequence. The papers [Miller (1993), Chen (1990), Yang (1990)] uses both the presence of MSIS and overlap between different test subsequences. The method given in [Miller (1993)] provides near-optimal test sequences in polynomial time.

Now we will briefly describe the algorithm given in [Miller (1993)]. A specification is said to be *nonconverging* if there are no two transitions entering the same state with the same input/output label. If the specification is nonconverging and has an *euler tour*, then the test sequence is any *euler tour* starting from the start state followed by $SIS(v_1)$, where v_1 is the start state. If the specification does not have an *euler tour* and it has an *euler path* starting and ending with the states v_k and v_l respectively, then the test sequence is any *euler path* of the specification from v_k to v_l followed by SIS_l. If the specification does not have an *euler path*, the specification is augmented with minimum number of edges to get an *euler path*. The test sequence is the *euler path* appended with the SIS sequence of the tail state of the path.

If there are converging edges in the specification, they are removed from the specification and edge-disjoint paths are computed from the resultant subgraph. For each edge-disjoint path, if the SIS sequence of the tail state of the path is appended to it, then all the edges that are in the disjoint path are verified. For all the converging edges, a test subsequence is generated by appending the SIS sequences of the tail state of the edges. These edge disjoint paths and the test subsequence corresponding to the converging edges are combined together in an optimal way by exploiting maximal overlap and minimal transfer subsequence. The detailed algorithm can be found in [Miller (1993)].

The lower bounds on the length of the test sequence based on D method *with* and *without overlap* between the test subsequences is given in [Ural (1993)]. An algorithm to obtain the test sequence using overlap between different test sequences are also given in [Ural (1993)].

In general, to obtain the shortest test sequence, *the length of the SIS should be minimum, multiple SISs should be used, the overlap between different test subsequences should be maximum and the transfer sequence between different test subsequences should be minimum.*

5.5 Regular checking method

The methods (D, W, U and E) combine the test subsequences without imposing any constraint on the ordering of the test subsequences. As a result, one fault may hide the presence of another fault and the faulty IUT may pass the test sequence. In this method (R-method) [Miller (1992)], a detailed study has been made on how the ordering of different test subsequences affects the fault detection capability. Different cases were identified, where one fault hides another fault. Care was taken so that all such hiding faults were detected.

A brief outline of the R-method is given here. For each edge e_{ij}, the effect of transfer faults on all $(n-1)$ states is identified. A test subsequence is generated which will detect all such errors. This step is repeated for all the edges in the specification. Finally all the test subsequences are combined in an optimal way to obtain the test sequence. Combining different test subsequences is made as per ordering constraints imposed by the structure of the specification graph. A detailed analysis of how the ordering of test subsequences affects the fault coverage and an algorithm for generating the test sequence can be found in [Miller (1992)].

5.6 Analysis of fault coverage

The U method may fail to detect some transfer faults if the U sequences are no longer the U sequences in the IUT [Chan (1989)]. Hence it was suggested that all the U sequences should also be verified to be unique in the IUT in addition to the state verification phase. Further testing can be continued only if there is no error detected during this phase. This added overhead is acceptable, as it provides better fault coverage. It was shown in [Lombardi (1992)] that UIO verification need not be done for all states and they can be computed only for those states where such extra testing is necessary.

The test sequences generated based on the methods given in [Aho (1988), Shen (1989), Chen (1990), Yang (1990), Miller (1993)] may fail to detect some faults since, the UIO sequences of the specification may not be the UIO sequences for erroneous IUT. Hence it was suggested in [Yao (1993)] to include state verification subsequence too while generating the test sequence for the IUT. The length of the test sequence generated based on this method is generally longer than the U method, as certain portions of a test cases are repeated several times in order to make sure that the same state is reached in different points of the test case.

The α, β, γ, approximations to protocol testing were proposed in [Sidhu (1990)]. For each approximation there is a corresponding test subsequence. A test subsequence is defined most generally as $L_{ij}.CS(v_j)$, where L_{ij} is a sequence of input/output labels that take the specification from state v_i to state v_j and $CS(v_j)$ is the characterization subsequence. If length of L_{ij} is zero then it is called an α subsequence and if the length is one then it called a β subsequence and so on. It was suggested that for better fault coverage, higher order approximation should be used while generating the test sequence.

It was pointed out that the test sequence generated based on the same method and also having the same length, has different fault detection capabilities [Sidhu (1990)]. This is because there is considerable arbitrariness in obtaining the test sequence from the test subsequences, which affect the fault detection capability. To get better fault coverage, the test sequence should have higher approximation in test subsequence. It was shown in [Miller (1992)] that only for some special classes of specifications one need to consider higher order of approximations, and in most of the cases, β subsequences will detect the faults. A survey of fault detection methods can be found in [Bosik (1991), Sarikaya (1982), Sidhu (1990), Sidhu (1989)].

6 FAULT DIAGNOSTIC METHODS

6.1 Fault resolution approach to protocol testing

This method (FR-method) assumes that for each transition e_{ij}, the label corresponding to e_{ij} should not occur in the SIS portion of the test sequence. This ensures that the erroneous e_{ij} does not mask the error in the SIS. Also, the IUT is assumed to have either output fault or transfer fault in at most one edge. This method assumes the presence of reliable reset in the IUT. The algorithm is given in [Ramalingam (1992)].

6.2 The diagnostic approach to protocol testing

In this method (DA-method), the initial testing is done by one of the following D, W, U methods. The basic assumption in this method is that all the faulty transitions are directly reachable by a test subsequence. The presence of *reliable reset* in the IUT is assumed. In this method, the observed outputs are compared with the expected ones. The difference, called *symptoms*, are identified for all the test subsequences. A set of edges where the fault might be present are identified based on the symptoms. Then, a set of *fault hypotheses* for each test subsequence is constructed. Fault hypothesis is the assignment of a specific fault to suspected edges. By changing the specification machine as suggested by fault hypothesis, if the observed output got for the original input sequence is the same, then the diagnostic candidate explains all the faults. This method may produce a number of diagnosis. Hence, further testing must be done in order to reduce the number of diagnosis. A detailed algorithm is given in [Bochmann (1993)].

6.3 Constraint satisfaction approach to protocol testing

This method generates a unique test sequence for the given specification which can distinguish any faulty IUT with the same number of states and the same number of inputs. The test sequence is incrementally generated from a set of test subsequences which represents the constraints imposed on the overall structure of the specification. In general, it is impossible to provide a unique test sequence for any specification [Kohavi (1978)]. But by making assumptions that, the specification is minimal, completely specified, strongly connected and the number of states in the IUT is equal to that of the specification (or within some upper bound) it is possible to generate a test sequence which is unique to the specification.

The test sequence generated based on the CSP method has two parts. They are (i) TS_o, the initial subsequence which is a transition tour, i.e., it covers all the edges that are in the specification and (ii) TS_d, the test subsequence which distinguishes the specification from all other erroneous specification. The algorithm to generate the test sequence for the CSP method is given [Vuong (1990)].

6.4 Multilevel method

By applying only a subset of inputs to the IUT, a subgraph of the specification will be taking part in the test. M-method uses this concept for testing the IUT. This method groups the inputs of the specification into disjoint sets like basic inputs, level-1 inputs, level-2 inputs and so on and tests them in that order. The ***Basic subgraph*** contains edges from ***basic inputs. Level-i*** subgraph contains inputs from ***basic subgraph, level-1 inputs*** etc., up to ***level-i inputs.*** The ***basic subgraph*** should have the following properties: (i) it is a Strongly Connected Spanning Subgraph (SCSS) of the specification (ii) it has the minimum number of inputs from the inputs of the specification and (iii) all the states in the basic subgraph have a SIS. The ***basic subgraph*** should be tested by a method which has better fault detection capability even in the presence of multiple faults in the IUT. Of all the fault detection methods, the R-method has better fault detection capability and it is used for testing the ***basic subgraph.*** For testing ***level-i*** inputs, the preamble and postamble should be used from the ***level-i subgraph*** which are already tested. If the preamble and postamble are error-free and testing starts from the state specified by the test sequence, fault diagnosis is possible for ***level-i*** inputs. Whenever an error is detected, homing and error recovery subsequences are applied before further testing can be continued. Multilevel method is described in detail in [Kripanandan (1994), Neelakantan (1994)].

7 COMPARISON OF TESTING METHODOLOGIES

If the IUT has a reliable reset in all the states and the SIS of the specification is also the SIS of the IUT, D, W and U methods can detect the faulty IUT. The test result analysis does not give the exact location of faults, as after the SIS verification phase, all the edges used in the SIS may not be error-free. As the test sequences are optimized, if any error occurs, it is not possible to distinguish between whether the error is due to a transfer fault in the previous transition or due to an output fault in the current edge under test. Also one fault may mask the presence of another fault in the IUT [Miller (1992)]. In general, the W-method will produce the longest test sequence and the U and D-methods will produce shorter test sequence of comparable lengths [Sidhu (1989)].

Fault coverage : The fault coverage or the effectiveness of the test sequence depends upon three factors: (i) for each edge in the specification the testing should start at the state specified by the test sequence. (ii) the preamble should be error-free and (iii) all the edges in SIS (postamble) should also be error-free. In the M-method, if the basic subgraph is error-free, all the three conditions are satisfied [Neelakantan (1994)]. The DA method assumes an error-free preamble for all the edges and errors can be present in the postamble. Hence to locate faults, additional testing is necessary. The CSP method can locate a single transfer fault easily. However, when multiple faults occur at successive edges, locating these faults can become difficult.

Number of subgraphs : In all the fault detection and fault diagnostic methods, the entire specification is taken for testing purposes. On the other hand, the M-method, splits the specification into different subgraphs and tests them one after another. In general D, W and U methods can be considered as special cases of the M-method where it has only two subgraphs namely the *basic subgraph* and *level 1 subgraph.* The edges in the SIS and state verification subsequence of the D, W and U methods can be considered as the basic subgraph. The transition verification phase can be considered as the level 1 subgraph. In D, W and U method, if the IUT succeeds the SIS verification phase, it does not guarantee that all the edges in the SIS are error-free. Whereas in the M-method, if *basic subgraph* succeeds the test, it guarantees that preamble and SIS are error-free.

Length of the test sequence : The length of the test sequence generated based on the FR approach will be comparable to D, W and U methods, as it assumes that only one fault can be present in the IUT. The length of test sequence based on the DA method increases with the number of faults present in the IUT and it will be comparable to the D, W and U method if there are no errors in the IUT. As the SIS of the *basic subgraph* of M-method is usually larger than the SIS used in other methods (D, W and U method), the length of test sequences of the M-method will be larger than the corresponding test sequence based on D, W, and U Method.

Error Reporting : Test result analysis and some cases, extra test sequences are required to identify the erroneous transitions in FR, DA, CSP methods, whereas in the M-method, the erroneous transitions are reported once they are detected.

Error recovery : The DA, FR, CSP methods uses *reliable reset* for error recovery in the IUT. On the other hand, the M-method uses homing and recovery sequences for error recovery.

Applicability to real life protocols : If *reliable reset* is present in the IUT, the DA, CSP and M methods are more attractive. If reliable reset is not present, the M-method can be used. If there is no basic subgraph which is error-free, the M-method may not be applicable. In that case, the DA method may be applicable if all the faults in the IUT are *directly reachable.*

8 SUMMARY AND CONCLUSION

In this paper we saw different aspects of protocol conformance testing. The testing methodology and framework provided by ISO has resulted in the development of standard test architectures and testing techniques and they were briefly explained. Different test sequence generation methods based on FSM models were also discussed. The testing methods can be grouped broadly into two categories: Fault detection methods, whose aim is to detect whether the IUT is faulty; Fault diagnostic methods, whose aim is to identify the erroneous transitions. We compared the test sequence generation methods in terms of its fault coverage, the availability of reliable reset and length of the test sequence. In general, the aim of all the test sequence method is to get minimal length sequence. But optimization introduces an undesirable side effect of reduced fault coverage. Thus, during test sequence generation effectiveness (fault coverage) should be given preference over the efficiency (length of test sequence). Most of the assumptions made by testing methods like completely specified, reliable reset, self loop at each node are not satisfied by most of the real life protocols. The possibility of overcoming these barriers should be explored.

REFERENCES

Aho A.V., Anton T. Dahbura, David Lee, and M. Umit Uyar (1991) An optimization technique for protocol conformance test generation based on UIO sequences and rural chinese postman tours. *IEEE Trans. on Communication*, 39:11, 1604-1615.

Bochmann G.V., R. Dssouli, A. Das, M. Dubuc, A. Ghedamsi, and G. Luo (1991) Fault models in testing. In *Protocol Test Systems*, pages 17–32.

Bochmann G.V., and A. Ghedamsi (1993) Multiple Fault Diagnostic Tests for Finite State Machines. In *IEEE Infocom 93*, pages 782–791.

Bosik B.S. and Uyar M.U. (1991) FSM based formal methods in protocol conformance testing : from theory to implementation. *CN and ISDN Systems*, 22:7–33.

Boyd S.C. and Hasan Ural (1991) On the complexity of generating optimal test sequences. *IEEE Transactions on Software Engineering*, 17:976–979.

Chen M, Choi Y., and Kershenbaum A. (1990) Approaches utilizing segment overlap to minimise test sequences. In *Protocol Spec. Testing, and Verification X*, pages 85–98.

Chow T. S. (1978) Testing software design modeled by finite state machines. *IEEE Transactions of Software Engineering*, SE-4:178–187.

Chan W. Y. L., S. T. Vuong, and M. R. Ito (1989) An improved protocol test generation procedure based on UIOS. In *SIGCOMM'89*, pages 283–294.

Fujiwara S., G. V. Bochmann, F. Khendek, M. Amalou, and A. Ghedamsi (1991) Test selection based on finite state models. *IEEE Trans. on Software Engineering*, 17.

Gonenc G. (1970) A method for the design of fault detection experiments. *IEEE Transactions of computers*, 19:551–558.

ISO (1989) *Information technology -OSI conformance testing methodology and framework - Part 1: General concepts.* International Organization for Standardization.

ISO (1989) *Information technology -OSI conformance testing methodology and framework - Part 2: Abstract test suite specification.* International Organization for Standardization.

ISO (1989) *Information technology -OSI conformance testing methodology and framework - Part 3: The Tree and Tabular Combined Notation.* International Organization for Standardization.

ISO (1989) *Information technology -OSI conformance testing methodology and framework - Part 4: Test realization.* International Organization for Standardization.

ISO (1989) *Information technology -OSI conformance testing methodology and framework - Part 5: Requirements on test laboratories and clients for the conformance assessment process.* International Organization for Standardization.

Linn R J. (1989) Conformance evaluation methodology and protocol testing. *IEEE Journal on selected areas in communications*, 7(7):1143–1158.

Kohavi Z. (1978) *Switching and Finite Automata Theory.* McGraw Hill.

Kripanandan R. S. and S. V. Raghavan (1994) Multilevel approach to protocol conformance testing. to appear in *CN and ISDN Systems.*

Kripanandan R. S. (1990) Multi-level approach to protocol conformance testing. Master's thesis, Department of Computer Science, Indian Institute of Technology Madras.

Lombardi F. and Y.N. Shen (1992) Evaluation and improvement of fault coverage of conformance testing by UIO sequences. *IEEE transactions on communications*, 40,8:1288–1293.

Miller R.E. and Sanjoy Paul (1992) Structural analysis of a protocol specification and generation of a maximal fault coverage conformance test sequence. Technical report, University Of Maryland.

Miller R.E. and Sanjoy Paul (1993) On the generation of minimal length conformance tests for communication protocols. *IEEE transactions on networking*, 1:116–129.

Neelakantan B. and S.V. Raghavan (1994) Scientific Approach to Multilevel Method. Communiated to *IEEE transactions on networking.*

Naito S. and M. Tsunoyama (1981) Fault detection for sequential machines by Transition tours. In *Proceedings of Fault tolerant computing systems*, pages 238–243.

Rayner D. (1987) OSI conformance testing. *CN and ISDN Systems*, 14:79–98.

Ramalingam T., Anindya Das, and K. Thulasiraman (1992) On conformance test and fault resolution of protocols based on FSM model. In *Networks 92*, pages 435–475.

Sarikaya B. and G.V. Bochmann (1982) Some experience with test sequence generation for protocols. In *Protocol Specification, Testing, and Verification II*, pages 555–567.

Sabnani K. and Anton Dahbura (1988) A protocol test generation procedure. *CN and ISDN Systems*, 15:285–297.

Sidhu D.P. (1990) Protocol testing: The first ten years, the next ten years. In *Protocol Specification, Testing, and Verification X*, pages 47–68.

Sidhu D.P. and Tingkau Leung (1989) Formal methods for protocol testing: A detailed study. *IEEE Transactions of Software Engineering*, 15(4):413–426.

Shen Y. N., F. Lombardi, and A. T. Dabhura (1989) Protocol Conformance Testing using MUIO sequences. In *Protocol Spec., Testing, and Verification IX*, pages 131–143.

Sidhu D.P. and Raghu Vallurupalli (1990) On arbitrariness in protocol conformance test generation. Technical report, University Of Maryland.

Ural H. and K Zhu (1993) Optimal length test sequence generation using distinguishing sequences. *IEEE transactions on networking*, 1:358–371.

Vuong S.T. and Kai C. Ko (1990) A novel approach to protocol test sequence generation. In *Globalcom'90*, pages 1880–1884.

Yao M., A. Petrenko, and G V Bochmann (1993) Conformance testing of protocol machines without reset. Technical report, University Of Montreal.

Yang Bo and Hasan Ural (1990) Protocol conformance test generation using multiple UIO sequences with overlapping. In *SIGCOMM'90*, pages 118–125.

Zeng H. X., S. T. Chanson, and B. R. Smith (1989) On Ferry clip approaches protocol testing. *CN and ISDN Systems*, 17:77–88.

9 BIOGRAPHY

B. Neelakantan received his B.E degree in Computer Science and Engineering from Thiagarajar College of Engineering Madurai, India in 1992. He is currently working towards his MS (by Research) degree at the Department of Computer Science and Engineering, Indian Institute of Technology, Madras. His research work is supported by Government of India research scholarship. His research interests are networks, protocol and graph theory.

S.V. Raghavan is on the faculty of the Department of Computer Science and Engineering, Indian Institute of Technology, Madras. He is also the Chief Investigator of the project on *Education and Research in Computer Networking* jointly sponsored by the Department of Electronics, Government of India and the United Nations Development Programme. He is a life member of the Computer Society of India, a member of the Institution of Engineers and a fellow of institute of Electronics and Telecommunication Engineers. He is presently serving on the Board of Editors of the journal of IETE for computers and control. He is also a member of the Editorial Advisory Board for Computer Communications, Butterworth-Heinemann Ltd. His research interests are networks, protocols, multimedia systems and performance.

PART FOUR

Distributed Systems and Applications

DISTRIBUTED SYSTEMS AND APPLICATIONS

Distributed Systems represent the modern computing environment where resources are physically spread out but are logically integrated, that too seamlessly. In such an environment, developing an application requires a language that is sensitive to the underlying system and its integration. Moreover, resources such as physical memory that is in each and every system, when viewed as one coherent piece of memory, presents challenging problems. Studying the performance of such a setup for understanding the configuration options leads to interesting results. After solving the problem of seamless access to resources, it is worthwhile examining the options available for layered application software such as X.500. There are four papers in this session that deal with these aspects and reveal some interesting results.

The first paper "EC: A language for distributed computing" by Ashok Kumar Naik and Gautam Barua deals with the issues that could be considered while designing a language for distributed systems. An important requirement is that programs written in the language should be portable. The details about the design of a new language, EC (Extended C) which is a superset of C and its support of process creation, message passing mechanism and exception handling are discussed. The second paper "A DSM-based portable distributed programming testbed " by Hariram and Gonsalves describes the design and implementation of a user level DSM (Distributed Shared Memory) on Unix and Linux. The implementation is tested with several parallel programs that are used to benchmark shared memory multiprocessors like matrix multiplication, linear equation solver and water.

The third paper "X.500 performance with large databases and OSInet" by Rajeev, Raghavan and Ashok Agrawala deals with the performance of X.500 on OSInet for large databases. The factors influencing the performance of the X.500 directory service, *viz.*, round trip time, name resolution time, database access time and protocol overhead time are discussed. Performance comparison of three different models: simple, chaining and referral are presented. Application of this model to the ERNET topology is illustrated. The fourth paper "A queuing network model Of distributed shared memory" by Venkatesulu and Gonsalves is concerned with the queuing models for distributed shared memory. Different schemes such as read replication, migration and full replication are analyzed using queuing network models. Performance comparison of various methods using MVA method are also presented.

12

EC : A Language for Distributed Computing

Ashok K. Naik
Gautam Barua
Department of Computer Science and Engineering
IIT Kanpur - 208016, India
email - {naik, gb}@iitk.ernet.in

Abstract

Computation intensive programs can utilise idle workstations in a cluster by exploiting the parallelism inherent in the problem being solved. A programming language for distributed computing offers advantages like early detection of type mismatch in communication and offers structured mechanisms to specify possible overlap in communication and computation, and exception handling for catching run time errors. Success of a language depends on its ease of use, expressibility and efficient implementation of its constructs.

EC is a super set of C supporting process creation, a message passing mechanism, and exception handling. The pipelined communication constructs and multiple process instances help in expressing concurrency between computation and communication. Data driven activation of EC processes is used for scheduling.

EC has been implemented in a Sun-3 workstation cluster. An inter-node message passing mechanism has been built on top of the socket interface using the TCP protocol, and intra-node message passing is done by passing pointers to improve efficiency. However, `message_type` variables hide the implementation details, improve type safety and location transparency of a program.

Keywords

Distibuted Languages, process, exception, pipelined communication

1 INTRODUCTION

With the decrease in the price of workstations along with their increase in computational power, workstation clusters are becoming an alternative to centralized computing platforms. But as the current generation operating systems allocate only one workstation to a user, most of the workstations are under-utilised. There have been attempts to distribute the load on a per task basis (Shivaratri *et al.*, 1992), but this does not reduce the computational time of large scientific and engineering applications unless the program is broken into smaller tasks. This requires considerable effort on the part of the programmer without any further support. Research in this area has been going on for quite sometime. The availability of fast, inexpensive systems, interconnected by fast networks, has given an added impetus to this area in recent areas. Commercial interest is being shown by a number of leading companies, such as Digital, HP-Convex, and IBM, in such clusters of workstations, as a low-cost alternative to vector machines. The typical commercial offering consists of a number of computer systems interconnected by a fast network such as FDDI. The software offering with such systems is the weak link, with the most common offering being a library of routines to help in the inter-node communication. What we have is a distributed system, with no sharing of memory, and with the need for a relatively coarse-grain level of parallelism to exploit the number of CPUs present.

Two main methods are being experimented with at various places. One is to use distributed shared memory to provide a shared memory model of multiprocessors to the users. The other is to provide primitives for explicit communication among processes in different processors, either by providing library functions or through distributed programming languages.

Some languages use sharing as an abstraction to hide issues related to communication. But, for efficient implementation, access to the lower level system software such as memory management unit, lower level communication primitives etc. is required, which is not possible in many cases. Also, the languages based on the shared memory paradigm provide support for synchronization, leaving the load distribution to the user and message traffic optimization is necessary to maximize the throughput. As problem specific optimization can not be done at runtime, performance depends upon the message volume generated by the heuristic used. This approach, being evolutionary in nature, may be suitable for using existing programs where portability rather than efficiency is of major concern. A variation of this is to use virtual distributed machines where a program includes definitions for data distribution mapping and program mapping. Fortran-D (Hiranandani *et al.*, 1991) uses data distribution mapping information to partition a sequential program and generate code for communication. DINO (Rosing *et al.*, 1990), PANDORE (Andre *et al.*, 1990), Spot (Socha, 1990) and Ensemble (Griswold *et al.*, 1990) also use these mapping information to generate a distributed program. Inefficiencies arise in dealing with irregular data like sparse matrices (Lu *et al.*, 1991). As the issues like data partitioning, communication optimization etc. are in NP-Complete, a programming language should provide facilities to allow programmers to exploit problem specific features.

UNITY (Mishra, 1991) is an abstraction mechanisms for data parallel programs. A UNITY program is essentially a declaration of variables and a set of assignment statements. Program execution consists in selecting nondeterministically some assignment statements, executing it and repeating it forever. Program developement is carried out in two basic steps: first a correct program is derived from specifications and this program is adapted to the target architecture by successive transformations of the original program in order to make the control structure explicit. UNITY uses abstract models for program development based on variant and invariant parts in the specification of a program. A naive implementation will result in inefficient programs. LINDA (Gelernter, 1991) is a language using an abstraction based on tuple space. All operations on tuple space are managed through predefined primitives. As access to the tuple space is based on key matching, all access to tuple space in LINDA programs are optimized by a compiler (Carriero *et al.*, 1991). The languages based on these abstraction mechanisms rely too much on compiler technology to generate efficient codes, which we believe is not yet mature enough to do this.

Programming in distributed systems using explicit communication primitives is difficult because of issues like data distribution, correct use of communication primitives, buffering, hiding communication latency etc. Moreover, when implementation dependent features like buffering are not hidden from programs, it becomes nonportable. It is possible to build an abstraction at the user level to provide portability. For example, PVM (Beguelin *et al.*, 1993) builds a virtual parallel machine model on top of existing systems at the user level to provide the portability. But being a library based approach, it lacks features like compile time type checking and language support for latency hiding. So developing distributed memory programs using libraries is still difficult and error-prone. In a language based method on the other hand, a virtual machine model can be adopted and by porting the compiler, programs can be run on a variety of architectures. A language for distributed systems aims to be expressive enough to program parallelism, to provide scalability and reusability of modules, and compile time and run time checks to detect any incorrect use of communication primitives. Process is the lowest level of abstraction used for expressing parallelism and it fits quite naturally with a distributed memory computer. Because of this, it can be implemented efficiently. In this abstraction, processes must be created and 'connections' must be established according to topology requirements of an algorithm used. Communication primitives may be of synchronous or asynchronous type. But synchronous primitives with static process creation limits the concurrency of local processes (Liskov *et al.*, 1986). This being the oldest abstraction in use, there are many languages based on this paradigm. The most recent proposals include Hermese (Strom *et al.*, 1991) and Darwin/MP (Magee *et al.*, 1992). In the past, many programming languages based on the message passing paradigm have been proposed. Many of them, i.e. OCCAM (Burns *et al.*, 1988), CONIC (Magee *et al.*, 1989) etc. are meant to be for Transputer based machines. Some of the early languages such as SR (Andrew *et al.*, 1988), Argus (Liskov *et al.*, 1983) etc. were more concerned about transaction oriented or system programming applications. Most of these languages have a common heritage, the guarded commands of CSP. One of the motivations for this was proving program correctness. Another motivation was to provide overlap in computation and communication through the nondeterminism in the guarded commands. Languages such as BSP (Gehani, 1984), DP (Hansen, 1978) are extension of CSP. We believe, these languages do not aid the compiler by providing appropriate information suitable to exploit the data parallelism present in scientific and engineering programs. For example, as guarded commands can not proceed until full data is available, the overlap of computation with communication can not take place.

Object based and object oriented languages for distributed systems use objects as an abstraction mechanism (Wegner, 1991; Wegner, 1992) In this abstraction, an object is used as a structuring mechanism for complex data structures. Concurrency in an object model has three different forms ie, sequential, quasi-concurrent and concurrent. In the first case there is only one thread of execution. In the second form, an object acts as a monitor; so all threads are queued except one. In the last case, multiple threads may be active at the same time. To express concurrency at the calling side, asynchronous method invocation is used if the result is not required immediately; otherwise future-based synchronization is used.

Declarative languages are based on the functional paradigm in which the order of execution is not specified. Instead, a declarative program specifies what to do. Thus execution order of an operation in dataflow style is data driven, that is execution starts once all operands are available. To reduce synchronization overhead for scheduling an instruction, a sequence of instructions called a thread, forms a unit of data driven computation (Gao, 1993; Grit, 1990) . In process based functional programming languages, a *future* (Halstead, 1989) type function is used to provide asynchronous features and *continuation* based function calls (Hieb *et al.*, 1990) provide coroutine semantics to a function call. To provide concurrent access to shared objects, objects with state called a *domain* (Kessler *et al.*, 1989) which is similar to a monitor, is used.

All distributed applications are derived from problems that are inherently distributed in nature. For example, differential equations representing the physical system do not have action at a distance. The interaction is always with the immediate neighbours. This allows the problem to be solved using divide and conquer, quite naturally. This feature makes object based or process based approach natural candidates. Our approach to the design of a language for distributed systems is evolutionary rather than revolutionary. We decided to use C as base language as it is popular and efficient and added the features that are orthogonal and easy to learn. So we have chosen the process abstraction as it is simple and easy to integrate with C. The proposed language, Extended C (EC), adds process and typed message variables to C and augments control structures by adding pipelines and process creation. Type checking of messages detect possible mistakes in communication. The process topology as required by the algorithm is specified at creation time.

2 BACKGROUND

Our experiments on numerical programs using DC (Subramaniam, 1991), a language for distributed computing, showed that performance depends very much on the amount of overlap in computation and communication. Sending large sized messages resulted in CPUs on remote nodes remaining idle, waiting for data transfer to be completed. So large sized messages had to be broken into smaller messages to improve the efficiency. But as this approach becomes cumbersome, some language support for specifying the concurrency in communication and computation was thought to be necessary.

From the user point of view, it was found that the point-to-point communication facility provided by the language was not convenient to use. Major portion of a typical program was for supporting communication, that is, connection set up and communication statements. It was felt that language level support for different communication patterns such as one-to-many and many-to-one type of communication will simplify programming.

3 THE LANGUAGE

3.1 Interprocess Communication

To allow a programmer to specify overlap in computation and communication, either asynchronous communication primitives or synchronous primitives with dynamic process creation has to be used. EC uses a hybrid of asynchronous and synchronous methods because synchronous versions are easy to understand and allow error handling to be specified as a part of the primitive. This form simplifies implementation of termination semantics of any exception raised during a receive. By specifying the computation to be performed while waiting for data, it allows the efficient use of CPU time. A similar construct has been proposed for RPC (Chang, 1989).

3.1.1 Message Variables

All communication is done through message variables. Message variables in EC are used as typed streams. A message variable is a tuple with two components: one component specifies the binding with the remote message variable, if any, and the other component specifies the buffer to be used for communication. No actual space is allocated for the buffer. This is done separately. A message variable is a first class variable, so a message variable can be passed from one process to another. This allows the connection topology to be changed after process creation. This allows a child process to do the processing on behalf of its parent. This is very useful in client server applications where there are multiple specialized servers and a client does not know which server to use.

Type checking in communication, like type checking of other variables, ensures that both sender and receiver communicate with messages of the same data type and size. EC defines message variables which act as typed streams, and like structures in C, name equivalence form of type checking is used for message variables. A **message_type** declaration creates a new message type whose components specify the message composition. The syntax for a message type declaration is similar to a struct as shown below.

```
message_type msg1 {
    int a, b;
    struct my_struct y[100];
} x;
```

But unlike a struct type, pointers are not allowed inside it because size information is not always present with a pointer (a pointer of type char is used in many C programs to point to arrays), and without this, type checking is not possible. Dynamic arrays can not be used either, because unlike a struct variable, scope of temporaries needed for message variables is within communication primitives and their sizes may vary from one primitive to another with dynamic arrays present, so some other linguistic mechanism must be provided to support this. Compile time check alone will not suffice because the sender and receiver sides may be separately compiled and size information is available at the receiving end only after data has been received. Further, EC allows a message type to be parameterized, with parameters specifying the size of variable sized components. So a run time check at the receiving end must be done to ensure that the correct amount of data has been received. The following example shows a declaration of a parameterized message type.

```
message_type msg2 (m) {
    double x[m];
} z;
```

We will discuss the syntax and semantics of send and receive using parameterized messages later. EC extends parameterization to **message_types** nested within a **message_type** declaration so that the number of such components is not fixed. This is useful for sparse data distribution in dynamic processor configurations and allows efficient implementation because it helps combine multiple communication primitives, thus saving run time overhead. Only one level of nesting is allowed, though. The following example illustrates the syntax.

```
message_type msg3 (p) {
    msg2 y[p];
} x;
```

message_type msg3 is composed of an array of **message_type** msg2, each of which refers to a vector of type double, and length m, which is a parameter of msg2. The number of msg2 elements, p, is a parameter of msg3. An instance of type msg3 has to now give values to both the parameters. This instance of the message variable is created either by a `receive` statement or by a C style assignment statement. For example, in the `for` loop shown below, variable x of type msg3 is used for communicating an upper triangular matrix of a square matrix a. As a pointer does not have any size information, and each component is of varying size, the assignment is type cast to type msg2 and the parameter of msg2 is used to specify the vector size.

```
for (i = 0; i < 100; i++)
    x.y[i] = (msg2(i)){&a[i][i]}
```

The final association of x with the parameter of msg3 is done in a `receive` statement, as for example,

```
receive x(100);
```

As the data area for message x is already associated inside the `for` loop, the above `receive` statement statement does not specify any data area. The parameter specified is checked against the previous association. (see Figure 1 for another form of the receive primitive).

In some applications, the message type to be communicated depends on the computation state at the sender side. A selective receive statement is usually used for this purpose. A selective receive is costly to implement because buffers for all alternatives must be specified, be enqueued for reception and be dequeued after a message has arrived. It is also error prone because the receiver side code must also be changed for any addition or deletion of message types at the sender side. So EC provides a **message_union** declarations similar to a union in C. This declares types of messages that are allowed in a selective receive statement using a message variable of message union type. A compile time check will detect if any message type is not allowed in the message union. Changes are limited to data structures, and existing code need not be changed. Also, this allows efficient implementation because a single message variable needs to be enqueued or dequeued and a "lazy" form of message initialization avoids unnecessary overhead for message types that have not arrived. The declaration syntax for **message_union** is the same as that of **message_type**.

3.1.2 Communication Primitives

EC supports an asynchronous **send** (although blocking, if there is no buffer space available) and a blocking **receive** primitive for interprocess communication. A **send** operation is said to be completed after all its data has been accepted by the kernel. As data may be buffered by the kernel at both sides, send and receives are not synchronous. A **receive** operation may specify its own buffer or ask the run time system to allocate buffer for its data. The latter form is suitable when data may arrive before it is requested. This is because EC processes share a connection to the same remote node to avoid exhaustion of system resources like buffers and sockets. So data must be removed from the kernel buffer to allow other EC processes sharing the same socket to read. Without allocating buffers at the user level, potential overlap of communication and computation is reduced. As specified earlier, a message variable is bound with another message variable, and this is done at process creation time. This binding sets up a connection between two or more processes. A local variable is specified (the user level buffer), to hold data in (for a receive), or to send data from (for a send), in a **send** or **receive** primitive. Figure 1 illustrates the simple use of **send** and **receive** primitives.

In the example, a receive on variable x is done, and the result is placed in the local variable a. In the send statement, data in variable c is sent out on the message variable z. There is no provision for overlap in communication and computation in Figure 1. So if the process blocks in the **send** or the **receive** statement, the run time system schedules another ready thread, if any.

A program can also exploit parallelism within a thread with the help of software pipelining. A **pipeline** construct specifies a code block to be run in parallel with communication. EC provides both interlocked and noninterlocked pipelines at the user level. Figure 2 shows an interlocked pipeline in a EC program.

In Figure 2, the program segment receives a two dimensional array of dimension m $\times$ 100 through message variable mat1, applies a function f to each each element in each row, and sends the output via message variable mat2. Without any pipelining, the program would have to wait for the entire array to be received before beginning computation, and would have to wait for the computation to complete before starting the sending. The code segment in Figure 2 pipelines the I/O statements, **receive** and **send** with the code in the *compute* block. The **forall** statement is like a parallel do statement, except, because of the **barrier** statement, each branch is executed only if the barrier can be passed. In the example, the barrier, mat1.b[j] can be passed, if the jth component of mat1 has been

```
message_type matrix {
   double p[10][10];
};

process matrix_multiply(x,y,z)
   in matrix x,y;
   out matrix z;
{
   double a[10][10], b[10][10], c[10][10];
   int i, j, k;

   receive x={a},y={b};
   for (i=0; i < 10; i++)
     for (j=0; j < 10; j++)
       for (k=0, c[i][j] = 0.0; k < 10; k++)
         c[i][j] + = a[i][k] * b[k][j];
   send z = {c};
}
```

Figure 1: A simple matrix multiplication program

received. The jth component is the jth row of the input array. The code within the **forall** statement applies the function f to each element of the jth row, puts the result in the jth row of variable y, and, when the computation for a row is complete, it assigns the jth component of y to the jth component of variable mat2 . This is a signal to the **send** statement, that the jth component of the output is ready for transmission, and this is then transmitted. So, as rows of input become available, computation on the row starts, and as a row of output is ready, it is sent. For machines with a single CPU, the **forall** is expanded to form another inner loop to avoid checking in every iteration.

The significance of declaring x and y inside the **pipeline** construct needs to be made clear. Statements within such a construct form a block, and the scope of these declarations are limited to this block. Further, since the run-time support has in all probability to allocate buffer space for the reception of messages (this has to be done, if data arrives before the **receive** statement can be executed), no space is allocated for these variables. Instead, they are aliased to the buffer space allocated by the run-time support routines. This helps reduce one copy from the run-time buffers to the variables themselves. Even though this is an implementation issue, and it should not be visible at the programming level, by providing such local variable declarations and encouraging programmers to use

```
message_type message_a {
   double arr[100];
};
message_type message_c(k) {
   message_a b[k];
}mat1, mat2;
int m, i, j;
        :
pipeline {
        double x[m][100], y[m][100];

        receive mat1(m) = {x};
} with {
        forall j = 0 to m barrier mat1.b[j] {
           for (i = 0; (i < 100); i++)
              y[j][i] = f(x[j][i]);
           mat2.b[j] = y[j][0];
        };
        send mat2(m);
} /*end pipeline */
```

Figure 2: Use of *interlocked pipeline* construct

them, optimization by the compiler is possible. When the end of the **pipeline** statement is reached, space for these variables is released.

A pipeline without interlocking is achieved by not using the **barrier** construct. This form of pipeline is useful when there is no dependency between computation and communication.

3.1.3 Process Declaration and Creation

Process declaration syntax is similar to that of a function. The **create** keyword, followed by the name of a process and its arguments, is used to create a new process. The location of the new process is specified by the keyword **on node**, that follows this. Figure 3 shows an example. A **create** statement has a body that specifies the processes to be created. In Figure 3 it is a simple statement with a repeater, **forall**. The body of a **create** statement can be a list of process instances, each with process name and its

```
main()
{
  matrix x[10], y, z[10];
  double a[10][10][10], b[10][10], c[10][10][10];

  /* Generate matrices a and b */
        :
  /* associate data buffers */
        :
  y = b;
  /* partition data for 10 processors */
  for (i= 0; i < 10; i++) {
    x[i] = {a[i]};
    z[i] = {c[i]};
  };
  create forall i = 0 to 9
    matrix_multiply(x[i],y, z[i]) on node i;
}
```

Figure 3: Create child processes for matrix multiplication

arguments, followed by node number on which the process is created. To create multiple instances of the same process, a `forall` repeater construct is used. Node 0 corresponds to the local node in this case. All available workstations are identified by an integer assigned by the run time system. Because of the synchronous nature, a `create` statement terminates when all its child processes are terminated. A process may have message variables and exceptions in addition to standard C style arguments. The messages and exceptions in process arguments get bound at the time of process creation. A message variable passed as an argument, is bound to the corresponding message variable that is created along with the process creation. Henceforth, communication sent on this message variable becomes available in the newly created message variable in the child. As the same message variable can be used as an argument in a number of create statements, multicasting becomes available. If a message variable has a buffer component as well, the data in such a buffer is also implicitly sent to the newly created message variable to which it is bound. It therefore acts like normal parameter passing. The need for an extra, explicit communication statement, is done away with. The child process starts as soon as it receives all its parameters. But it does not wait for the data bound to the message variable in the process argument to arrive. This allows pipelining of communication by using the pipeline construct. Figure 1 shows the syntax for process declaration. Figure 3 shows the syntax for process creation. Variables x, y and z are message types of type matrix (see Figure 1 for a declaration). In order to allow implicit sending of messages using these variables, on process creation, variables have to be associated with each message variable. The statements following the `for` loop and the one preceding it do this. Thus, the statement,

x[i] = {a[i]};

associates the variable a[i] with the message variable x[i]. Now, when the process matrix_multiply is created a few lines hence, the values in variable a are sent to the newly created process through the message variable x due to this association. Similarly, the contents of b are sent via the message variable y. Nothing is sent via z since it is declared as an `out` variable (Figure 1), even though it is associated with the variable c. Only a binding between z and the corresponding parameter in matrix_multiply is created. When a send occurs from the child process on this message variable, it will be received via z and thus placed into variable c. Even though the message variable y is being used in a number of `create` statements, no multicasting takes place at this time, as there is a separate send for each `create`. However, y is bound to a number of other message variables. So, if later, there is a send on y, then it will be multicast to all these variables in the child processes.

The process topology to be used depends upon the algorithm used. Figure 3 uses a tree topology because of the worker-coordinator paradigm used. In solving linear equations, a completely connected graph is required to implement pivoting. Similarly, FFT requires a hypercube topology. Different topologies for different problems is achieved by appropriate association of message variables in the argument.

3.2 Other Features

3.2.1 Process Synchronization

EC processes are normaly nonpreemptive thus eliminating the need for any critical sections. But a new process may be invoked by a `create` statement or by raising an exception from a remote node. This requires some form of synchronization primitives. Also, sometimes it is required to define a process execution order for correct execution. A conditional synchronization facility is provided by `block` and `activate` primitives which take a process variable as argument. A critical section is implemented by using the `barrier` construct described earlier. By separating synchronous and asynchronous process activation, use of critical sections are further reduced.

3.2.2 Exceptions

Though operating systems like Unix support signals to implement limited interprocess communication and exception handling, we decided against using this as it is expensive and also unsuitable for multithreading environment supported by EC. Exceptions in EC serve two main purposes: handling error conditions, and modelling asynchronous activities such as load balancing. Another possible use is to implement a user level distributed shared memory scheme. EC supports both the termination model as well as the resumption model for its exceptions. The resumption model is included in the language because the termination model is not suitable for modelling asynchronous activities. The resumption model in EC provides a masking facility for recoverable asynchronous faults. User specifies a model of execution for the exception handler in an exception declaration. An exception declaration is similar to a function prototype of type **exception** with optional modifier **resumption_type** specifying type and, **in**, **out**, and **inout** specifying the direction of exception propagation from the block it is defined.

In EC, exceptions are raised by the **raise** statement and handlers are specified by **try** statements similar to those in C++. Apart from user initiated exceptions, the EC run-time system raises two exceptions **io_break** and **process_exited** to handle abnormal situations arising when an EC process terminates prematurely. In the case of **io_break**, the system rejects any data pending in queues and synchronizes the message streams appropriately.

4 IMPLEMENTATION

The implementation consists of a compiler front end and run-time system library providing an interface independent of the underlying Operating System or architecture. The compiler front end is a modification of the GNU C compiler, in which, grammar, typechecking and code generation for new features of EC is added. As with the GNU C compiler, the resulting compiler is CPU independent. The run-time system provides a library for light weight processes, and communication support. The current implementation is on a cluster of Sun workstations and uses the socket library interface.

4.1 Implementation of EC Process

EC does not use the light weight library interface in SunOS because language semantics demand nonpreemptive threads. So there is no need for critical sections in nonpreemptive EC processes. It also allows the compiler to optimize the context switching overhead. The language allows preemption of a process at the time of handling **resumption_type** exceptions. So a user writes code for critical sections only if a handler may concurrently modify sharable data structure. As the compiler knows a context switch may occur only at

send, **receive**, **barrier** or **create** primitives, it can determine the life time of registers and thus good register allocation algorithms using constructs as hints for determining the lifetime of register variables (Gupta *et al.*, 1989; Davidson *et al.*, 1991) can limit context switch overhead to saving and restoring of just the instruction pointer and frame pointer registers. A low context switching latency helps in programs involving frequent context switches such as distributed discrete event simulation. Another advantage of using nonpreemptive threads is that debugging is as easy as conventional single threaded applications because there is no need for a separate debug window for each thread. Other reasons for not using Sun threads are portability and also the fact that there is no direct support for creation of threads on remote nodes, which is essential for our applications. Also, EC implementation of threads use a single stack per processor which reduces memory requirements in contrast to preemptive threads such as Sun's light weight process library, and reduces the error due to stack overflow in threads with multiple stacks.

All EC processes are created by a stub residing on the run-time system. The compiler just generates calls to this stub with appropriate parameters to create a EC process either on the local or a remote node. Binding of message parameters of an EC process is also done by this stub procedure.

4.2 Communication Support

This part of the run time system is actually system dependent. Our implementation provides a communication interface on the top of the socket library of Unix. Any call through the compiler interface builds up the list of data to be sent or received and passes this information to an asynchronous communication routine which is driven by interrupts. Under Unix, SIGIO and SIGALRM signals are used to drives this routine. This allows the overlap of communication and computation.

The sending of data is regulated according to the capacity of the network. This is done by sending a fixed amount (which is 8KB in the current implementation) every time and then waiting for a SIGALRM signal to occur. When this occurs, another segment is sent. This way, the buffer requirements in the kernel are kept within limits as otherwise this adversely affects the data transfer rate.

The data reception is done through a handler routine for the SIGIO signal. If there is no receive pending, then buffer space is allocated, as explained in section 3.1.2, otherwise the buffer designated by the receive is used. As TCP/IP does not support multicasting, this is implemented through an intermediate layer which sends separate messages to all remote nodes. The composition of a number of message variables in one variable results in the use of the scatter-gather facility of the socket library. This reduces copying overhead. Reducing the number of copy operations is very important for performance as large volumes of data are transferred in numerical problems, and the implementation takes special care on this.

All intra-node interprocess communications using message variables do not pass through the usual path. The run-time system passes these data through pointer manipulation whenever a construct does not explicitly specify that data be copied to a to different area. The use of message variables ensures that type safety is not sacrificed for efficiency. Also, this makes a program location transparent.

The code generated by an EC compiler will be dependent not only on the CPU type, but also on the machine architecture and the communication software and hardware. Thus, if a node is a multiprocessor system, separate stacks for each thread will have to be present, to allow simultaneous execution. If the hardware supports multicasting, then this can be used. The flow control implemented on **sends** will be influenced by the buffer sizes available and the speed of the network. While an attempt has been made to make the code generated sensitive to a number of such parameters, the issues seem to be too complex and not understood clearly enough, to be able to make the compiler code generation phase portable. Manual tuning will be required.

5 CONCLUSION

EC, an extension of C for distributed programming in a network of computers, has been described. The language provides primitives for fast communication, and for overlap between computation and communication. The compiler generates code for a network of SUN-3 systems interconnected by an Ethernet, running SUNOS 4.1. Some example problems have been run on the system. These include, Matrix Multiplication, solving a set of linear equations and FFT. Preliminary results have shown that a speedup of 3.6 has been achieved on four nodes, with matrix multiplication. A number of other problems are proposed to be run and evaluated. The experience gained from this will be used to fine tune the implementation.

A question arises as to whether the language will be useful in the environment of newer systems that are faster, and that have faster communication links. This question can be answered fully only after gaining experience, but our assertion is that the basic assumptions behind the current design will continue to hold. The ratio between communication and computation speeds is essentially the same in the newer architectures using fast RISC CPUs and FDDI network links. So there is a need to provide overlap between the two activities. The grain of parallelism will remain the same, as software overheads remain relatively the same.

Programming using EC is clearly more difficult than writing a Fortran program. The comparison, however, should be made with other parallel programming languages. Even High Performance Fortran (Loveman, 1993) has many of the complexities that can be seen in EC. The current state-of-art in Distributed Programming is such that such complexities seem difficult to remove. It is hoped that EC is a small step forward, allowing a programmer to write code reasonably easily without sacrificing on the performance. As in other language proposals, libraries of frequently used routines can be programmed using EC which can then be called from a Fortran or C program.

References

[1] Andre, F., Pazat, J.L. and Thomas, H.(1990) Data Distribution in PANDORE. In *The Fifth Distributed Memory Computing Conference*, IEEE Computer Society Press.

[2] Andrew, G.R., Olsson, R.A. , Coffin, M., Elshoff, I., Wilson, K., Purdin, T., and Townsend, G..(1988) An Overview of the SR Language and Implementation. *ACM Transactions on Programming Languages and Systems*, **1**,51–86.

[3] Beguelin, A., Dongarra, J., Geist, A. and Sunderam, V.(1993) Visualisation and Debugging in a Heterogenous Environment. *Computer*, **6**,88–95.

[4] Burns, A.(1988) Programming in OCCAM-2. Addison- Wesley Publishing Company.

[5] Carriero, N. and Gelernter, D. (1991) A Foundation for Advanced Compile time Analysis of LINDA programs. in *Languages and Compiler for Parallel Computers(LNCS-589)* (ed. U. Banerjee et.al.), Springer Verlag.

[6] Chang, C.(1989) REXDC: A Remote Execution Mechanism. in *SIGCOM'89 Symp. Communication Architecture and Protocols*, ACM Press.

[7] Davidson, J.W. and Whalley, D.B.(1991) Methods for Saving and Restoring Register values across Function Calls. *Software Practice and Experience*, **2**,149–165.

[8] Gao, G.R.(1993) An Efficient Hybrid Dataflow Architecture Model. *Journal of Parallel and Distributed Computing*, **4**, 293–307.

[9] Gehani, N.H.(1984) Broadcasting Sequential Processes. *IEEE Transactions on Software Engineering*, **4**,343–351.

[10] Gelernter, D.(1991) Current Research on LINDA. in *Research Directions in High level Parallel Programming Languages (LNCS-574)*, (ed. J. P. Banatre and D. Le Metayer), Springer Verlag.

[11] Griswold, W.G., Harrison G.A., Notkin, D., and Snyder, L.(1990) Scalable Abstraction for Parallel Programming. in *The Fifth Distributed Memory Computing Conference*, IEEE Computer Society Press.

[12] Grit, D.H.(1990) A Distributed Memory Implementation of Sisal. in *The Fifth Distributed Memory Computing Conference*, IEEE Computer Society Press.

[13] Gupta, R., Soffa, M.L. and Steel, T.(1989) Register Allocation via Clique Separator. in *The Proceedings of the ACM SIGPLAN'89 Conf. on PLDI*, ACM Press.

[14] Halstead, R.H.(1989) New Ideas in Parallel Lisp: Language Design, Implementation and Programming Tools. in *Parallel Lisp:Language and Systems(LNCS-441)*, (ed. Robert H. Halstead Jr. and T. Ito) Springer Verlag.

[15] Hansen, P.B.(1978) Distributed Processes: A Concurrent Programming Concept. *Communications of the ACM*, **11**,934–941.

[16] Hieb, R., Dybvig, R.K. and Bruggeman, C.(1990) Representing Control in Presence of first class Continuation. in *The Proceedings of the ACM SIGPLAN'90 Conf. on PLDI*, ACM Press.

[17] Hiranandani, S., Kennedy, K., Koelbel, C., Kremer, U. and C. W. Tseng.(1991) An Overview of Fortran D Programming System. in *Languages and Compiler for Parallel Computers(LNCS-589)* (ed. U. Banerjee et.al.), Springer Verlag.

[18] Kessler, R.R. and Swanson, M.R.(1989) Concurrent Scheme. in *Parallel Lisp:Language and Systems(LNCS-441)*, (ed. Robert H. Halstead Jr. and T. Ito) Springer Verlag.

[19] Liskov, B. and Scheifler, R.(1983) Guardians and Actions: Linguistic support for Robust Distributed Programs. *ACM Transactions on Programming Languages and Systems*, **3**,381–404.

[20] Liskov, B., Herlihy, M. and Gilbert, L.(1986) Limitations of Synchronous Communication with Static Process Structure in a Language for Distributed Computing. in *The Thirteenth ACM Symp on POPL*, ACM Press.

[21] Loveman, D.B.(1993) High Performance Fortran. *IEEE Parallel and Distributed Technology*, **1**,25–42.

[22] Lu, L.C. and Chen, M.(1991) Parallelizing Loops with indirect array reference or Pointers. in *Languages and Compiler for Parallel Computers(LNCS-589)* (ed. U. Banerjee et.al.), Springer Verlag.

[23] Magee, J., Kraumer, J. and Sloman, M.(1989) Constructing Distributed Programs in CONIC. *IEEE Transactions on Software Engineering*, **6**,663–675.

[24] Magee, J., Kraumer, J. and Dulay, N.(1992) Darwin/MP: An Environment for Parallel and Distributed Programming. in *Proceedings 1992 Hawai International Conference on System Sciences*, IEEE Computer Society Press.

[25] Mishra, J. (1991) A Perspective in Parallel Program Design. in *Research Directions in High level Parallel Programming Languages (LNCS-574)*, (ed. J. P. Banatre and D. Le Metayer), Springer Verlag.

[26] Rosing, M. and Weaver, R.P.(1990) Mapping Data to Processors in Distribution Memory Computations. in *The Fifth Distributed Memory Computing Conference*, IEEE Computer Society Press.

[27] Shivaratri, N.G., Krueger, P. and Singhal, M.(1992) Load Distributing for Locally Distributed Systems. *Computer*, **12**,33–44.

[28] Socha, D.G.(1990) An Approach to Compiling Single-Point Iterative Programs for Distributed Memory Computers. in *The Fifth Distributed Memory Computing Conference*, IEEE Computer Society Press.

[29] Strom, R.E., Bacon, D.E., Goldberg, A.P., Lowery, A., Yellin, D.M. and Yemini, S. A.(1991) *Hermes : A Language for Distributed Computing.* Prentice Hall International, 1991.

[30] Subramaniam, O.S.M.(1991) Enhancements to DC : A Distributed Programming Language. Master's thesis, Indian Institute of Technology, Kanpur, 1991.

[31] Wegner, P.(1991) Design Issues in Object based Concurrency. In M. Tokoro, O. Nierstrasz, and Peter Wegner, editors, *Object based Computing (LNCS-612)*, Springer Verlag, 1991.

[32] Wegner, P.(1992) Dimensions of Object Oriented Modeling. *Computer*, 21(10):12–19, 1992.

BIOGRAPHY: Dr. Barua is Head of Computer Science and Engineering Department and Computer Center. His areas of interest include Operation Systems and Distributed Computing.

Mr. Naik is a Ph.D. candidate in the department. His areas of interest include Compilers and Parallel Computing.

13

A DSM-based portable distributed programming testbed

R.K. Hariram *T.A. Gonsalves*
Department of Computer Science and Engineering
Indian Institute of Technology, Madras, India.
E-mail address : tag@iitm.ernet.in

Abstract

For interprocess communication, Shared Memory is a simpler and more transparent programming paradigm than Message Passing. Distributed Shared Memory (DSM) is an abstraction of shared memory on distributed computers having no physically shared memory. This paper presents the design and implementation of a completely user-level DSM system on top of Unix. Our testbed uses only System V shared memory IPC, Berkeley sockets and Unix's signal handling facility. The developed testbed is highly portable because all these facilities are now supported by almost all Unix systems. As the testbed is fully user-level it can be used for testing various DSM algorithms. Currently, it supports two consistency algorithms. We have also implemented distributed semaphores and distributed barriers for synchronization of distributed processes. We evaluate our testbed using a set of applications taken from different fields including applications from the SPLASH (Stanford Parallel Applications for Shared Memory) suite. These experiments show close to linear speedup in many cases.

1 INTRODUCTION

Recent developments in microprocessor technology and experiences in computer networks have increased the interest in distributed computers or loosely-coupled multiprocessors. Distributed computers provide good scalability, reliability, fault-tolerance and cost-performance ratio. However, programming on distributed computers had traditionally been using message passing technique as distributed computers do not share memory physically. Programming with shared memory, which is naturally supported by tightly-coupled multiprocessors, provides a better interface than message passing. It simplifies sharing complex data structures among parallel processes. Hence, recent researches are oriented towards supporting shared memory paradigm on distributed computers. An abstraction of supporting the notion of shared memory on distributed computers is referred to as Distributed Shared Memory (DSM) [Stumm (1990), Nitzberg (1990)].

One of the approaches in supporting DSM is Shared Virtual Memory [Li (1986), Fleisch (1989), Zhou (1992)], where a virtual address space is shared among all processors in the distributed computers. DSM address space can be structured either into pages of the size of physical memory pages or into segments of user-defined size. Our DSM testbed follows the latter scheme which is called "Paged Segmentation". Other approaches base sharing not on pages but on more software-oriented approaches like tuple spaces [Carriero (1989)] or objects [Ramachandran (1991), Tanenbaum (1992), Carter (1991)].

Basic issues in the design of DSM are granularity, heterogeneity [Zhou (1992)], scalability, recoverability, level of implementation, consistency and synchronization. They are addressed in [Nitzberg (1990)]. Consistency of DSM regulates the access of shared data so as to maintain the correctness of execution. Strict consistency requires that a read should always return the most recent write. Sequential consistency [Lamport (1979)] is slightly weaker but most widely assumed. It requires the execution of a parallel program to appear as some interleaving of the execution of the parallel processes on a sequential machine. Recently, weaker consistencies like Processor Consistency [Bisiani (1990)], Weak Consistency [Dubois (1986)] and Release Consistency [Carter (1991)] have been analyzed. They impose less restriction on data accesses and hence exploit more parallelism. However, they complicate programming. Our testbed implements Sequential Consistency.

Shared Virtual Memory can be supported either at the hardware level [Bisiani (1990)], kernel-level [Fleisch (1989)] or at user-level [Zhou (1992), Li (1986)]. User-level implementations, even though slightly less efficient than their kernel counterparts, make the system highly portable. Also, experimentation can be carried out without rebooting the system and hence such systems are suited for testing multiple protocols. So, we chose user-level for our implementation. Many of the user-level implementations reported so far either use distributed kernels of Unix or require some changes to kernel [Zhou (1992), Li (1986)]. However, we design and implement DSM on Unix completely at the user-level without any changes to kernel. Our testbed uses System V Shared Memory IPC, Unix signal handlers and Berkeley's UDP/IP sockets, which are now available on almost all Unix systems.

We also implement distributed semaphores and barriers at user-level which can be used to synchronize the distributed processes. The programming interface supported is similar to that of System V IPC. This simplifies porting of various parallel applications written for shared memory multiprocessors onto our system.

Our testbed currently runs on SunOS and Linux, a public domain Unix OS. We have conducted experiments on a network of Sun workstations connected by 10 Mb/s Ethernet.

Here is the road map to the rest of this paper. Section 2 describes various design aspects of our DSM testbed and Distributed Semaphores. Section 3 explains the implementation details. In section 4, we present the performance of our testbed. A survey of related work is in section 5. Section 6 concludes this paper with a look to the future.

2 DESIGN

Our testbed consists of three independent functional subsystems: Distributed Shared Memory (DSM), Distributed Semaphores (D*Sem*) and Distributed Barriers (D*Bar*). All three use a ***Naming Subsystem*** to manage object names globally. In this section, we present the design of these subsystems. As the design and implementation of D*Bar* is very similar to that of D*Sem*, we will discuss only D*Sem* in detail.

2.1 The Naming subsystem

Naming or binding high level names to communication objects is used by all IPC systems. Our naming scheme follows that of System V IPC. It uses two types of names: keys and handles. A key is a 32-bit integer that the user associates with the object (DSM segment or distributed semaphore). When the user creates or queries an object using a key, a handle is returned. This handle is subsequently used to operate on the object.

The separation of the naming subsystem from the other functional components is a natural one. Creating and deleting an IPC object is considerably less frequent than the functional operations on that IPC object. So, we implement the naming component in a centralized, yet reliable, manner. On one of the sites, a Name Server daemon is run. As there is no dependence between the names of DSM segments and distributed semaphores, our testbed uses different Name Servers, one for each. These daemons can be located on different sites so as to distribute the naming load. The Name Server daemon maintains a name database keeping all keys and handles. In addition, it also records the address of the first creator of the object as it will be useful to locate the server for that object. The DSM and D*Sem* subsystems send requests to their respective Name Servers for inserting, querying and deleting the names of their objects.

2.2 The DSM subsystem

2.2.1 DSM algorithms

Consistency of DSM is achieved by a combination or variation of one or more of the following four basic algorithms [Stumm (1990)].

Central Server Algorithm The entire DSM data is held in a single server. Every access to DSM data by a non-server causes a request to be sent to the central server.

Migration Algorithm In this *Single Reader Single Writer* algorithm, DSM is divided into fixed-sized pages. Each page resides on the machine which is currently using it. When a fault occurs on any location within a page, the entire page is transferred from the current holder of the page to the faulting processor. After that, all accesses to that page by the new holder become local, until the page is taken away by a fault on another processor. This algorithm makes use of locality of accesses to minimize network traffic.

Read Replication Algorithm This *Multiple Readers Single Writer* algorithm allows either multiple readers or a single writer to a page. A read access to a page just replicates the page whereas a write invalidates all other copies leaving only one writable copy.

Full Replication Algorithm This *Multiple Readers Multiple Writers* algorithm allows every processor to read and write into a page. A read is done directly from the local copy whereas a write is sent to a *sequencer* which then broadcasts it to all processors.

No single algorithm performs well under all conditions. Performance depends on the application's read/write characteristics. A queuing network model to analyse these algorithms is presented in [Venkatesulu (1994)].

Our DSM testbed currently implements Read Replication and Migration. The user can choose one of these by passing proper flags to the *dsmat()* call. He can also dynamically change the consistency algorithm by reissuing the *dsmat()* with proper flags. The user can choose Migration if the data is migratory. Data is said to be migratory if a write is going to occur soon after a read. For migratory data, Read Replication will cause read and write faults whereas Migration causes only one fault. Full replication and centralized server algorithms are not attractive as they require every write of the user process to be intercepted which is difficult to achieve and very costly in any user-level implementation.

In Read Replication and Migration algorithms, for every DSM page, one process is termed as owner. In Migration algorithm, owner is a node that has an exclusive read and write access to the page. In Read Replication algorithm, owner is either a reader or the only writer to the page. The owner maintains the *copy set*, the list of readers to that page. A process that wants to write to a page acquires the ownership first and then writes. Thus the owner of a page changes dynamically. Owner of a page can be located in either a centralized or fixed distributed or dynamic manner [Li (1986)]. Our testbed adopts the dynamic approach. In this approach, each process keeps track of the ownership of all pages in a local table. The ownership field in the table points either to itself, if it is the true owner of the page, or to another processor which may be the true owner or in turn point to a sequence of processors through which the true owner can be found.

2.2.2 OS requirements

In order to implement Shared Virtual Memory on top of an existing operating system, we need the ability to (a) install page fault handlers for DSM, (b) change the access protection of individual pages, and (c) send messages from one processor to another.

We designed our testbed for systems that meet all these requirements at user-level. In Unix, the signal SIGSEGV (Segmentation Violation) is issued to a process whenever it tries to access an unmapped memory location or when it attempts to write to a location mapped only for read. The signal handler for SIGSEGV is given the address of fault as a parameter.* Such a signal handler acts as the page fault handler in our system.

Even though Unix does not allow a user-level process to change its page access protections, this could be done in two indirect ways. One method is to let the process issue memory map, unmap and protect calls to the virtual page that corresponds to the DSM page. The other way is to let the process issue System V Shared Memory IPC calls to a segment that corresponds to the DSM page. We have chosen the latter approach as System V IPC is widely used. In this indirect approach, every DSM page is mapped to one local shared memory segment of the same size. If a process has to invalidate its access to a DSM page, it issues *shmdt()* system call to the shared memory segment corresponding to that page. Similarly to change access privilege from/to read-write to/from read-only, it issues *shmdt()* followed by *shmat()* with the required permission flags set.

In order that the DSM processes communicate with each other for servicing page faults, we implement our own communication layer using Berkeley's UDP/IP sockets. This layer adds reliability to unreliable UDP/IP socket communication and provides request-response mode of communication with forwarding capability.

2.2.3 DSM architecture

Figure 1 shows the organization of DSM implementation on a site. Each dotted box denotes a process. Descriptions within a dotted box denote modules. On every site, users can create any number of processes that belong to either the same distributed application or different applications. The user program is linked to a library which provides our DSM subsystem. This user process is hereafter referred to as the *DSM process*. The DSM library contains the following major modules:

Page Fault Handler To handle local faults by sending requests to the the page owner.

Segment Interface Handler To handle user's call to create, delete, attach or detach DSM segments by communicating with the Name Server and Segment Manager.

Memory Controller To degrade or invalidate the process's access privilege to a page, on a remote fault.

Page Server and Segment Manager are the two server daemons that run on each node using DSM. The Page Server maintains the copy sets for pages that are owned by local DSM processes. Any fault requests from remote processes are serviced by the page server

*Not all Unix systems provide this address to the signal handler.

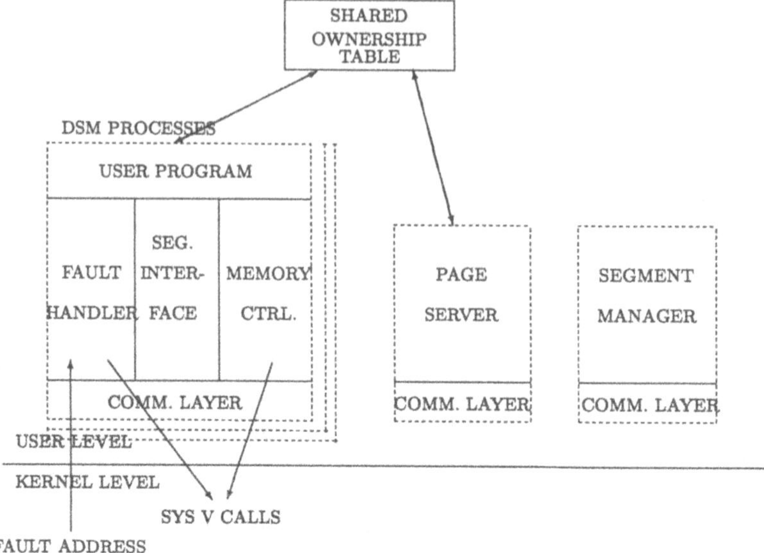

Figure 1 Organization of the DSM subsystem

with or without the help of local owner process. We have designed the functions of the Page Server and its co-ordination with the DSM process such that the Page Server takes up time-consuming jobs with little or no intervention of the DSM process. Having a Page Server daemon gives us the following advantages.

- After a user process has finished its job and exits, the pages owned by it need not be flushed out to any other site, as all remote requests are directed only to Page Server and it can still service them.
- Some of the remote faults can be serviced by the page server without disturbing the DSM process. For example, if a read fault occurs at a remote DSM process and the local owner process is a reader of that page, the page server replies with a copy of the local page and updates the copy set maintained by it.
- Some remote faults can be serviced by the page server with less disturbance to the DSM process. For example, if the local DSM process is owner of a page with write permission and a remote read fault request arrives to the Page Server, the DSM process is just requested to reduce its access permission to that page to read-only and the remaining task of sending entire page is carried out by the Page Server.

The page ownership information is stored in a locally shared segment shared by DSM processes and Page Server. So, a DSM process can communicate with page owners directly without the help of the local Page Server thereby avoiding an otherwise context switch.

The Segment Manager maintains the status of all segments that are created by local DSM processes. Even though pages and their ownership may move from site to site, the segment management always remains with the Segment Manager on the site on which the segment was created.

2.3 The D*Sem* subsystem

Distributed semaphores can be implemented either on top of DSM or on top of messages independent of DSM. In the first approach, the data structures for distributed semaphores are stored in DSM space. Consistency of DSM does not itself ensure atomicity as a page may be pre-empted by a remote faulter at any time during operations on distributed semaphores residing in that page. Atomicity can be ensured in two ways:

1. By an option to wire-in DSM pages to a process until they are released. DSM should delay servicing of remote faults until the local process releases such pages.
2. By using a lock in DSM space, the processes can use atomic test-and-set instruction on that lock before entering into operations on distributed semaphore. This approach will cause large network traffic due to thrashing of the DSM page containing the lock, when many processes test the lock simultaneously.

Because both the ways of ensuring atomicity are not satisfactory, we implement D*Sem* on messages. In this approach [Tanenbaum (1987), Yuan (1992)], the data structures for the semaphores are kept in private address space and atomicity is ensured by queuing the requests. As in DSM, semaphores can also be replicated, migrated or centralized. We observe that semaphore operations exhibit poor locality compared to operations on DSM. So, we take a simple way of distributing semaphores in a fixed manner.

2.3.1 D*Sem* architecture

The user program is linked to a library that provides our D*Sem* subsystem. The D*Sem* module accepts requests to create, delete and operate on a distributed semaphore. It handles create and delete requests by sending request messages to the D*Sem* Name Server. To service requests to execute *up* or *down* on a semaphore, it locates the Semaphore Server owning that semaphore using a table and then communicates with that Semaphore Server.

On every site there is a Semaphore Server daemon. It owns all semaphores that are created by local processes. The data structures of the distributed semaphores owned by this server are stored in its private address space. Since semaphores do not migrate, their distribution among different sites is determined by the pattern of semaphore creation by the user processes.

2.4 Programming model

The programming semantics supported by our DSM system are similar to those of System V shared memory IPC [Bach (1986)]. It provides *dsmget(), dsmat(), dsmdt() and dsmctl()* calls to create, attach, detach and remove a segment respectively. In addition to the above four IPC calls, as our testbed is at the user-level it also provides routines InitDsm() and ExitDsm() for initializing and finishing up operations. Similar set of calls are supported in case of D*Sem* and D*Bar* subsystems.

3 IMPLEMENTATION

3.1 Table handling

The modules of the DSM subsystem linked to the user program maintain the status of segments used by the user program in a table called *SegTable*, shown in Figure 2.(a). Each entry in *SegTable* contains information about one segment, consisting of one or more pages. The table is hashed by the segment id. Any page within a segment is identified by the tuple <segmentId, offset>, where the *offset* is in number of pages within the segment.

Each valid entry of *SegTable* has segment identifier, segment size, flags for type of attaching, address of segment manager, a pointer to a list of attached addresses and a pointer to a page table. Each entry in the page table contains the id of the System V shared memory segment corresponding to that page and the permission with which this System V segment is currently attached to the process.

The segment table of the Page Server contains entries with fields segment id, size and a pointer to page entries. These page entries contain more information including the copy set and a pointer to the queue of processes waiting for the page. The segment table of the Segment Manager is very simple and contains only segment id, segment status and the list of processes attached to that segment.

As both the Page Server and DSM processes need the ownership information of the DSM pages, it is kept in a locally shared space, *shTable* (Figure 2).(b). This table has two parts: a pointer array and an ownership table. Each entry in the pointer array corresponds to one DSM segment and points to the ownership table entries for that segment. Each entry in the ownership table corresponds to one DSM page. All the entries of one segment are linked together. All the free entries are linked into a free list. Each entry contains a lock to control access to that entry, a pointer to the list of processes waiting for that lock, access permission of the page owner, address of the owner and links to the next page entry. Using *shTable*, given <segmentId, offset>, the owner of the page can be found.

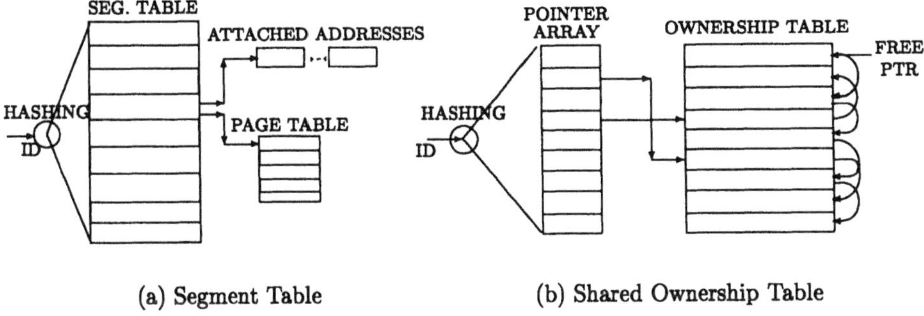

(a) Segment Table (b) Shared Ownership Table

Figure 2 Structure of Tables

3.2 Page fault handling

Local faults

When a page fault occurs the Page Fault Handler module of the DSM process checks whether the faulting address lies in attached DSM address space. If so, it maps the address to the <segmentId, offset> tuple that identifies the DSM page in which the fault address lies. The fault handler then finds the address of the owner of the page from the shared ownership table and sends a request to the Page Server of the owner's node. In case of a Read Fault, it receives the page as reply and then returns. In case of a Write Fault, it receives the page and the copy set. It then sends invalidation messages to all processes listed in the copy set, receives acknowledgements, updates the shared ownership table by setting the ownership of the page to itself and then returns. Figure 3 illustrates handling of Read Fault and Write Fault.

When a process writes to a read-only page, it generates an Upgrade Fault, sends an upgrade request to the Page Server. If the upgrading is successful, the reply contains copy set alone. If the upgrading fails, the reply contains copy set and the page. As in case of Write Fault, it sends invalidation messages to all readers in the copy set. Support of upgrading avoids transfer of a page when the reader already has a valid copy. In order to check whether the copy of the page held by the reader is valid, a version number is used for each page. Whenever an upgrade or write fault occurs, the version number is incremented. On a read fault, the version number is also received along with the page.

Remote faults

When Page Server receives a Read Fault or Write Fault or Upgrade Fault request for a DSM page <segmentId, offset> it checks the entry in the ownership table to see whether the owner of the page is a local client. If not, it forwards the request to the owner listed in the ownership table. If the owner is local, the Page Server maps <segmentId, offset> to

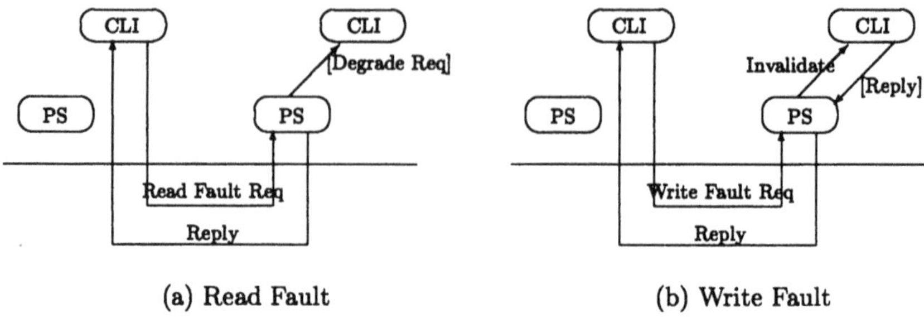

(a) Read Fault (b) Write Fault

Figure 3 Page Fault Handling

an address at which the Page Server has attached the System V shared memory segment corresponding to that DSM page. In case of a Read Fault, it sends degradation message to the local owner process and all other local processes that have write access to that page, adds the requestor to the copy set of that page and replies with the page. In case of a Write Fault, it sends invalidation message to all local writers and readers using that page. Only the writers reply to this invalidation message. After receiving reply from all local writers, the Page Server replies to the remote faulter with the page and copy set. In case of an Upgrade Fault, it invalidates local clients as in Write Fault servicing. It compares the version number of the page held by the requestor to the current version number. If both are same, it replies with copy set alone. If both are different, it replies with copy set and the page.

When the Page Server receives an invalidation request, it checks whether the DSM process is alive and if so, it forwards the request to that process and sends an acknowledgement to the requestor. When a DSM process receives an invalidation request or degradation request for a page, its Memory Controller module detaches or reattaches the System V segment corresponding to that page with new permissions.

3.3 Communication substrate

The implementation of DSM, D*Sem* and Naming requires a reliable communication protocol. They typically operate in request-response mode. For example, on a page fault, the fault handler sends a request to the probable owner of the page which either replies with the page or forwards the request to another host. The most suitable protocol for these remote operations is therefore a request-response protocol, with forwarding capability.

While Sun's RPC package supports request-response mode it does not provide forwarding and would reduce portability. Hence we selected the more widely available Berkeley sockets. TCP/IP sockets are unsuited as the pattern of communication between DSM processes varies dynamically according to the access to the DSM data and a process can communicate with multiple processes. So, it is expensive to keep multiple connections alive or to open and close a single connection to suit the pattern of communication. Hence, we opted for the UDP/IP connectionless protocol. Unfortunately, UDP/IP does not provide reliable communication. Hence, our protocol implements timeout and retransmit mechanism to ensure reliability.

The communication layer provides *SendReqRecvReply()*, *SendReply()*, *ForwardReq()* and *SendAck()* calls to the upper layer. *SendAck()* is used by a replier to prevent the requestor from retransmitting if replying may take a long or arbitrary time. The upper layer should provide functions *ServiceReq()* and *ServiceDupReq()* which are called by the communication layer on receiving a new or a duplicate forwarded request respectively.

4 PERFORMANCE EVALUATION

We now examine the performance of our DSM testbed. First we describe a series of experiments for evaluating the costs of important operations. Besides providing insight into our implementation, these can be used as inputs to analytical or simulation models of DSM. Next we discuss the performance on our DSM of several parallel programs that are commonly used to benchmark shared memory multi-processors. Experiments were performed on machines interconnected by a 10 Mb/s Ethernet. We used *gcc* compiler. During experiments, the facility was devoted solely to us.

4.1 Costs of faults

On Sun 3/60+, the average end-to-end delays in handling a read fault (without forwarding the request), a write fault (without invalidations) and an upgrade fault (with reply containing no page) are 36.6 ms, 35.6 ms and 20.6 ms respectively. The cost of sending and receiving a 30 byte fault request packet by UDP is 3.3 ms and of an 8 KB fault reply packet is 18.3 ms. Thus, our implementation adds an overhead of 15 ms, 14 ms and 14 ms for read, write and upgrade faults respectively.

Our times of 35–37 ms/fault compare favourably to the 40 ms/8KB transfer time using SunRPC on a 10 Mb/s Ethernet reported by Van Renesse et al. [Renesse (1988)]. Ramachandran et al. [Ramachandran (1991)] report 43.4 ms/fault in their DSM system, about 16-20% poorer than ours.

The read and write fault costs in the case of Linux, with an i386 node generating the faults and an i486 servicing them, are 26.0 ms and 26.3 ms respectively. The time between sending a 30 byte request packet and receiving 4KB reply packet is 20.2 ms. So, the overheads for read and write fault are 5.8 ms and 6.1 ms respectively.

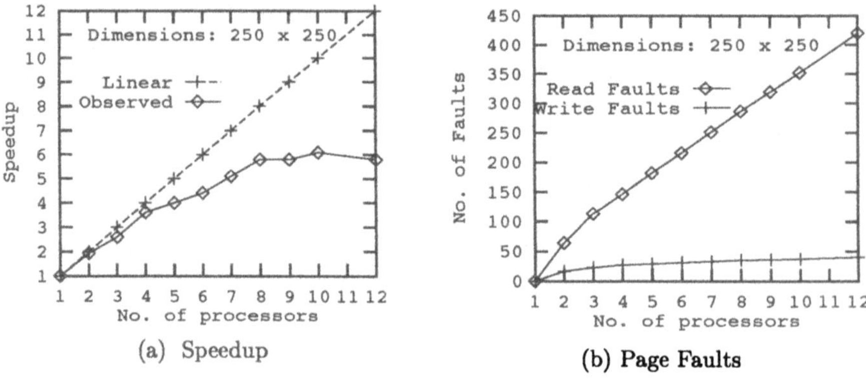

(a) Speedup

(b) Page Faults

Figure 4 Performance of Matrix Multiplication

4.2 Performance of applications

We have currently implemented three applications with different shared access behaviours and synchronization requirements on our testbed. The input data size is chosen not to be too large so as to test the performance of our testbed for medium granularity also. All measurements were on Sun 3/50 machines.

4.2.1 Matrix multiplication

Matrix multiplication is inherently parallel with minimal sharing. The result matrix is divided into a number of sets of adjacent rows equal to the number of processes p, and assigned to the processes. A master process initializes the two input matrices, creates one process on each remote machine and then computes its share of the result matrix. A barrier is used to synchronize the termination of all processes. The input matrices are read-shared and the result matrix is write-shared. During computation there will not be any communication between worker processes except for any page containing rows assigned to different processes. This *false sharing* becomes negligible for large matrices.

Figure 4 illustrates the results. For matrices of 250×250, near linear speedup was observed for upto 4 processors (Figure 4.(a)). As the number of processors increases, the computation assigned to each processor becomes small. Also, the read faults for the input matrices increase (Figure 4.(b)) as each processor needs one of the input matrices fully. So, for $p > 4$ the speedup rises slowly, levels off and then starts to fall for $p > 10$.

4.2.2 Linear equation solver

Linear Equation Solver implements a parallel version of Jacobi's[Bertsekas (1989)] iterative algorithm to solve a set of n linear equations, $Ax = b$, where A is $n \times n$ matrix and

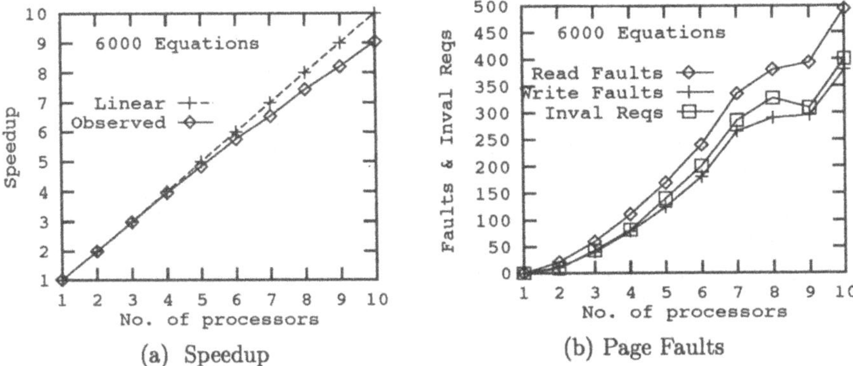

(a) Speedup (b) Page Faults

Figure 5 Performance of Linear Equation Solver

x and b are $n \times 1$ vectors. Starting with some initial vector $x(0)$, a new vector $x(t+1)$ is computed in the $(t+1)^{th}$ iteration using,

$$x_i(t+1) = \frac{1}{a_{ii}}\{b_i - \Sigma_{i \neq j} a_{ij} x_j(t)\}$$

This algorithm is parallelized by dividing the result vector x into a number of sets of adjacent elements equal to the number of processes, p, and assigning the task of evaluating that portions to the processes. Every iteration has two phases: computation phase and update phase. In the computation phase, each process calculates its portion of x into a temporary array. In the update phase, every process copies the new values from the temporary array into the globally shared array x. Processes synchronize at the end of each phase using barriers.

We coded A and b into our program as the number of DSM pages available in our testbed is limited by the number of available Sys V segments. As A and b are only read shared, this does not affect the overall performance of the application appreciably. x is stored in DSM space. The entire vector x is read by all processes in the computation phase and portions of it are written by each process in the update phase.

Graphs in Figure 5 show the speedup and page faults for solving a system of 6000 equations using 5 iterations. We observe linear speedup upto $p = 6$. For $p > 6$, speedup is only marginally below linear.

4.2.3 Water

Our third test program, *Water*, is taken from the SPLASH [Singh (1992)] suite of realistic benchmarks for shared memory multiprocessing systems. *Water* is an N-body molecular dynamics program that evaluates forces and potentials in a system of water molecules

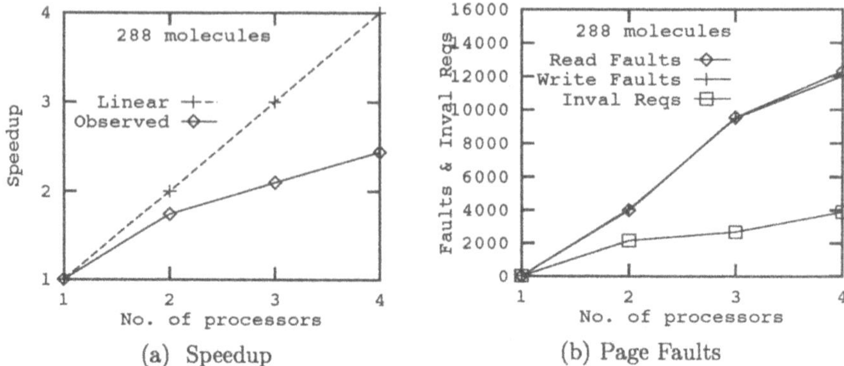

(a) Speedup

(b) Page Faults

Figure 6 Performance of *Water*

in the liquid state. The computation is performed over a user-specified number of time-steps. Every time-step involves setting up and solving the Newtonian equations of motion for water molecules in a cube with periodic boundary conditions. The total potential is computed as the sum of intra- and intermolecular potentials.

The main data structure is a large array of structures, VAR. Each structure holds all the data for one molecule, about 600 bytes. For intra-molecular computation, parallelism is maintained by assigning every processor a set of molecules that are located next to each other in the VAR array. There is no communication in the intra-molecular computations, except in the small number of accumulations to a global sum every time-step. For intermolecular computations, a processor computes the interactions for every molecule in its partition with the $n/2$ molecules following it in the VAR array. Then a processor communicates with exactly half the total number of processors.

Distributed Barriers are used to maintain dependencies before and after computing intermolecular interactions. Distributed semaphores are used to accumulate the private sum of each process into the shared sum once every time-step and also to make updations to the force locations of all atoms to be mutually exclusive.

The problem size used in our measurements is 288 molecules. Process creation, segment creation and semaphore creation are omitted in time measurements as we simulate only 2 time-steps. The speedup obtained is shown in Figure 6.(a). As the communication involved in intermolecular computations is much larger than that in other applications, the number of faults is substantially higher as seen from Figure 6.(b). Therefore, the speedup is not as linear as in the other two applications. However, reasonable speedup is observed for upto 4 processors.

5 RELATED WORK

A number of software DSM systems have been developed in the recent past. Ivy [Li (1986)] was the first page-based DSM system developed on an Apollo ring using memory map calls to attach and detach pages. Ivy-II is a user-level implementation but requires addition of some system calls to the kernel. It implements consistency using ***Single Writer Multiple Readers*** protocol. It supports page-based sharing whereas our testbed supports paged-segmentation which is more flexible.

Mirage [Fleisch (1989)] is a kernel-level paged-segmentation DSM system on Locus operating system. It locks a page with a certain processor for a certain time to avoid page thrashing due to false sharing. Mermaid [Zhou (1992)] is a heterogeneous page-based DSM system developed for Sun and Firefly workstations. It is also at user-level, but requires addition of a new system call to SunOS so as to set the page access permissions from user-level.

Ramachandran *et al.* [1991] have implemented an object based DSM system on Clouds, a distributed operating system, and also on Unix. Their Unix implementation uses Sun's Network Interface Tap for communications and Unix file system to store DSM segments whereas our testbed uses Berkeley sockets and System V shared memory IPC for respective needs. Their design involves more context switches in servicing page faults as each request and reply has to go through a special process meant for communication. In our design we use asynchronous sockets to handle asynchronous events. Also, we minimise context switches by using a locally shared table between DSM user processes and Page Server. Hence, our system provides 16-20% better performance.

Munin [Carter (1991)] is an object-based software DSM system. It supports multiple coherency protocols including Release Consistency. The user annotates the shared data with the type of its expected access pattern and these annotations are then used by the run-time system to choose a consistency protocol best suited to that access pattern. A.S.Tanenbaum *et al.* [Renesse (1988)] implement distributed shared objects using Full Replication over a reliable broadcast layer. They have also designed a new parallel programming language, called Orca [Bal (1992)] which supports parallel processes and distributed shared objects.

6 CONCLUSIONS

We have designed and implemented a distributed programming testbed for a network of Unix machines based on the shared memory communications paradigm. In addition to a distributed shared memory interface similar to the Unix System V shared memory, our testbed provides distributed semaphores and barriers. The implementation is entirely at the user-level and is hence portable. The testbed currently runs on SunOS and Linux.

The testbed could be used for evaluating various DSM algorithms. Currently it supports two consistency algorithms, Read Replication and Migration. It also supports Upgrading for write faults onto a read-only page.

The fault handling times of our user-level implementation (35 ms/8 KB page) compare favourably with other implementations. Several realistic applications exhibit near linear speedups for upto 8 processors. However in an application with substantial interprocess communication, the observed speedup is considerably less than linear.

In our testbed currently the user specifies whether Read Replication or Migration algorithm is to be used for a segment. The future work will be to design an automatic switching by the DSM subsystem between these algorithms depending on the shared access behaviour of the application.

REFERENCES

Bach, M.J. (1986) *The Design of the Unix Operating System*. Prentice-Hall of India, New Delhi.

Bal, H.E., Kaashoek, M.F. and Tanenbaum, A.S. (1992) "ORCA: A Language for Parallel Programming of Distributed Systems". *IEEE Trans. Soft. Engg.*, 18(3):190-205.

Bertsekas, D.P. and Tsitsiklis, J.N. (1989) *Parallel and Distributed Computation - Numerical Methods*. Prentice-Hall Int'l Editions.

Bisiani, R. and Ravishankar, M. (1990) "PLUS: A Distributed Shared Memory System". In 17th *Annual Int'l Symp. Comp. Arch.*.

Carriero, N. and Gelernter, D. (1989) "How to Write Parallel Programs: A Guide to the Perplexed". *ACM Computing Surveys*, 21(3):323–357.

Carter, J.B., Bennet, J.K., and Zwaenepoel, W. (1991) "Implementation and Performance of Munin". In *Proc. of* 13th *ACM Symp. Op. Sys. Principles*, pp. 152–164.

Dubois, M., Scheurich, C. and Briggs, F.A. (1986) "Memory Access Buffering in Multiprocessors". In 13th *Annual Int'l Symp. Comp. Arch.*, pp. 434–442.

Fleisch, B.D. (1989) *Distributed Shared Memory in a Loosely Coupled Environment*. PhD thesis, University of California, Los Angels.

Lamport, L. (1979) "How to Make a Multiprocessor Computer that Correctly Executes Multiprocess Programs". *IEEE Trans. Comp.*, C-28(9):690–691.

Li, K. (1986) *Shared Virtual Memory on Loosely Coupled Multiprocessors*. PhD thesis, Department of Computer Science, YALE University.

Nitzberg, B. and Lo, V. (1990) "Distributed Shared Memory: A Survey of Issues and Algorithms". *IEEE Comp.*, 24(8):52–60.

Ramachandran, U. and Khalidi, M.Y.A. (1991) "An Implementation of Distributed Shared Memory". *Software-Practice and Experience*, 21(5):443–464.

Renesse, R.V., Staveren, H.V. and Tanenbaum, A.S. (1988) "Performance of the World's Fastest Distributed Operating System". *Operating Systems Review*, 22(4):25–34.

Singh, J.P., Weber, W.D. and Gupta, A. (1992) "SPLASH: Stanford Parallel Applications for Shared Memory". *Comp. Arch. News*, 20(1):5–44.

Stumm, M. and Zhou, S. (1990) "Algorithms Implementing Distributed Shared Memory". *IEEE Comp.*, 23(5):54–64.

Tanenbaum, A.S. (1987) *Operating Systems: Design and Implementation.* Prentice-Hall of India, New Delhi.

Tanenbaum, A.S., Kaashoek, M.F, and Bal, H.E. (1992) "Parallel Programming using Shared Objects and Broadcasting". *IEEE Comp.*, pp. 10–19.

Venkatesulu, D. and Gonsalves, T.A. (1994) "A Queuing Network Model of Distributed Shared Memory". In *Proc. of NETWORKS '94 - Int'l Conf. Computer Networks, Architecture and Applications*, Madras, India.

Yuan et al, S.M. (1992) "Design and Implementation of a Distributed Semaphore Facility". In *Third Workshop on Future Trends on Distributed Comp. Sys.*, pp. 180–184.

Zhou, S., Stumm, M., Li, K. and Wortman, D. (1992) "Heterogeneous Distributed Shared Memory". *IEEE Trans. Parallel and Distributed Sys.*, 3(5):540–554.

R.K. Hariram received a B.E. (Computer Science) from Government College of Engineering, Tirunelveli, Tamil Nadu, India in 1992. He was the University top rank holder in B.E. (Computer Science). He is currently an M.S. research scholar in the Department of Computer Science and Engineering, I.I.T., Madras. His research interests include distributed systems, parallel processing and computer networks.

Timothy A. Gonsalves received a B.Tech.(Electronics) from I.I.T., Madras in 1976, an M.S. (EE) from Rice University, Houston, USA in 1978, and a Ph.D.(EE) from Stanford University, USA in 1986.

From 1986 to 1988 he was an Assistant Professor in the Computer Science Department, Worcester Polytechnic Institute, USA. In January 1989, he joined the Department of Computer Science and Engineering, I.I.T., Madras where he is currently an Associate Professor. His research interests include distributed computer systems, computer networks, and performance evaluation. He is active in technology/product development and consultancy for a number of industries. Dr.Gonsalves is a member of CSI, ACM and IEEE.

14

X.500 Performance with Large Databases and OSInet[1]

P.V. Rajeev[2] *S.V Raghavan*[3]
Department of Computer Science and Engineering
Indian Institute of Technology, Madras, India
rajeev%hss@lando.hns.com
svr@iitm.ernet.in
Ashok K. Agrawala
Department of Computer Science,
University of Maryland,
College Park, MD USA.
agrawala@cs.umd.edu

Abstract

The X.500 directory provides a powerful mechanism for storing and retrieving information about objects of interest in the networking world. The rapid increase in the deployment of the directory has resulted in a very large amount of data in the global directory. Hence, the need arises for the development of a model which predicts the performance of the directory in an internet environment with large amount of data.

In this paper, we present a deterministic model for predicting the performance of X.500 directory operations. Actual measurements are carried out in a real life environment, with the directory having a large amount of data distributed across the internet. The model is validated by comparing the results of these actual measurements with the predicted values. The application of the model for the selection of DIT structure in a wide area network is explained.

Keywords:

X.500, Directory Services, Performance, Deterministic Modelling.

[1]Network Software developed at Indian Institute of Technology, Madras.
[2]Currently with Hughes Software Systems, New Delhi
[3]Further correspondence may please be addressed to Dr. S.V. Raghavan.

1 INTRODUCTION

With the rapid expansion in the electronic communication, there is a growing need for an organized global information service. The X.500 directory provides such a powerful mechanism for storing and retrieving information about objects of interest in the networking world[CCITT (1988), Huitema (1989), Bumbulis *et al.*(1993)]. X.500 directory defines an architecture whereby the directory can be distributed over an arbitrary number of systems. Each system manages a portion of the global directory. Typically, in the X.500 directory, the information is distributed and managed across the network of many autonomous administrations. These systems are integrated into a single directory by a set of protocols. The directory stores information about *objects* and defines a hierarchical relationship between *objects* stored in the directory. The hierarchical structure reflects the national and organizational structure of the real world objects.

The X.500 directory was originally designed for use with X.400 message handling systems (MHS)[CCITT (1988)]. Message handling system uses directory for a variety of purposes including: retrieval of O/R address when O/R name is given, user friendly naming of users in the MHS, expansion of distribution lists, e-mail routing, etc. To provide these services to the MHS, the directory must store information about all e-mail users in the network. The number of e-mail users in the network itself will be of the order of millions. A recent survey on the usage of internet indicated a very high rate of increase in the number of users. A large amount of information has to be stored in the directory to cater to the needs of MHS.

Apart from MHS, directory is now being used for a variety of other applications. For example directory is now used to store the topology and service related information about communication networks [Mansfield *et al.*(1993), Hong *et al.*(1992a), Hong *et al.*(1992b)], to map object names to internal object identifiers in object based distributed operating systems [Rennesse *et al.*], to store bibliographic information [Barker (1992)], to store the information about ftp archives [Barker (1993)], and so on. Once all these applications start using the directory to store information, directory will be used extensively and an enormous amount of data will be stored in the directory.

The X.500 directory deployment in the internet is now progressing at a very fast pace. Presently, there are more than 450 Directory System Agents (*DSAs*) spanning 35 nations, covering a total of more than a million entries [Goodman (1992)]. The growth rate is found to be significant, with about 150 *DSAs* added every year since November 1990, with a proportional increase in the number of entries. With this rate of increase in the deployment of directory in the internet, the internet directory will soon be populated with enormous amount of information. The information about the e-mail users itself will be substantial. The internet directory is also used for storing information about various services available in the internet. Directory helps in giving an integrated view of the internet and the services available in the internet.

The concern of the designers is to estimate the critical database size and interconnection topologies among the *DSAs* and understand the implication of chaining and referral and guarantee response time whenever possible. This necessitates the study of the performance of the X.500 directory system, with special emphasis on large databases, wide area distribution and variable interconnections through different topologies.

In this paper, we present the performance measurements carried out in an internet environment with the directory distributed over two sites in the internet. A model for estimating the performance of the directory is also described.

The rest of the paper is organized as follows. Section 2 describes the X.500 and its implementation over OSInet. A deterministic model for X.500 is presented in the section 3. Section 4 validates the model with the help of actual measurements. The application of the model for the selection of DIT topology in a wide area network is explained in section 5.

2 X.500 OVER OSInet

The X.500 directory used for the performance measurements presented in this paper, is implemented [Sriram (1991), Sriram *et al.*(1994), Raghavan *et al.*(1992)] over OSInet network kernel. OSInet [Raghavan (1990), Raghavan (1993a)] is a network protocol software which provides a subset of services defined in ISO-OSI reference model. OSInet has a network kernel which provides the basic networking support and user applications that run over the kernel. The kernel provides the services required for a reliable networking environment. OSInet follows the layered architecture of the ISO-OSI reference model. The interfaces to upper and lower layers are implemented using queues and the layer specific actions are realized through finite state machines. The OSInet architecture is shown in Figure 1a.

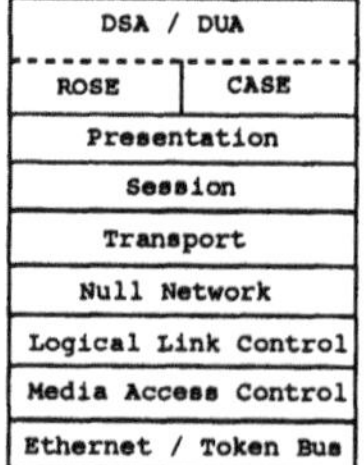

a. OSInet Architecture For LANs

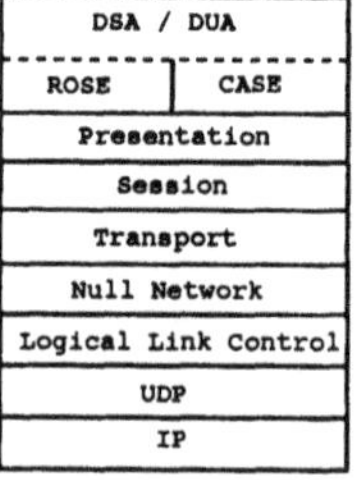

b. OSInet Architecture For Internet Experiments

Figure 1: The OSInet Architecture.

The X.500 implementation makes use of ROSE and ACSE of OSInet. The binding and unbinding to *DSAs* are done by using the ACSE. ROSE is used to invoke the directory operations. For the internet experiments the lower layers of OSInet stack is modified to include UDP as shown in Figure 1b.

The X.500 standard [CCITT (1988)] does not specify any database mechanism for storing the directory information. It is upto the implementor to choose an appropriate storage mechanism. In the case of OSInet X.500, B-Tree++ indexing mechanism is used [Comer (1979), Raghavan *et al.*(1993b)] for the database access. The tree structure of the directory is mapped onto the underlying file systems' directory structure. The knowledge information regarding the location of various naming contexts known to *DSAs* is stored in a sub-directory called *tree directory*, which is the working directory of a *DSA*. The information about an entry is stored as pairs <attribute type, attribute value> in the directory information files. The knowledge information file is indexed using the distinguished name as the key and the directory information file is indexed using the relative distinguished name (RDN) as the key.

3 A DETERMINISTIC MODEL FOR X.500 DIRECTORY

A deterministic model for distributed directory services allows one to investigate the effect of various design choices and mode of interaction between the *DSAs* on the cost of directory operations. Very few attempts have been made so far in the literature to quantify the performance of a distributed directory[Neufeld (1992)].

The major factors that affect the performance of the directory in a large internet environment are,

- the communication delay between the Directory User Agents (*DUAs*) and *DSAs*, and between *DSAs*,
- the database access time, and
- the name resolution time.

In this section we discuss a model for the performance estimation of the directory based on these factors.

Typically, a directory service consists of a number of *DSAs* distributed throughout the internet. In our model, we assume that the *DSAs* are always available and there are no communication link failures. A *DSA* is run as a single process in the system.

The *DUAs*, or the clients of the directory services, may be a specific program, or it may be a part of another program which uses the directory for some lookup operation. For example, it may be a MHS UA which uses the directory to provide user friendly names to the MHS users. A *DUA* need only know the address of its closest *DSA*.

If a *DSA* is unable to service a request, it may chain the request to another *DSA* which may be able to answer the request. This may be repeated many times to resolve the name and to locate the requested entry. The *DSA* may also return a referral back to the requested *DUA* which in turn uses the referral to contact another *DSA* whose address is returned in the referral. This *DSA* can resolve the name if the requested object is present in its local database or can give a referral to another *DSA* which holds the entry or is more closer to the requested information. These modes of interactions are called chaining and referral respectively.

3.1 Communication

The roundtrip time between the *DUA* and the DSA_i is given by C_i. C_i strongly depends on the sites at which the *DSA* and the *DUA* are executing. This value depends on the number of gateways between the *DSA* and *DUA* and also on the speed of the intermediate transmission links. We assume that the number bytes per packet for different operations are the same. Thus the different operations have the same roundtrip time. We also ignore the variations due to the network congestion. The round trip time between a DSA_i and a DSA_j is given by C_{ij}.

3.2 Name resolution

The *DSA* resolves the name by scanning through its knowledge information base. If the requested object is located in locally held naming context, the *DSA* goes down the tree and locates the requested object. If the *DSA* finds that the requested object is not held locally, it finds out the address of the *DSA* which stores the object or a *DSA* which can give a more precise location of the *DSA* which stores the object. Then it chains to the that *DSA* or returns a referral back to the requested *DUA*. Let N_j be the average amount of time required to perform the name resolution in the DSA_i.

3.3 Database Access

The second factor that affects the performance of the directory is the database access time. As we are storing enormous amount of data in the directory, the database access time will contribute to a significant portion of the response time as seen by the client. The database access time depends on two factors: the object accessed and the type of operation requested. Let the time required to perform the database operation be D_i to perform an operation in the DSA_i. This is taken as an average amount of time taken to access different objects in the directory.

3.4 Other factors

Another factor that affect the performance of the directoty is the protocol overhead due to the lower layers. The time spent in the protocol stack, P_i, is assumed to be constant for all objects and all operations. This will be true in most of the cases because the amount of data transferred by the directory operations will be almost the same. The protocol overhead depends mainly on the size of the PDUs transferred.

Partitioning of the name space also influences the performance of the directory. If enough cross references are stored in the *DSAs*, then the *DSAs* will be able to resolve the name by contacting fewer number of *DSAs*. Also by suitably selecting the higher level *DSAs*, the performance of the system can be improved.

3.5 Scenarios

Now, we proceed to find out the response time of the directory as seen by the user of the directory. We consider the following three different cases:

i The local *DSA* itself can answer the query.

ii The local *DSA* is unable to resolve the query and uses chaining mode of interaction.

iii The local *DSA* is unable to resolve the query and uses referral.

We consider each of these three cases separately and find out the response time required in each case. In all these cases, we assume that the *DUA* is already bound to its closest *DSA* (primary *DSA*).

3.6 Case I - The Simple Case

In this case the interactions that are taking place is shown in the Figure 2. The *DUA* sends a request to the *DSA* and *DSA* reads its local database and responds to the query.

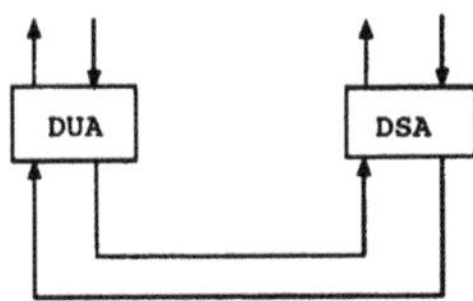

Figure 2: Single *DSA*, Single *DUA*

The response time as seen by the users of the *DUA* for performing an operation can be given by:

$$\boxed{T = P_c + C_1 + P_1 + N_1 + D_1}$$

Where,

$P_c =$ Protocol overhead in the client (DUA).

$C_1 =$ Round trip communication delay between the DUA to DSA_1.

$P_1 =$ Protocol overhead in the DSA_1 (including queuing delay).

$N_1 =$ Name resolution time in DSA_1.

$D_1 =$ Database access time in DSA_1.

3.7 Case II - Chaining

In this case, the primary *DSA*, DSA_1, decides to chain the request to a remote *DSA* which can answer the query. The interactions that are taking place are shown in the Figure 3. The primary *DSA* has to bind to the remote *DSA* before invoking the remote operation on the second *DSA*. Thus the total time required to perform the operation includes the time required to bind to the remote *DSA* + the time required to chain the request to the remote *DSA* + the time required to merge the results obtained from the remote *DSA* and send the result back to the *DUA*.

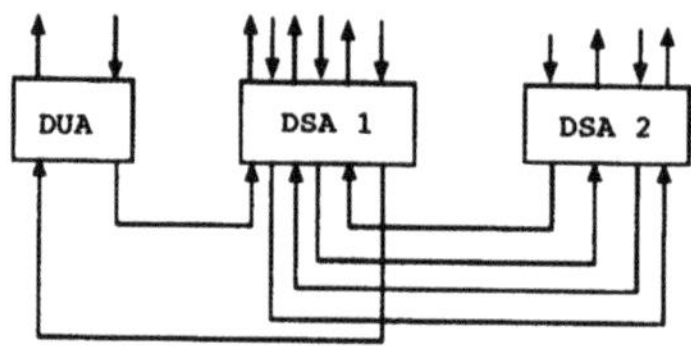

Figure 3: Multiple *DSAs*, Chaining mode of interaction

Thus the response time as seen by the user of the system is

T = time taken by the *DUA* to send the request + time taken by the *DSA* to perform the complete operation (include the time to bind to the second *DSA* and to invoke remote operation on the remote *DSA*) + time required to merge the results and return the result to the *DUA*

$$T = P_c + C_1 + T_{o1} + rm_1 + P_1$$

$P_c =$ Total protocol overhead in the client (DUA).

$C_1 =$ Round trip communication delay between DUA and DSA_1.

$T_{o1} =$ Total time required by the DSA_1 to complete the operation including chaining.

$rm_1 =$ Time required to merge the results.

$P_1 =$ Total protocol overhead in the DSA_1 (including the queuing delay).

T_{o1} consists of three parts. Name resolution time in the first *DSA* + time required to bind to remote *DSA* + time required to perform the operation on the remote *DSA*.

$$T_{o1} = N_1 + T_b + T_o$$
$$T_b = bf_1 + C_{12} + P_2 + b_2 + P_1$$
$$T_o = qf_1 + C_{12} + P_2 + N_2 + D_2 + P_1$$

$N_1 =$ time required for name resolution in DSA_1.

$bf_1 =$ Time taken to form the bind request.

$C_{12} =$ round trip communication delay between DSA_1 to DSA_2.

$P_2 =$ protocol over head in DSA_2 (including the queuing delay).

$b_2 =$ time required to perform bind operation.

$qf_1 =$ time required to form the query.

$N_2 =$ time required for name resolution in DSA_2.

$D_2 =$ Time required to perform database operation o in DSA_2.

Thus the total time is given by:

$$T = P_c + C_1 + 3P_1 + N_1 + bf_1 + 2(C_{12} + P_2) + b_2 + qf_1 + N_2 + D_2 + rm_1$$

Now consider a general case with the chaining goes to an arbitrary depth, say n *DSAs*. All but the last *DSA* act as relays to pass the query to the next *DSA* and all but the first *DSA* requires binding. It should be noted that the unbind time will not affect the total response time as unbinding is done after sending the result to the *DUA* or to the previous *DSA* in the chain. Thus the following actions are repeated in each *DSA* except the last one:

- resolve the name,
- bind to next *DSA*,
- chain the operation on the next *DSA* in the chain, and
- merge the result obtained and pass it back to the previous *DSA*.

Thus the total time taken is given by:

$$T = P_c + C_1 + 3P_1 + N_1 + bf_1 + 2C_{12} + qf_1 + \sum_{k=2}^{n}(P_k + b_k) + \sum_{k=2}^{n-1}[3P_k + N_k + bf_k + 2C_{k(k+1)} + rm_k] + P_n + N_n + D_n$$

3.8 Case III - Referral

In this case the *DSA* returns a referral to the *DUA*, which in turn uses this referral to contact another *DSA* which can answer the query. The interactions that are taking place are given in Figure 4.

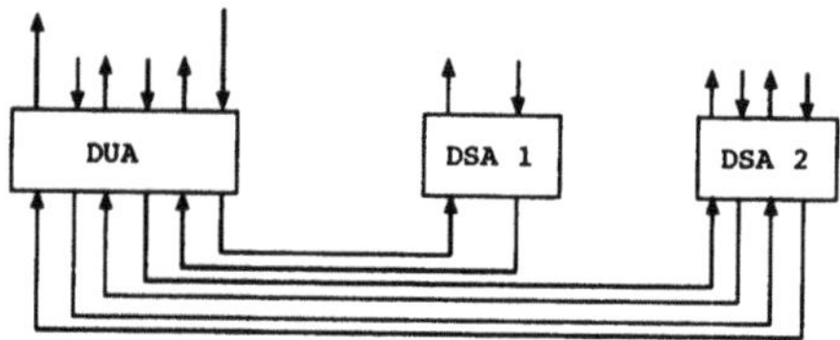

Figure 4: Multiple *DSAs*, Referral mode of interaction

The operation can be split into the following sub-operations:

- The *DUA* sends the request to DSA_1 which returns a referral.
- *DUA* uses these referral and binds to a *DSA* whose address is given in the referral.
- *DUA* sends the operation request to the second *DSA* which services it and returns the result.

Timing for each of these operations are given below.

$T_1 = P_c + C_1 + P_1 + N_1$
$T_{2b} = bf_c + P_c + C_2 + P_2 + b_2$
$T_{2o} = qf_c + P_c + C_2 + P_2 + N_2 + D_2$
Thus the total time T is given by:

$T = 3P_c + C_1 + P_1 + N_1 + qf_c + bf_c + 2(C_2 + P_2) + b_2 + N_2 + D_2$
If the referral extends to n *DSAs*, the total time required is given by:

$$T = (2n-1)P_c + (n-1)bf_c + n * qf_c + \sum_{k=2}^{n}(C_k + P_k + b_k) + \sum_{k=1}^{n}(C_k + P_k + N_k) + D_n$$

Hitherto, we developed a deterministic model to predict T, the time required to perform the directory operations, in terms of the parameters P_c, P_1, P_2, N_1, N_2, C_1, C_2, C_{12}, D_1, D_2, b_1, bf_c, bf_1, qf_c, qf_1 and rm_1. The completion of the model requires estimation of these parameters.

4 PARAMETER ESTIMATION OF THE MODEL

A series of experiments were designed to measure the parameters identified in the section 3. Routines for logging the time are included at various points in both *DUA* and *DSA*. A set of ten experiments are conducted and the average values are taken. The parameters required by the model is estimated from these measured values. Figure 5 shows the architecture of *DSA* and *DUA* in the experimental setup. The shaded area indicates the routines added to measure the timings.

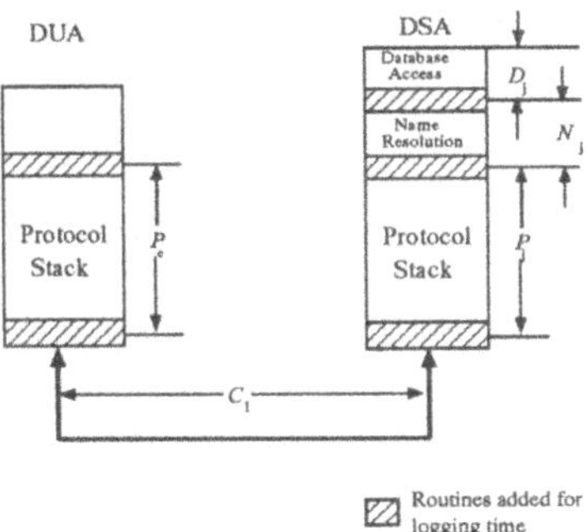

Figure 5: Experimental Setup

In the *DUA*, the time is logged in just before the query is passed onto the protocol stack and once again before the packet is passed to the network. Similarly, in the *DSA*, time is logged when a packet is received from the network, before and after the name resolution and during the database access. That way, database access time and the name resolution time can be calculated. As the clocks in the two systems are not synchronized with each other, it is not possible to measure the oneway communication delay. But by measuring the packet transmission time and packet reception time at both ends, it is possible to measure the round trip communication time.

Experiment Number	DSA_1	DSA_2	DUA	Where reported	Mode
1	IITM	N/A	IITM	Table 2	N/A
2	UMD	N/A	IITM	Table 3	N/A
3	IITM	UMD	IITM	Table 4	Chaining
4	IITM	UMD	IITM	Table 5	Referral

Table 1: Experiments Conducted.

The DIT structure used for the measurements is shown in the Figure 6. Experiments were conducted with different network topologies and different modes of interactions. The experiments conducted are listed in the Table 1.

The database used for the experiments is generated from a subscriber database used by telecommunication department. The entries in the database are duplicated to create bigger databases. An unique identifier is appended so as to make the name of every subscriber unique. This name is then used as the RDN of the entry. Subordinates of a given entry is stored in a separate file and this file is indexed using the B-Tree++ algorithm, with RDN as the key.

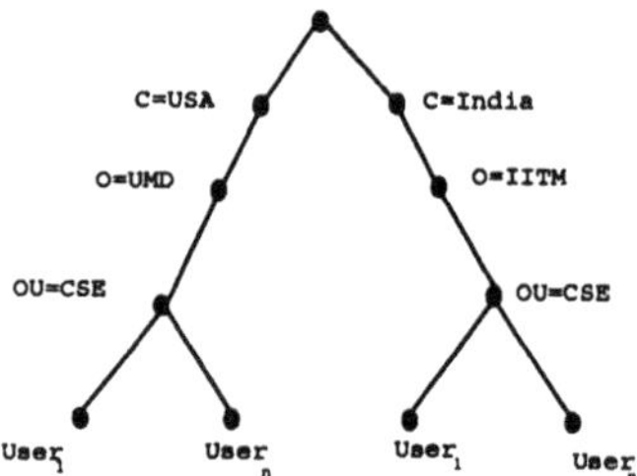

Figure 6: DIT structure

Tables 2 and 3 present the measurements that are taken between three different machines in the internet. Table 2 gives the measurements with *DSA* running in the Sparc machine in IIT, Madras, India and *DUA* running in the same LAN in a Tektronix machine in IIT, Madras, India.

Table 3 shows the measurements with the *DSA* running in the Ultrix machine in University of Maryland, USA, and the *DUA* running in the Sparc machine in IIT, Madras, India. All values given in the tables are the average over a set of ten measurements.

Table 4 shows the measurements when the directory is distributed among two systems in the internet, using *chaining mode* of interaction between the *DSAs*. The *DSAs* were running in the Ultrix machine in University of Maryland, USA and in the Sparc machine in IIT, Madras and *DUA* in the tektronix machine in IIT, Madras. Table 5 shows the measurements done with the same configuration using the *referral mode* of interaction.

From the tables, the following observations can be made:

- The time required to perform the database operation depends on the type of operation. From the tables, it can be seen that the most time consuming directory operation is *modifyrdn* operation. This is because the *modifyrdn* operation requires searching the database twice to perform the operation - once to check the existance of the entry whose modification is requested and then to confirm that no other entry with the same name exists.
- The database operation time and the name resolution time increases with the increase in the number of entries in the directory database. This is because of the increase in the depth of the B-Tree.
- The time required to perform the *list* operation does not depends on the database size. As the number of subordinates of the base object is always larger than the sizelimit parameter, the *list* operation always returns the first few entries and returns the *sizelimit exceeded* error. This is a limitation of the 1988 standard of X.500[CCITT (1988)], which has been overcome in 1993 version[CCITT (1993)].

The estimated values of model parameters are used in predicting the directory operation time T. In all the experiments, the directory operation time T was also measured. Table 6 gives the parameters related to directory operations. The values D_1 is the database operation time and N_1 the name resolution time in the DSA where the entry is located. Table 7 gives the $DSA - DUA$ interaction parameters. The values presented include the bind query formation time, the bind operation time, the query formation time and the result merging time. Table 8 gives the parameters for communication delay and protocol overhead. In the table, C_l is the communication delay in a LAN environment and C_i is the communication delay in an internet environment.

The variations in the predicted value and the actually measured value are computed. The error rate is calculated as *(predicted value - measured value) / predicted value * 100.* The error rate is plotted against database size for all directory operation for all the four experiments in the Figure 8, 9, 10, and 11. The graph given in the Figure 9 shows the variation in the percentage error rate with the number of entries for experiment 2, with no referral or chaining in an internet environment. From this graph, it is evident that the percentage error rate decreases as the number of entries increase. The estimated database access time is more accurate than the round trip communication time. The round trip communication time is considerably smaller when compared to the database access time, when the database size reaches 50,000. Thus the percentage variation in the predicted response time will become smaller with the increase in the number of entries in the directory. This can be predicted accurately. But the round trip communication time across the internet cannot be predicted accurately [Sanghi *et al.*(1993)]. When the database size is small the round trip communication time will be small and thus the percentage variation in the predicted value from the measured value will be high.

Figure 10 shows the percentage variation of error with increase in database size for chaining mode of interaction. The variation in this case is also small for large database size. In our experiments, the *DUA* and one *DSA* was in the same LAN. Thus the communication delay between the *DUA* and the first *DSA* will be very less. Figure 11 in the appendix gives the percentage variations of the error with increase in database size for referral mode of interaction. Again, the variation in the predicted value from the actual measurements decreases with increase in the database size.

5 APPLICATION OF THE MODEL FOR ESTIMATING THE NAME LOOKUP TIME

We now use the deterministic model to predict the expected response time for read operation in a national network. Two different DIT structures are considered and the response time are computed for each structure with the help of the model.

We consider the *ERNET*[4] topology, and estimate the response time using the analytical model discussed in the section 3.

The network topology of *ERNET* is shown in the Figure 7a. *ERNET* consists of 8 different sites, connected by point-to-point links of capacity 9.6 kbps. We estimate the average response time for a query given to any one of the site in the *ERNET* to access the data in any other site. The label on the arcs connecting the sites are the round trip time between the nodes. The actual communication delay between the sites is measured using the `ping` command.

Figure 7b shows the logical tree structure used to estimate the response time for a directory operation. The average response time estimated using the analytical model for the *ERNET* directory for the DIT structure shown in the Figure 7b is shown in the Table 9. Figure 7c shows another logical directory tree structure with the first level *DSA* located at a different site. The response time with this DIT configuration is given in the Table 10. The values given are for a single lookup (read) operation with 500 entries in each *DSA*. The value in column i and row j indicates the time required to access an object in site i by a *DUA* in site j. We assume that *DSAs* stores no cross references to any other *DSAs*, thus the name resolution is done by strict tree traversal.

The Table 9 and 10 show the response time for directory operations with different DIT structures. The readings given are the estimated time for directory read operations. The difference in the readings clearly show the advantage of selecting a proper DIT structure

[4]Education and Research NETwork, setup by Department of Electronics and UNDP.

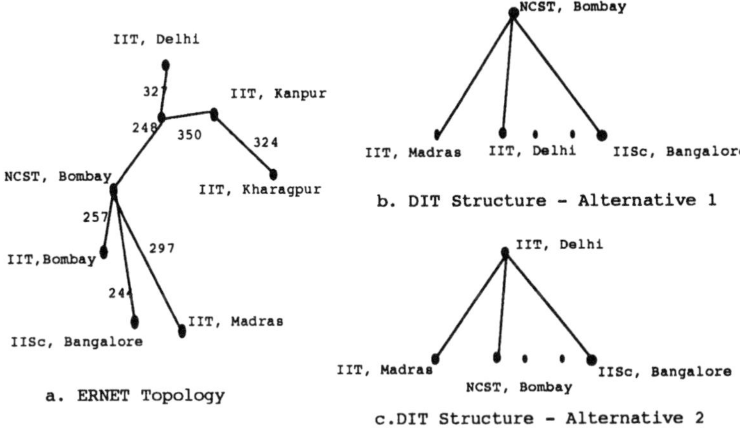

Figure 7: ERNET directory

for a given topology. The directory structure given in Figure 7b is best among the two topologies as it gives the minimum response time.

6 CONCLUSION

In this paper, we presented a simple deterministic model for predicting the performance of the X.500 directory. This model will be helpful for the X.500 directory managers to tune the directory to get a better performance. The model assists the directory manager in estimating the critical database size and the optimal topology to be used. The model is validated by comparing the results predicted using the model and the actual measurements carried out in the internet with a large database. The percentage deviation in the predicted results from the measured results was found to be very less thereby confirming the usefulness of the model. Also, the deviation of the predicted results from the measured value decreases with increase in the database size because the variations in the communication delay, which is an unpredictable factor, will be a small percentage of the total time required to perform the directory operation. The model is used to select the optimal topology for DIT, when the physical interconnection topology of a large national network is given.

References

[CCITT (1988)] CCITT. *Data Communication Networks, Message Handling Systems, Recommendations X.400 - X.420.* 1988.

[CCITT (1988)] CCITT. *The Directory - Overview of Concepts, Models and Services, Recommendations X.500 - X.521.* 1988.

[CCITT (1993)] CCITT. *The Directory - Overview of Concepts, Models and Services, Draft Recommendations X.500 - X.521.* 1993.

[Goodman (1992)] David Goodman. *PARADISE International Report 4.* PARADISE Pilot Group, University College London, November 1992.

[Sanghi *et al.*(1993)] Dheraj Sanghi, Olafur Gudmundsson, Ashok Agarwala. Study of network dynamics. *Computer Networks and ISDN Systems*, 26:371–378, 1993.

[Comer (1979)] Douglas Comer. The Ubiquitous B-trees. *ACM Computing Surveys*, 11(2):121–137, June 1979.

[Huitema (1989)] G. Huitema. The X.500 Directory Services. *Computer Networks and ISDN Systems*, 16:161–166, September 1989.

[Neufeld (1992)] Gerald Neufeld, Barry Brachman, Murray Goldberg, Duncan Stickings. The EAN X.500 Directory Services. *Internetworking : Research and experience*, 3:55–81, 92.

[Mansfield *et al.*(1993)] Glenn Mansfield, K Jayanthi, K Higuchi, Thomas Johannsen, Y Nemoto. Mapping communication networks in the directory. *Computer networks and ISDN Systems*, 26:327–335, 1993.

[Hong *et al.*(1992a)] James W Hong, Micheal A Bauer, J Michael Bennet. Integration of Directory Services in the network management framework. In *Third International Symposium on Integrated Network Management*, April 1992.

[Hong *et al.*(1992b)] James W Hong, Micheal A Bauer, J Micheal Bennet. Integragion of the Directory Services in the Distributed systems management. In *International Conference on Parallel and Distributed Systems*, 1992.

[Barker (1992)] Paul Barker. DIT Structures to support searching for Bibilographic information and information resources. Technical report, University College London, December 1992.

[Barker (1993)] Paul Barker. Implementing FTP archive searching using X.500. Internet Draft : OSI-DS 40, March 1993.

[Bumbulis *et al.*(1993)] P.J. Bumbulis, D.D. Cowan, C.M. Durance, T.M. Stepien. An Introduction to the OSI Directory Services. *Computer Networks and ISDN Systems*, 26:239–249, 1993.

[Sriram (1991)] R Sriram. Studies in OSInet Directory Services. Master's thesis, Indian Institute of Technology Madras, January 1991.

[Sriram *et al.*(1994)] R Sriram, S V Raghavan. Example implementation of Name Services using Directory Services. To appear in Computer Networks and ISDN Systems, 1994.

[Raghavan (1990)] S.V. Raghavan. *Solutions to Local Area Networks, The Indian Context.* Tata McGraw Hill Publishing Company Limited, 1990.

[Rennesse *et al.*] Robbert Van Rennesse, Andrew S Tanenbaum. A Directory service supporting avaliability and consistency. Vrije Universiteit, Amsterdam.

[Raghavan *et al.*(1992)] S. V. Raghavan, R. Sriram. Implementation of X.500 Directory Services in OSInet. *IETE Technical Review*, 9(3), May - June 1992.

[Raghavan (1993a)] S.V. Raghavan. OSInet: Open Systems Interconnection network Software:: Description of its Protocol Architecture, Design, Implementation, Application and Performance. Technical Report ERNET/06/93/IITM, CSE/NSL/08/93/IITM, Indian Institute of Technology, Madras, April 1993.

[Raghavan *et al.*(1993b)] S.V. Raghavan, B. Prabhakaran, K. Shivkumar. An enhanced B tree design for databases. Technical Report ERNET/10/93/IITM CSE/NSL/12/93/IITM, Indian Institute of Technology, Madras, May 1993.

7 AUTHOR INFORMATION

Rajeev P.V. obtained his Bachelor's degree in 1991 from the University of Calicut, India in the field of Computer Science and Engineering. He completed his Master's in Computer Science and Engineering from Indian Institute of Technology, Madras, India. He was working as a research fellow in Networks Systems Laboratory, Department of Computer Science, Indian Institute of Technology, Madras, India. At present, he is with Hughes Software Systems, Delhi. His research interests are networks and multimedia systems.

S.V. Raghavan is on the faculty of the Department of Computer Science and Engineering, Indian Institute of Technology, Madras. He is also the Chief Investigator of the project on *Education and Research in Computer Networking* jointly sponsored by the Department of Electronics, Government of India and the United Nations Development Programme. He is a life member of the Computer Society of India, a member of the Institution of Engineers and a fellow of institute of Electronics and Telecommunication Engineers. He is presently serving on the Board of Editors of the journal of IETE for computers and control. He is also a member of the Editorial Advisory Board for Computer Communications, Butterwoth-Heinemann Ltd. His research interests are networks, protocols, multimedia systems and performance.

Ashok K. Agarwala is a Professor of Computer Science at the University of Maryland, College Park. For the past twenty years he has been actively involved in the research in various aspects of computer system design, implementation and performance. He received his the B.Sc. degree from Agra University in 1960, the B.E degree in Electrical Technology, and the M.E. degree in Applied Electronics and Servomechanisms from Indian Institute of Science in 1963 and 1965, respectively. He earned the A.M. and Ph.D. degrees in Applied Mathematics from Harward University, Cambridge, MA in 1970. Dr. Agrawala is fellow of the IEEE, and a member of the Association for Computing Machinery, American Association for the Advancement of Science Xi.

DSA running in Sparc, IIT Madras and DUA in Tektronix, IIT Madras							
Number of Entries = 5000							
Time	Read	Compare	List	Add	Modify	Remove	ModifyRDN
D_1	531	337	301	1427	1011	210	2203
N_1	1096	1118	781	1219	1137	1060	1256
P_1	33	31	20	31	19	29	26
C_1	38	29	27	38	25	31	38
P_c	18	19	17	29	17	22	17
T	1710	1534	1146	2744	2209	1352	3540
Number of Entries = 10000							
D_1	593	392	1610	1213	232	1932	1496
N_1	1461	1460	798	1432	1472	1503	1496
P_1	34	25	25	22	30	25	27
C_1	38	38	37	32	26	33	29
P_c	25	25	18	18	21	17	27
T	2151	1940	1250	3114	2762	1810	3511
Number of Entries = 25000							
D_1	581	410	363	3793	2478	241	3975
N_1	2262	2145	728	2453	2320	2413	2417
P_1	23	32	32	34	34	31	33
C_1	26	35	37	38	31	36	29
P_c	19	30	20	20	30	18	25
T	2911	2652	1201	6338	4893	2739	6479
Number of Entries = 50000							
D_1	512	409	410	6781	4923	303	7013
N_1	3813	3623	813	3713	3691	3686	3594
P_1	29	25	32	19	34	34	24
C_1	26	30	35	27	33	34	26
P_c	22	18	30	18	25	26	22
T	4402	4105	1320	10558	8706	4083	10679

All values in milliseconds

Table 2: DSA and DUA on LAN.

DUA running in Sparc, IIT, Madras and DSA in Univ. of Maryland							
Number of Entries = 5000							
Time	Read	Compare	List	Add	Modify	Remove	ModifyRDN
D_1	531	320	343	1461	1075	180	2304
N_1	1220	1180	712	1207	1230	1141	1162
P_1	26	29	27	31	27	26	29
C_1	1290	1214	1195	1175	1108	1178	1187
P_c	16	21	18	29	22	23	23
T	3126	2764	2293	3903	3262	2558	4695
Number of Entries = 10000							
D_1	532	393	327	1772	1246	210	2802
N_1	1400	1364	719	1767	1879	1364	1782
P_1	24	27	23	31	29	20	32
C_1	1239	1206	1195	1223	1156	1259	1150
P_c	23	20	27	22	18	23	21
T	3318	3010	2270	5015	4528	2886	5067
Number of Entries = 25000							
D_1	546	403	308	3632	2590	230	3815
N_1	2561	2232	738	2762	2260	2123	2974
P_1	27	23	29	33	29	23	29
C_1	1319	1294	1299	1226	1213	1302	1264
P_c	21	18	21	29	27	19	23
T	4464	3952	2395	7682	6122	3897	7805
Number of Entries = 50000							
D_1	573	420	380	6585	4833	287	6952
N_1	3905	3685	790	3917	3729	3519	3923
P_1	29	23	31	28	26	29	25
C_1	1278	1123	1145	1292	1240	1153	1231
P_c	21	18	27	21	23	22	19
T	5808	5226	2363	11841	9851	5010	12150

All values in milliseconds

Table 3: DSA and DUA Across Internet.

DSAs in Maryland and Sparc, IIT, Madras and DUA in Tektronix Chaining mode of interaction Number of Entries = 5000							
Time	Read	Compare	List	Add	Modify	Remove	ModifyRDN
D_2	618	526	428	1198	1063	230	1839
N_2	1003	1029	721	1102	1121	1017	1482
N_1	307	314	418	472	381	238	325
P_1	26	22	29	30	22	28	24
P_2	20	21	28	21	30	20	24
P_c	19	23	19	20	23	22	20
C_1	34	34	36	26	29	28	31
C_{12}	1162	1227	1241	1181	1286	1280	1143
bf_1	38	32	39	30	31	34	35
qf_1	45	38	35	42	39	45	48
rm_1	50	36	49	46	38	39	30
b_2	122	110	112	142	124	106	146
T	4678	4704	4482	5572	5547	4443	6362
Number of Entries = 10000							
D_2	597	387	441	1780	1312	348	1899
N_2	1492	1480	892	1471	1398	1501	1496
N_1	318	329	393	403	307	321	481
P_1	28	29	30	20	33	20	34
P_2	30	33	26	27	21	33	22
P_c	22	18	24	23	20	18	18
C_1	32	35	35	25	32	31	30
C_{12}	1286	1156	1108	1259	1257	1295	1291
bf_1	33	39	33	33	35	30	35
qf_1	45	49	35	48	49	44	42
rm_1	48	38	54	49	35	43	36
b_2	117	107	143	100	120	114	125
T	5420	4947	4408	6564	5963	5166	6890
Number of Entries = 25000							
D_2	588	476	480	3578	2489	480	3857
N_2	2378	2247	760	2782	2490	2389	2415
N_1	321	370	280	420	338	264	407
P_1	33	32	31	32	22	31	33
P_2	23	32	27	23	28	27	27
P_c	18	19	24	20	19	24	18
C_1	25	38	37	34	36	37	30
C_{12}	1304	1212	1162	1257	1222	1162	1264
bf_1	38	30	37	39	34	37	33
qf_1	37	39	47	36	43	47	38
rm_1	53	52	51	32	36	51	33
b_2	115	122	106	107	114	106	133
T	6326	5977	4293	9704	8165	5906	9645
Number of Entries = 50000							
D_2	621	508	405	6817	4293	421	6983
N_2	3873	3692	821	3985	3961	3796	3975
N_1	298	264	269	347	469	256	442
P_1	20	28	31	32	22	34	22
P_2	26	33	20	31	33	24	26
P_c	21	22	24	23	21	19	21
C_1	30	36	27	26	38	27	26
C_{12}	1142	1198	1313	1235	1323	1243	1248
bf_1	37	30	31	30	37	30	31
qf_1	38	35	46	41	35	48	36
rm_1	50	40	44	40	48	35	52
b_2	138	135	147	124	117	108	128
T	7502	7308	4573	14061	11797	7376	14308

All values in milliseconds

Table 4: DSAs Across Internet with Chaining + DUA.

DSAs in Maryland and Sparc, IIT, Madras and DUA in Tektronix Referral mode of interaction Number of Entries = 5000							
Time	Read	Compare	List	Add	Modify	Remove	ModifyRDN
D_2	639	570	400	1247	1036	242	1793
N_2	1027	1001	658	1051	1150	1073	1407
N_1	329	360	394	463	414	217	334
P_1	25	34	32	34	26	23	23
P_2	34	27	33	33	23	21	23
P_c	20	22	18	22	24	23	22
C_1	38	33	25	37	30	33	30
C_2	1228	1222	1233	1154	1136	1164	1194
bf_c	32	38	34	32	39	39	38
qf_c	43	36	41	48	46	40	38
b_2	115	117	138	116	138	138	112
T	4832	4753	4308	5468	5269	4244	6275
Number of Entries = 10000							
D_2	607	384	448	1785	1296	299	1850
N_2	1430	1512	918	1445	1391	1483	1510
N_1	284	369	396	432	349	349	496
P_1	25	32	31	22	33	21	24
P_2	32	33	29	29	24	29	30
P_c	22	23	21	23	24	23	18
C_1	39	37	26	25	38	28	31
C_2	1223	1281	1154	1149	1167	1120	1130
bf_c	33	32	38	36	36	32	32
qf_c	38	41	49	47	38	35	44
b_2	100	133	135	141	134	139	104
T	5132	5237	4470	6358	5769	4753	6465
Number of Entries = 25000							
D_2	591	449	444	3533	2486	494	3836
N_2	2336	2211	726	2711	2465	2348	2417
N_1	303	407	302	383	332	222	372
P_1	25	34	21	25	23	21	32
P_2	23	23	26	24	26	34	26
P_c	20	19	19	24	20	23	22
C_1	25	26	33	25	35	25	37
C_2	1114	1343	1100	1165	1157	1211	1134
bf_c	35	36	34	37	33	36	37
qf_c	48	38	40	44	42	39	35
b_2	128	114	145	113	111	117	120
T	5825	6104	4054	9321	7953	5861	9272
Number of Entries = 50000							
D_2	670	499	386	6816	4265	432	6960
N_2	3918	3624	760	4025	3920	3737	3998
N_1	264	302	244	361	483	271	461
P_1	21	23	30	32	30	27	25
P_2	23	30	32	25	29	33	30
P_c	22	21	23	24	18	19	24
C_1	26	31	29	39	31	30	27
C_2	1285	1114	1282	1240	1289	1127	1243
bf_c	31	30	34	35	38	38	39
qf_c	37	46	47	39	42	35	45
b_2	104	116	146	145	104	116	116
T	7753	7022	4373	14094	11603	7063	14289

All values in milliseconds

Table 5: DSAs Across Internet with Referral + DUA.

Variable Estimated	Read	Compare	Add	Modify	Remove	ModifyRDN
Number of entries = 5000						
D_1	531	320	1500	1075	200	2300
N_1	1220	1180	1200	1230	1140	1160
Number of entries = 10,000						
D_1	532	393	1772	1246	210	2802
N_1	1400	1364	1767	1879	1364	1782
Number of entries = 25,000						
D_1	546	403	3632	2590	230	3815
N_1	2561	2232	2762	2260	2123	2874
Number of entries = 50,000						
D_1	573	420	6585	4833	287	6952
N_1	3905	3685	3917	3729	3519	3923

Table 6: Estimated Values of Model Parameters Related to Directory Operations.

Estimated Variable	Value
Time to search knowledge information (N_1)	550
Bind request formation time (bf_1)	40
Bind time (b_i)	130
Query Formation time (qf_1)	45
Result merging time (rm_1)	45

Table 7: Estimated Values of *DSA* - *DUA* interaction parameters.

Estimated Variable	Minimum Value	Maximum Value	Mean	Standard Deviation
P_1	19	34	27	0.427
P_2	20	34	26	0.588
P_c	16	30	21	0.380
C_l	25	39	30	0.860
C_i	1100	1343	1198	7.020

Table 8: Estimated Values of Protocol Overhead and Communication Delay Parameters.

	ncst	iitm	iitb	iisc	doe	iitkgp	iitk	iitd
ncst	1667	3282	3090	3086	3109	4759	3880	3850
iitm	3230	1607	4244	4280	4311	5511	5204	5154
iitb	3174	4336	1549	4152	4199	5685	5146	5054
iisc	3084	4304	4098	1458	4205	5657	5198	5144
doe	3159	4330	4088	4180	1499	4665	4062	3958
iitkgp	4723	6928	7048	7022	6109	1582	5496	6774
iitk	3908	5964	5850	5986	4887	4861	1498	5574
iitd	3763	5792	5666	5810	4641	6005	5446	1542

All values in milliseconds.

Table 9: Predicted Read Operation Timing in DIT Alternative 1.

	ncst	iitm	iitb	iisc	doe	iitkgp	iitk	iitd
ncst	1667	4534	4342	4338	4361	6011	5132	3850
iitm	5089	1607	5180	5216	5247	6447	6140	4231
iitb	4989	5300	1549	5116	5163	6649	6110	4203
iisc	5079	5434	5228	1458	5335	6787	6328	4279
doe	3893	4258	4016	4108	1499	4593	3990	3152
iitkgp	6709	6620	6740	6714	5801	1582	5188	4480
iitk	5509	5948	5834	5970	4871	4845	1498	3957
iitd	3763	4182	4056	4200	3031	4395	3836	1542

All values in milliseconds.

Table 10: Predicted Read Operation Timings in DIT Alternative 2.

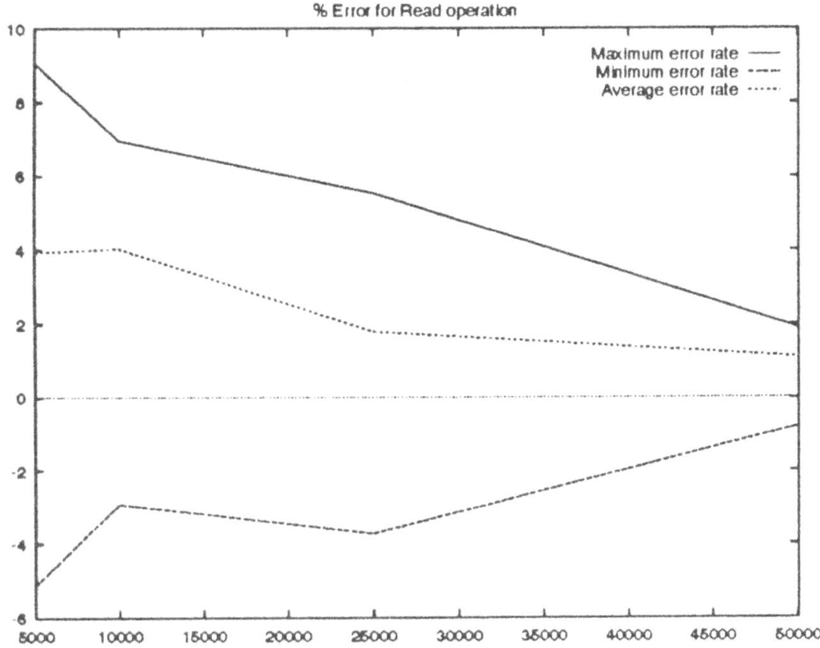
% Error for Read operation
Maximum error rate
Minimum error rate
Average error rate
10
8
6
4
2
0
-2
-4
-6
5000
10000
15000
20000
25000
30000
35000
40000
45000
50000

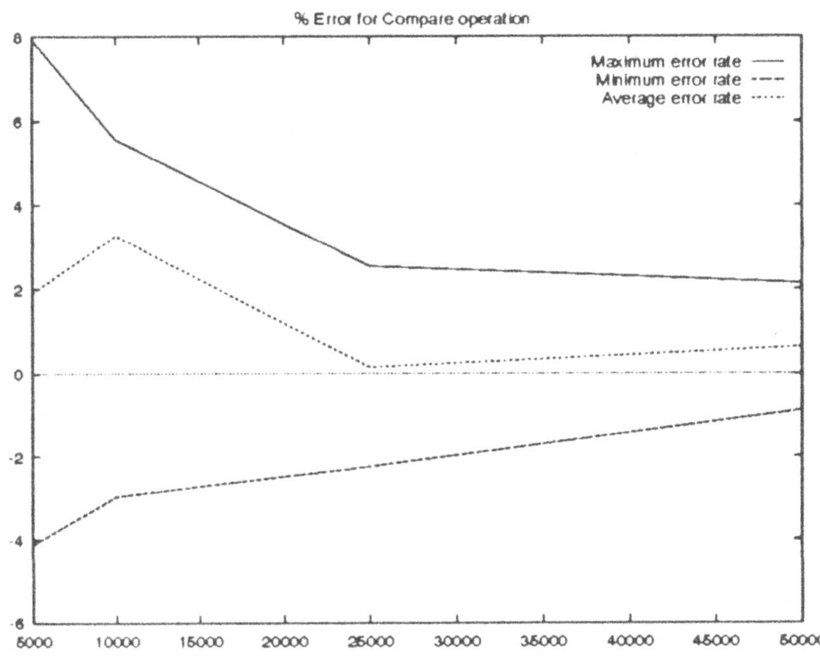
% Error for Compare operation
Maximum error rate
Minimum error rate
Average error rate
8
6
4
2
0
-2
-4
-6
5000
10000
15000
20000
25000
30000
35000
40000
45000
50000

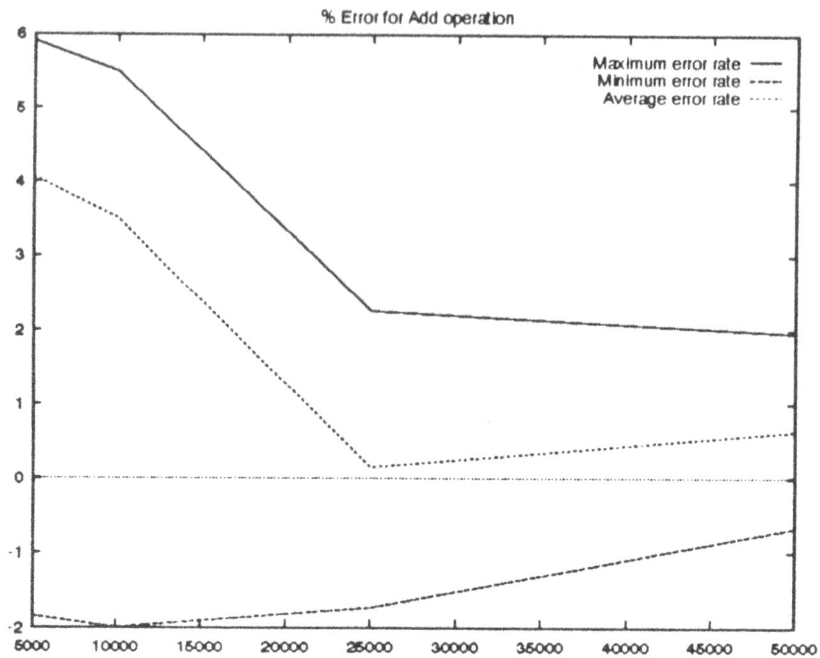
% Error for Add operation
Maximum error rate
Minimum error rate
Average error rate
6
5
4
3
2
1
0
-1
-2
5000
10000
15000
20000
25000
30000
35000
40000
45000
50000

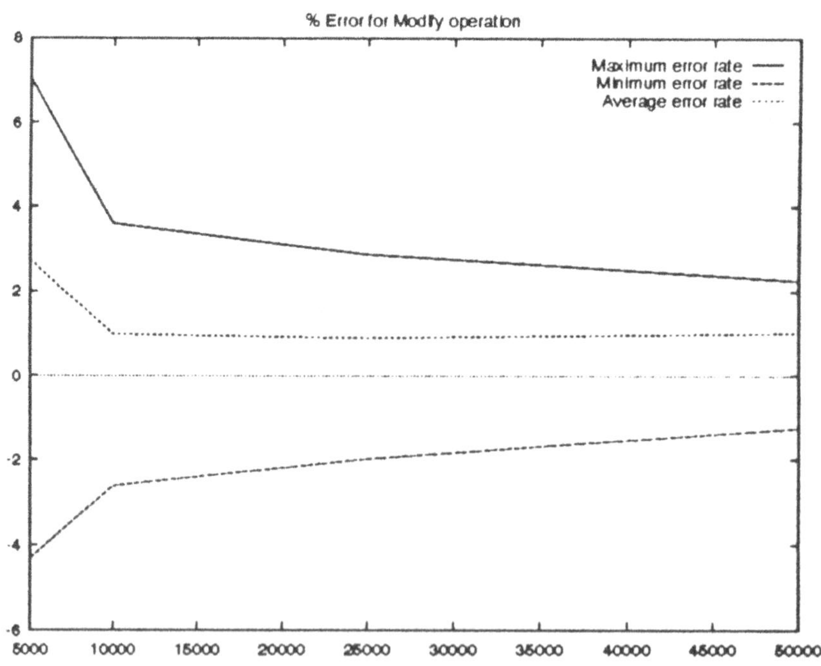
% Error for Modify operation
Maximum error rate
Minimum error rate
Average error rate
8
6
4
2
0
-2
-4
-6
5000
10000
15000
20000
25000
30000
35000
40000
45000
50000

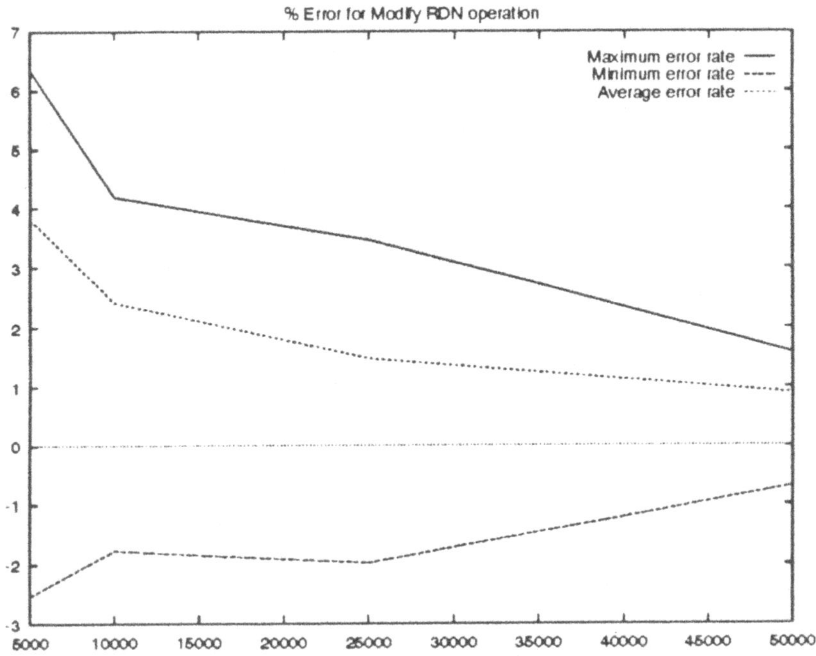

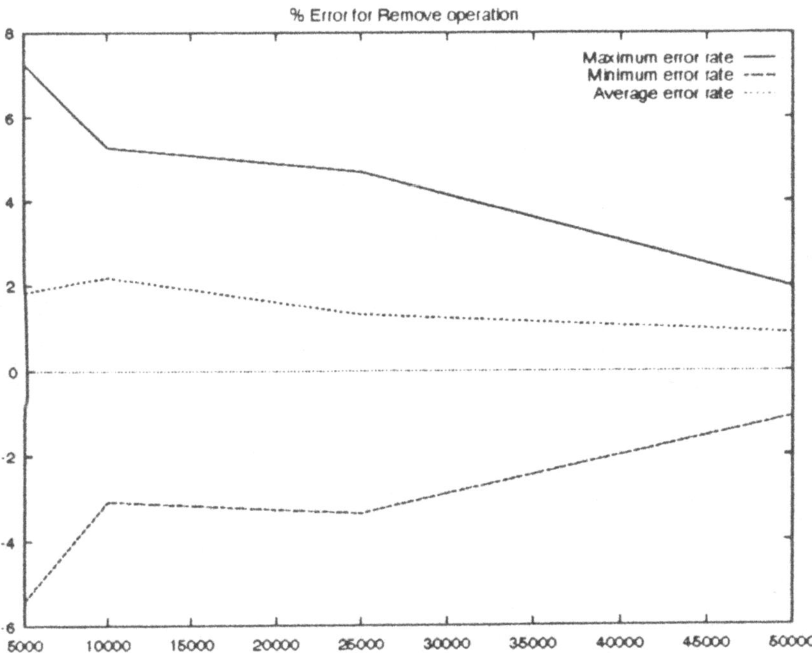

Figure 8: Percentage Error rate Vs Database size, Experiment 1.

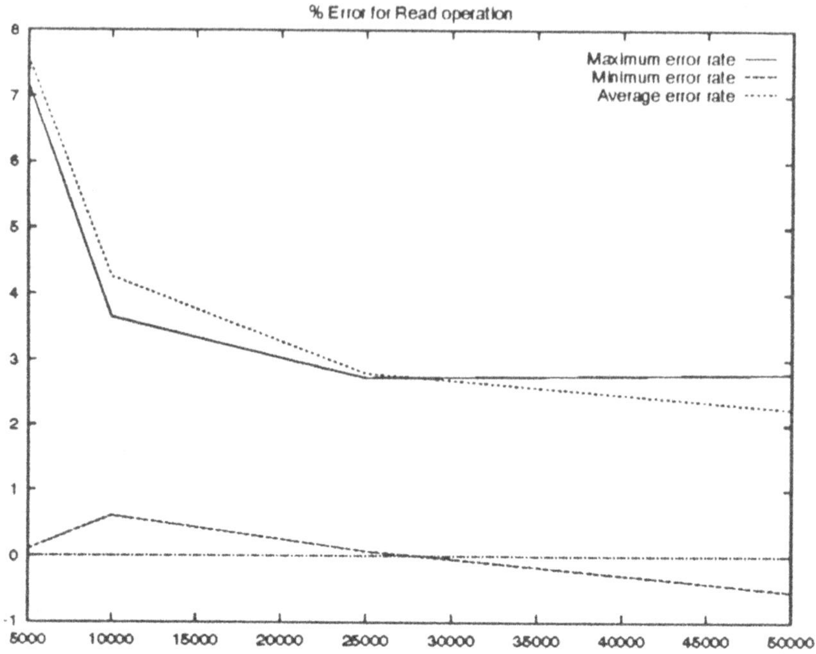
% Error for Read operation
Maximum error rate
Minimum error rate
Average error rate
8
7
6
5
4
3
2
1
0
-1
5000
10000
15000
20000
25000
30000
35000
40000
45000
50000

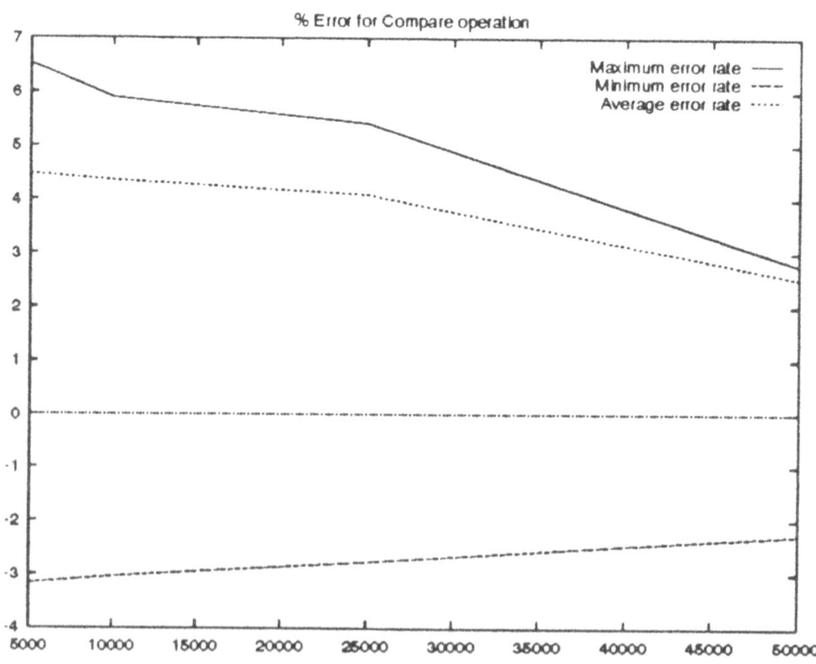
% Error for Compare operation
Maximum error rate
Minimum error rate
Average error rate
7
6
5
4
3
2
1
0
-1
-2
-3
-4
5000
10000
15000
20000
25000
30000
35000
40000
45000
50000

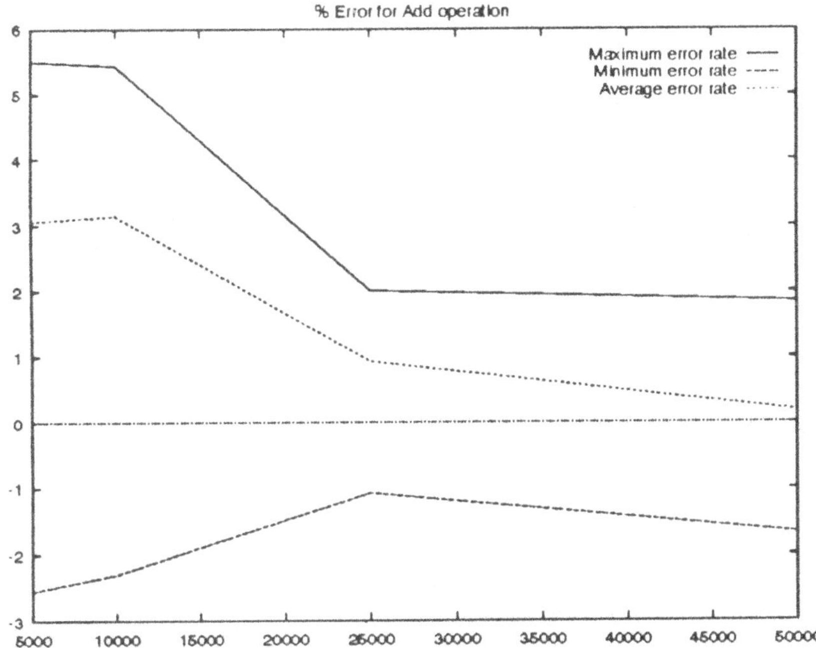
% Error for Add operation
Maximum error rate
Minimum error rate
Average error rate
6
5
4
3
2
1
0
-1
-2
-3
5000
10000
15000
20000
25000
30000
35000
40000
45000
50000

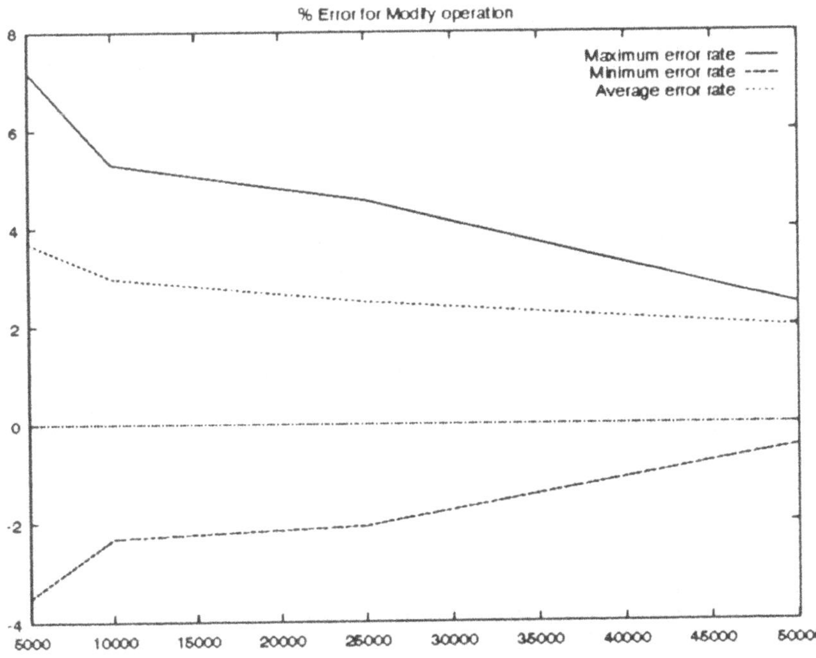
% Error for Modify operation
Maximum error rate
Minimum error rate
Average error rate
8
6
4
2
0
-2
-4
5000
10000
15000
20000
25000
30000
35000
40000
45000
50000

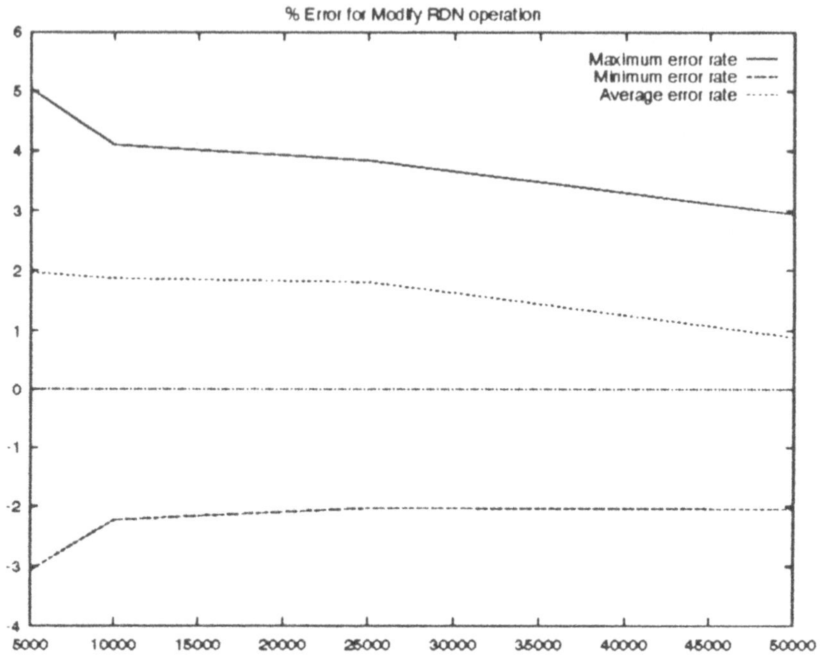

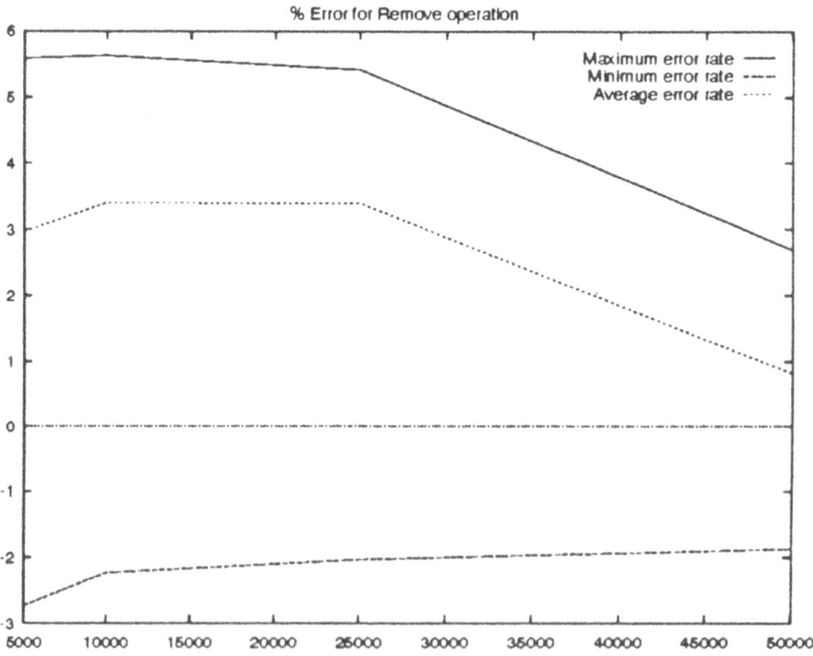

Figure 9: Percentage Error rate Vs Database size, Experiment 2.

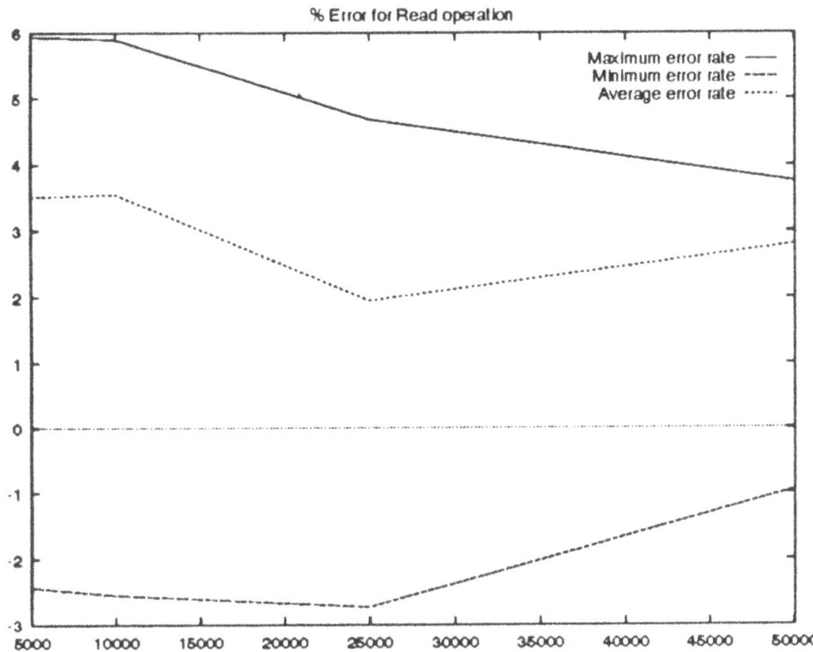
% Error for Read operation
Maximum error rate
Minimum error rate
Average error rate
6
5
4
3
2
1
0
-1
-2
-3
5000
10000
15000
20000
25000
30000
35000
40000
45000
50000

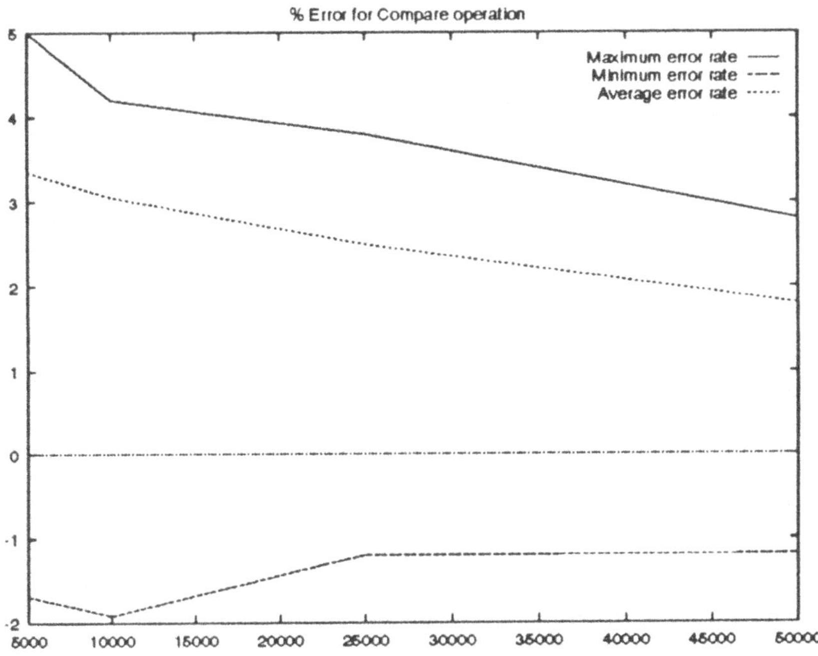
% Error for Compare operation
Maximum error rate
Minimum error rate
Average error rate
5
4
3
2
1
0
-1
-2
5000
10000
15000
20000
25000
30000
35000
40000
45000
50000

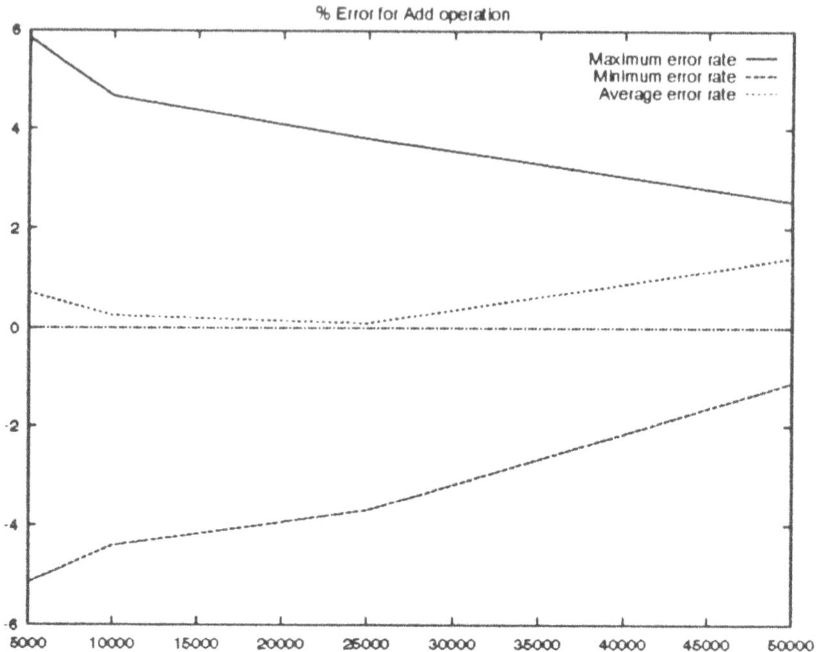
% Error for Add operation
Maximum error rate
Minimum error rate
Average error rate
6
4
2
0
-2
-4
-6
5000
10000
15000
20000
25000
30000
35000
40000
45000
50000

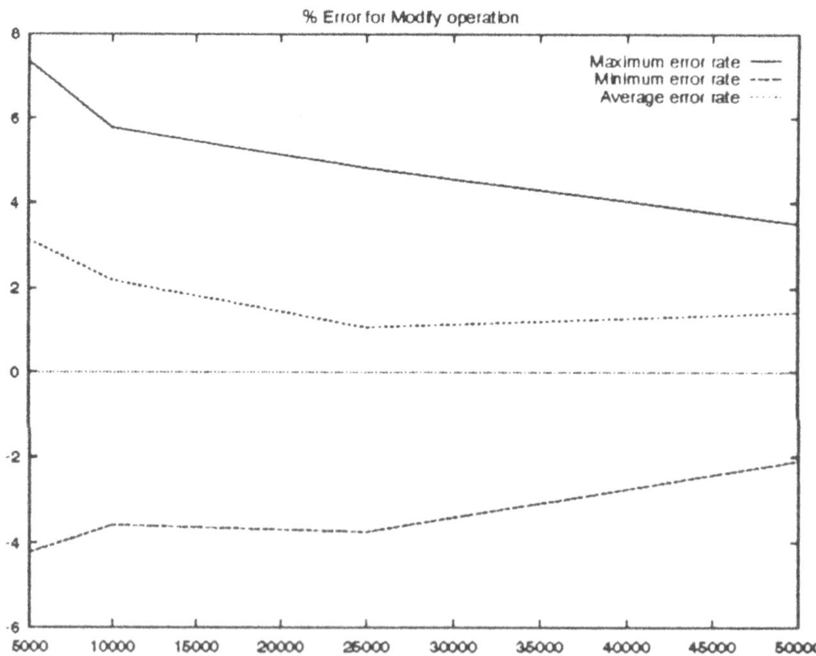
% Error for Modify operation
Maximum error rate
Minimum error rate
Average error rate
8
6
4
2
0
-2
-4
-6
5000
10000
15000
20000
25000
30000
35000
40000
45000
50000

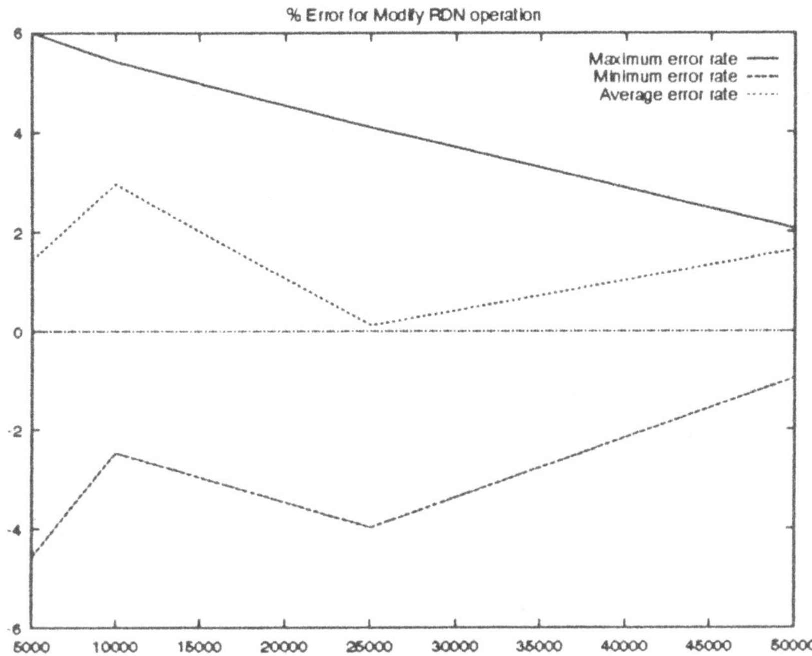

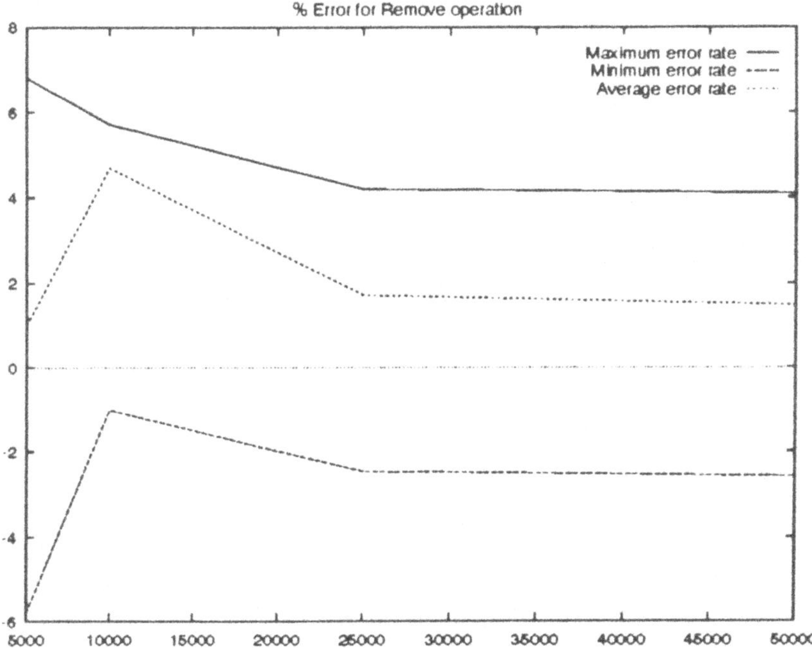

Figure 10: Percentage Error rate Vs Database size, Experiment 3.

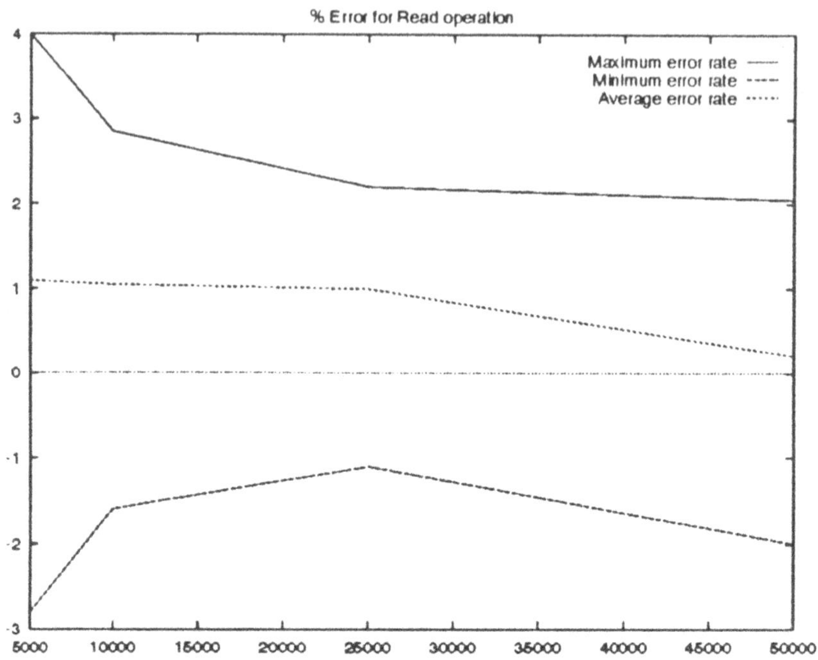
% Error for Read operation
Maximum error rate
Minimum error rate
Average error rate
4
3
2
1
0
-1
-2
-3
5000
10000
15000
20000
25000
30000
35000
40000
45000
50000

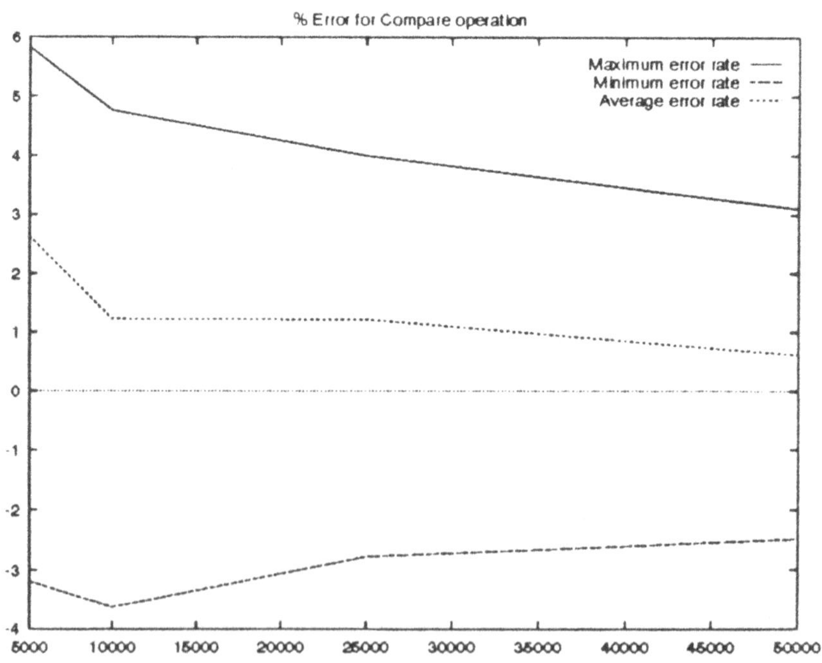
% Error for Compare operation
Maximum error rate
Minimum error rate
Average error rate
6
5
4
3
2
1
0
-1
-2
-3
-4
5000
10000
15000
20000
25000
30000
35000
40000
45000
50000

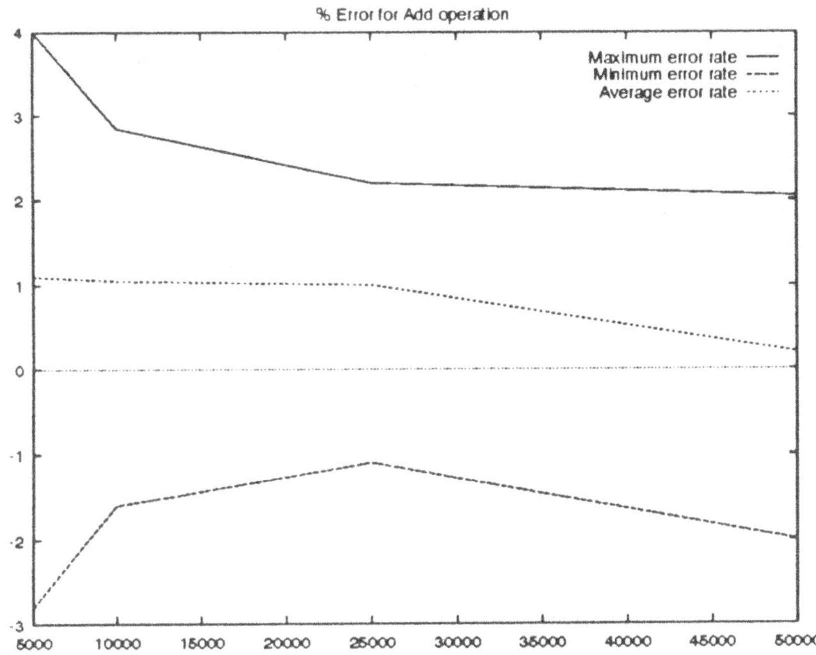
% Error for Add operation
Maximum error rate
Minimum error rate
Average error rate
4
3
2
1
0
-1
-2
-3
5000
10000
15000
20000
25000
30000
35000
40000
45000
50000

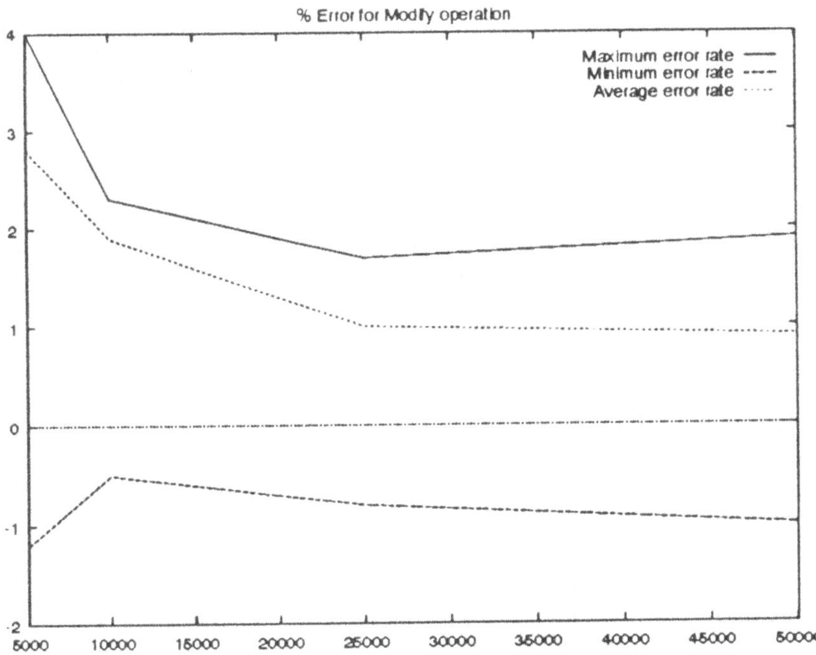
% Error for Modify operation
Maximum error rate
Minimum error rate
Average error rate
4
3
2
1
0
-1
-2
5000
10000
15000
20000
25000
30000
35000
40000
45000
50000

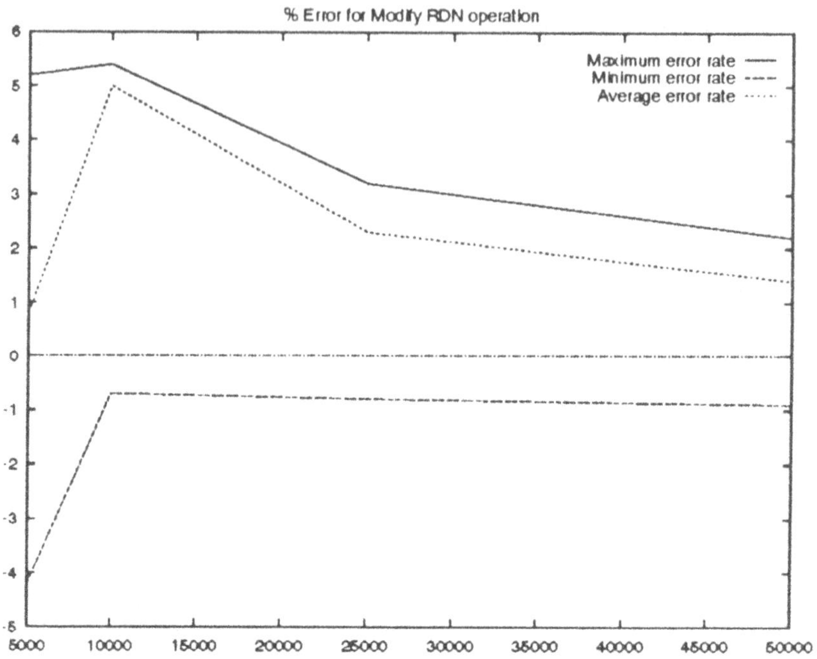

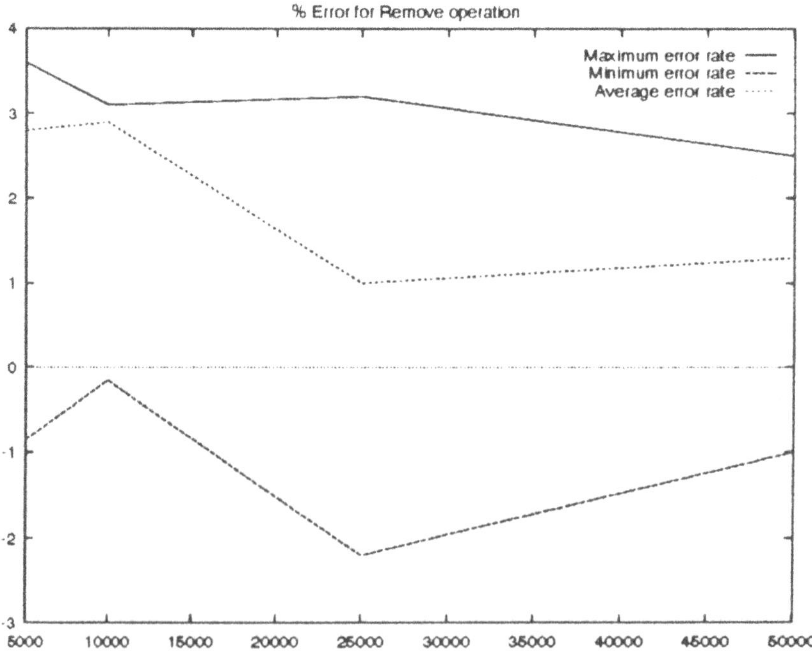

Figure 11: Percentage Error rate Vs Database size, Experiment 4.

15

A Queueing Network Model of Distributed Shared Memory

*D. Venkatesulu * & T.A. Gonsalves*
Department of Computer Science and Engineering,
Indian Institute of Technology,
Madras 600036, India.
E-mail: dvn%kurinji@iitm.ernet.in
tag@iitm.ernet.in

Abstract

Distributed Shared Memory (DSM) is a mechanism which provides logical shared memory in a distributed system. This simplifies the task of distributed programming. We develop simple queueing network models that can be used to compare the performance characteristics of several DSM algorithms for location of shared pages. Our models can be evaluated efficiently using the MVA (Mean Value Analysis) technique. Using these models, we can explore the regions of applicability of various DSM algorithms. By modelling contention at shared resources such as the network and memory servers, our model is able to provide more accurate results than the earlier analysis by Stumm and Zhou.

Keywords

Distributed Shared Memory, Computer Networks, Inter-Process Communication, Queueing Network Model, Mean Value Analysis.

1 INTRODUCTION

In a distributed system each node has its own processor and memory. Memory is not physically shared between any nodes. Traditionally, the I/O-based communication paradigm using message-passing is used in distributed programming. This is significantly more complex than the shared memory communication paradigm used in shared-memory multiprocessors. *Distributed Shared Memory (DSM)* [Cheriton (1986), Li (1986), Fleisch (1988), Kessler *et al.* (1989), Stumm *et al.* (1990)] is a mechanism which provides an abstraction of logically shared memory on the physically distributed memory of a distributed system. By making the separation of memories transparent to the programmer, DSM supports data sharing directly and passing of complicated data structures by reference, and simplifies process migration. Several issues need to be considered in a DSM. These include [Nitzberg *et al.* (1991)]:

----

*Supported by a Govt. of India Research Scholarship.

- *Structure and granularity of Memory*
  Structure refers to the layout of the shared data in memory, it could be a variable length segment (object) or fixed length linear array of words, and Granularity refers to the size of the unit of sharing: byte, word, page, or complex data structure.
- *Algorithms for Location of Shared Memory*
  There are several ways of locating shared data [Li (1986), Fleisch (1989)]. First one is the *Centralized Manager* which holds the information about the location of shared data at one place. The second one is the *Broadcast* of every request, which broadcasts every request to know the location of shared data. The third one is the *Distributed Manager* which reduces the load at the centralized manager and every node holds approximate information about the location of shared data.
- *Coherence Semantics and Protocols*
  The most appealing semantics for memory coherence is the *strict consistency*. By strict consistency we mean that a read operation to a memory location should return most recently written value into that location. There are several algorithms to provide strict memory coherence [Cheriton (1986), Li (1986), Fleisch (1988), Kessler *et al.* (1989), Stumm *et al.* (1990)].

While DSM can be implemented in hardware, Memnet, [Delp (1988)], the use of software mechanisms akin to virtual memory is more common. Our interest is in such software mechanisms since these can be implemented on commercially available computers. Our basic objective is the efficient comparative evaluation of the performance of various DSM algorithms. The most promising approach is the use of approximate queueing network models which can be evaluated efficiently.

In the next section, we describe the various DSM algorithms that we study. Section 3 explains the workload characterization, System characteristics, queueing network models for these algorithms, Stumm & Zhou's performance model, and validation of our model. Performance results and conclusions are presented in sections 4 and 5 respectively.

2 DSM ALGORITHMS

Stumm and Zhou have proposed four algorithms for data access in a DSM [Stumm *et al.* (1990)]. Of these, two are non-migratory and two are migratory. The latter algorithms migrate a shared data page from the owner node to the requesting node on demand, whereas the former hold shared data statically at each node.

The *Central Server* algorithm keeps only one copy of shared data in the system. The *Migration* algorithm allows only one read-write copy of a page to exist in the system at a time and migrates the page on demand to the requesting node. The *Read-replication* algorithm allows either one read-write copy or multiple read-only copies of a page to exist at a time. All read-only copies of a page are invalidated on a write to the page. Thus it is a *write-invalidate* algorithm. The *Full-replication* algorithm maintains read-only copies of each shared page at all nodes. It updates all copies of a page on a write operation, i.e it is a *write-broadcast* algorithm.

2.1 Central Server Algorithm

The simplest DSM algorithm has a single memory server that holds the entire shared memory of the system. Hence, every access to the shared data from client processes is remote which requires network access. Client processes on other nodes send requests explicitly to the server for every read or write of shared data. Each node also has local memory. The server serializes all the remote memory access requests ensuring strict consistency. However, there is no concurrency in processing the shared data access requests.

The advantage of this algorithm is that it is easy to implement and it is quite useful in client-server model applications. The disadvantage of this algorithm is that the central server will become the bottleneck of the system as the number of nodes increases.

One way of reducing the load at the central server is to distribute the shared data statically among several servers. The resultant increase in concurrency in processing shared memory requests will reduce the average waiting time. In the extreme case, every node acts as both a memory server and runs client processes. This could slow down the client processes.

2.2 Migration Algorithm

When a node references a page, it is likely to make additional references to it in the near feature. We can exploit this *locality of reference* by migrating the page to the requesting node. This node now becomes the new *owner* of the page. Initially the owner of a page will be the node at which it has been created. It is necessary to keep track of the location of pages. This can be performed by having a separate *manager* process running on one node which keeps track of the current owners of every page. The manager forwards the request to the owner of the page. Alternatively, the request could be broadcast.

The migration algorithm allows only one read/write copy of a page to exist at a time. The local copy of a page is invalidated when it is migrated to another node. The advantage of this algorithm is that it is similar to the virtual memory management system. It performs well when application exhibits a high degree of locality in its shared data accesses. It performs very poorly when there are interleaved read accesses to the same page by different nodes. This causes a page transfer between the two nodes for every access which increases the page fault rate.

The size of the page to be transferred also plays major role in determining the performance of this algorithm. As the size of the page increases, the transmission time at the network increases but the page fault rate could decrease.

2.3 Read-replication Algorithm

If a page has many readers with only occasional writes, the migration algorithm will experience considerable overhead as the page moves back and forth between readers. In such cases, the *Read-replication* algorithm can be advantageous. This allows either multiple read-only copies of a page or only one read-write copy of the page to be present in the system at a time. It solves the problem of memory coherence by invalidating all the read-only copies of a page on a write access to that page. The new owner of a page is the node at which the write fault occurred most recently. The owner holds either a read-write or a read-only copy of a page, and it maintains a list of nodes having read-only copies of a page, i.e the *copy set*. The owner of a page changes only on write faults. The manager keeps track of the owners of different pages.

On a read fault, the requesting node sends a message to the manager, which forwards the request to the owner. The owner adds the faulting node to the copy-set of the page, it then transfers a read-only copy of the page to the requesting node and changes its own access rights to read-only, if necessary.

A write fault is similar to the read fault except that the owner transfers ownership and the copy-set of the page to the faulting node. The faulting node sends invalidation requests to the copy-set members. The faulting node will resume its write operation only after sending invalidation messages.

2.4 Full-replication Algorithm

To further reduce overhead when reads are predominant and there is a high degree of contention for a page between readers and writers, the *Full-replication* algorithm is more suitable. It maintains read-only copies of all pages at all nodes. On a write fault, the faulting node broadcasts the modification to all other nodes.

For consistency, the broadcast must be reliable and writes should be processed at all nodes in the same order. This can be achieved via a *reliable broadcast protocol* that uses one *sequencer* node. Each write operation is sent to the sequencer node which adds a sequence number to the request. It then broadcasts the request. It also maintains a history of recently received requests to handle retransmission requests. The sequence number of the request allow the server process at each node to determine whether the update request is a duplicate or whether it has missed any broadcasts. In the latter case, it requests retransmission from the sequencer.

3 SYSTEM MODEL

We now develop an approximate queueing network model to evaluate the performance of the DSM algorithms described in the previous section. First, we characterize the workload of a DSM system. Next, we present our model. Finally,we describe an earlier model [Stumm *et al.* (1990)].

3.1 Workload Characterization

The performance of DSM depends on the memory access behaviour of the processes running on the various nodes. We assume that a single process runs on each node. The memory accesses of the process are divided into two types: *local* access which does not cause a fault, and *remote* access which causes a fault. Local access includes all accesses to private memory, and accesses to shared memory that is locally resident. Only access to shared memory that is resident on another node such that a read or write fault is generated, is remote access.

Thus, a program is characterized by the lengths of the local *burst*, Z, between successive remote accesses [kessler *et al.* (1989)]. This is a random variable that is a complex function of both the interaction between different processes and also the DSM algorithm. Consider, for example, a series of accesses to the same shared page. In the Central server algorithm, every one will be a remote access. In the migration algorithm, on the first fault, the page is migrated and hence the remaining accesses become local ones. Thus, Z_{mig} is expected to be greater than Z_{cs}.

In algorithms with multiple read-only copies of a page, the read and write accesses may have different effects. Hence, the read-to-write ratio, r, is an important characteristic. For most programs, $r > 1$.

3.2 System Characteristics

The system consists of N nodes interconnected by a broadcast network of capacity C. Each node has some private memory and some shared memory. We assume that there are two sizes of packets: *short packets*, typically about 64 bytes, are used for control, requests and individual data objects; *large packets* each contain a memory page (a large packet may actually consists of several maximum size LAN frames back-to-back). The processing cost of sending or receiving a small packet is p and of a large packet is P.

Requests for shared memory access received from remote nodes are given higher priority than local processing. In addition to the N nodes, some algorithms require a manager and/or sequencer. These are dedicated nodes that do not participate in user computation or hold shared memory. The service demand of each request at the manager and sequencer are denoted by D_m and D_{seq} respectively.

3.3 Queueing Network Model

We have chosen analytical modelling to evaluate the different DSM algorithms, because it is less expensive than *measurement* or *simulation* and provides quick insight into the performance of a system. In order to efficiently evaluate models with large numbers of nodes, we resort to approximate mean value analysis (MVA) [Lazowska *et al.* (1984)]. We validate our models by comparison with published results (section 3.5).

Trace-driven simulation or measurement with actual distributed applications will certainly yield more accurate results than analytical models. However, these results would be applicable to only specific workloads. Distributed workloads are highly variable in their memory reference behaviour and communication patterns. Hence, the results of simulation or measurement may not yield insight into the system.

The workloads in our models do not correspond to any particular application. However, by focussing on *relative* performance of different algorithms for a wide range of parameters, we are able to obtain insights into DSM behaviour that could not be easily obtained otherwise. We do not, in general, ascribe especial importance to the absolute values of performance predictions.

We use closed network of queues to represent the system [Lazowska *et al.* (1984)]. Servers correspond to nodes, managers, sequencers and the network. Local access bursts and local packet processing are modelled by *think time* using the *interactive workload*. When the same node serves both local processing and remote requests, the remote requests get absolute priority. To model this, we use the technique of *service demand inflation* [Lazowska *et al.* (1984), Chap. 11]. We first find the utilization of the server, U_r, due to remote requests. The local processing time, Z, is then inflated to $Z' = 1/(1-U_r)$.

3.3.1 Central Server Model

We assume that the shared data is statically distributed among all N nodes. That is, each node acts as a client and also as a memory server. Accesses to shared data are uniformly distributed. Hence, the probability that a shared access results in a fault is $(N-1)/N$. Each remote access takes two short packets. The access cost is $2p$ at the requesting node and $2p$ + memory access time at the memory server.

In our queueing network model for the Central server algorithm, Figure 1, each node is represented by two servers: a *delay* service centre to represent the client functions, and a *queueing* service centre that represents the memory server function. The network is represented by a queueing service centre.

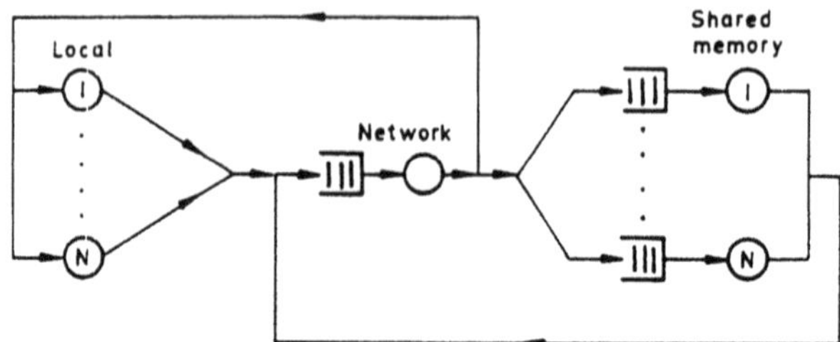

Figure 1 Model of Central server algorithm. Node i is represented by local server i and shared memory server i.

Each job corresponds to a local access burst followed by one remote access. Hence, we have the following service demands:

Local server, l — local burst + two short packet events. $D_l = Z + 2p$

Memory server, k — One memory access + two short packet events, multiplied by visit ratio, $1/N$. $D_k = (d + 2p)/N$, where d is the local memory access time.

Network server, n — Two short packet transmissions. $D_n = 2l/C$, where l is the length of a short packet.

This allows concurrency between client and memory server processing at each node that does not exist in reality. To account for this, since by assumption the memory server runs at higher priority, we use demand inflation at the local server: $D'_l = D_l/(1- U_k)$, where U_k is the utilization of the memory server. Initially, D'_l is set to D_l and we iterate until convergence. Finally, the average shared memory access time, $T_{cs} = (N\text{-}1)/N$ * average remote access time.

3.3.2 Migration Model

To extend the Central server model to migration, we add a manager service centre which is a queuing centre with service demand D_m. Next, the service demands at the other centres are adjusted to account for the cost of one request packet(small) plus one data packet(large) for each page fault. Hence, we have the following service demands:

Local server, l — local burst + one short packet event + one large data packet event. $D_l = Z + p + P$

Memory server, k — one short packet event + one large data packet event, multiplied by visit ratio, $1/N$. $D_k = (p + P)/N$.

Manager server, m — two short packet events. $D_m = 2p$.

Network server, n — two short packet transmissions + one large data packet transmission. $D_n = (2l + L)/C$, where L is the length of the large data packet.

As above, servicing of incoming remote requests is given higher priority and hence local processing is subjected to inflation. To compute the average shared memory access time, we assume a page fault rate (with out replication), f, for references to shared memory. As discussed earlier, f would typically be smaller than the corresponding rate in Central server algorithm and is a complex function of program behaviour and memory page size. The average shared memory access time, $T_m = f$ * average remote access time.

3.3.3 Read-replication Model

The model of this algorithm is same as the migration algorithm with one exception. In this algorithm, the service demands at the memory servers vary with the read/write ratio r, and the number of nodes having replicated copies of a page. This is due to the fact that on every write operation, all read-only copies in the system are invalidated. We have the following service demands:

Local server, l	local burst + one short packet event + one large data packet event. $D_l = Z + p + P$
Memory server, k	one short packet event + one large data packet event + average cost of invalidation, multiplied by visit ratio, $1/N$. $D_k = (p + P + (1/(r + 1)) * Np)/N$
Manger server, m	two short packet events. $D_m = 2p$
Network server, n	three short packet transmissions + one large data packet transmission. $D_n = (3l + L)/C$.

The average shared memory access time, $T_{rr} = f' *$ average remote access time, where f' is the average fault rate with replication.

3.3.4 Full-replication Model

Since every node has read-only copies of all pages, read operations on shared data are local accesses. Every write operation to shared memory requires that the writer to send the update request in a short packet to the sequencer. The sequencer then broadcasts the packet which is received and processed by all other nodes. Hence, we have the following service demands:

Local server, l	local burst + one short packet event. $D_l = Z + p$.
Memory server, k	one short packet event, multiplied by visit ratio, 1. $D_k = p$.
Sequencer server, seq	two short packet events. $D_{seq} = 2p$.
Network server, n	two short packet transmissions. $D_n = 2l/C$.

Demand inflation and iteration is used as in the other models. The average shared memory access time, $T_{fr} = (1/(r + 1)) *$ average remote access time

3.4 Stumm and Zhou's Model

Stumm and Zhou have derived expressions for average access time using a simple deterministic model of the DSM algorithms [Stumm *et al.* (1990)]. They ignore network transmission time, local memory access time and queueing delays at servers. The average access time is given by the probability that an access to shared data is remote multiplied by the average remote access time. The first term depends on the number of nodes, N, in the case of Central server algorithm, on the fault rate, f and f', in Migration and Read-replication, and read/write ratio, r, in Full-replication. The second term depends on the number of packets involved and the packet processing times. They have been derived expressions for the average access time:

Central Server	$T_{cs} = (1 - 1/N) * 4p$
Migration	$T_m = f * (2P + 4p)$

Read-replication $T_{rr} = f' * (2P + 4p + (1/(r+1) * Np)$

Full-replication $T_{fr} = (1/(r+1)) * (N + 2) p$

3.5 Validation

Since queueing network models are approximate models, it is necessary to validate these models by comparison with earlier models or with the results of measurement or simulation. First, we validate our model by comparing it with the results of Stumm & Zhou. By setting the service demands to zero at the network and manager servers in the Migration model, and at the network server in Central server model, and by assuming the local processing time between successive remote requests is negligible, we are able to get the equal performing curve almost the same as Stumm & Zhou's model as shown in Figure 2. Also in Figure 2, we have shown that the effect of increase in local processing time, when $D_n = 0$ and $D_m = 0$ ms.

We have also validated our model by comparing it with simulation results [Kessler *et al.* (1989)] for *Hot Potato* algorithm. We found that the difference is with in the acceptable limits (3-9%).

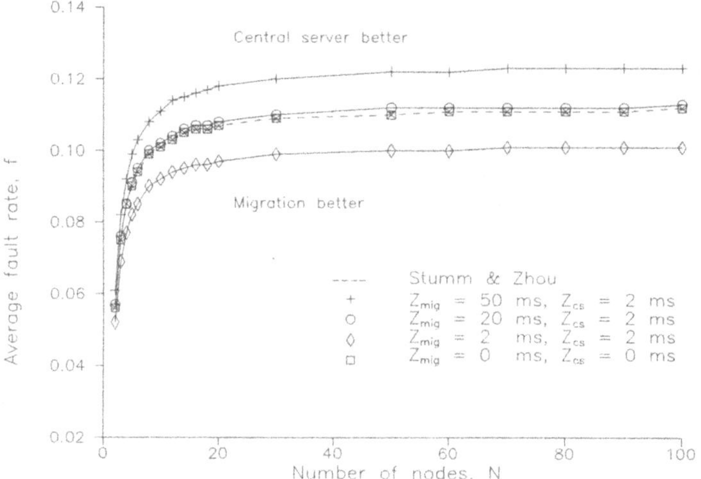

Figure 2 Validation of the Queueing Network Model

4 PERFORMANCE RESULTS

We now use our queueing network models to examine the trade-offs between the different DSM algorithms described in section 2. Since the memory reference behaviour of a distributed program is not well known, the absolute magnitudes of performance measures such as average memory access time are likely to be of limited use. However, the relative performance of different algorithms can be gauged. To this end, we present *equal performing curves* for each pair of algorithms. For two independent parameters, the average memory access time of both algorithms is the same along the equal-performing curve. Above the curve, one algorithm has higher performance, while below the other is better.

Table 1 Parameter values used in the evaluations

Parameter	Value
Request/Response message size:	64 B
Memory page size:	8 KB
Time to send/receive a message, p:	1 ms
Time to send/receive a Memory page, P:	16.75 ms
Network Bandwidth, C:	10 Mb/s - 1 Gb/s
Service demand at the manager, D_m:	0 - 2 ms
Service demand at the Sequencer, D_{seq}	2 ms
Local processing time, Z:	0 - 100 ms
Number of nodes, N:	2 - 100

We assume that the memory page size is 8 KB and that the CPU time to send or receive a request message (short packet), and a memory page (large data packet) is 1 ms and 16.75 ms respectively. We have assumed that the CPU is faster than the one assumed by Stumm & Zhou. The network service demand is the message transmission time dependent on the network bandwidth. In the case of Central server and Full-replication, one data item is contained in each request/response message. Allowing for overheads, these messages are assumed to be of size 64 B. The local processing time between memory accesses in the central server algorithm, Z_{cs}, is set to 2 ms. In the case of the other algorithms, this varies depending on the locality of reference. Hence, Z_{mig} is varied between 2 to 50 ms. Table 1 is a list of the values used for each parameter.

4.1 Central Server versus Migration

To compare the Central server and Migration algorithms, for different number of nodes, N, we find the fault rate in the migration algorithm, f, at which the two algorithms perform equally.* In Figure 3 equal-performing curves are plotted for $C =$ 10, 100 Mb/s, $Z_{cs} = 2$ ms and $Z_{mig} =$ 2, 20 and 50 ms. $Z_{mig} = 2$ ms $= Z_{cs}$ represents a program with very little locality of reference. $Z_{mig} = 20$ ms and 50 ms ($>> Z_{cs}$) represent programs with greater locality of reference. For each case, migration performs better below the curve and central server is better above.

At Ethernet speeds, 10 Mb/s, the central server algorithm performs better always. This is because the network is the bottleneck in the case of migration except for small N when the contention for a page is reduced. For $N > 40$, the degree of locality makes no difference. For $N < 40$, performance of migration improves somewhat with higher locality (i.e Z_{mig} = 20 and 50 ms).

If the network bandwidth is increased to 100 Mb/s, performance of migration improves further and we see that program locality has a larger impact except for large N. Further increase of C to 1 Gb/s does not improve performance (the curves overlap the 100 Mb/s ones and hence are not shown).

----

*Note that since we assume that all nodes act as servers, the fault rate in the central server algorithm is fixed at (1-1/N)

Also shown in Figure 3 is the curve using Stumm & Zhou's analysis. Since they ignored the network overhead, and the queuing delay at the manager their model considerably over-estimates migration performance especially for large systems. For small N (typically for Z_{mig} = 50 ms), their model underestimates the performance of migration because they assumed that the local processing time between successive remote accesses is negligible.

In migration, every page fault requires service from the manager. Increasing the CPU speed of the manager node, i.e. reducing D_m results in improvement of migration performance (Figure 4). At 10 Mb/s, since the network is the bottleneck, a faster manager has minimal effect. At 100 Mb/s, on the other hand, with $D_m <= 1$ ms, performance improves substantially in large systems.

Sensitivity Analysis

Based on the equal performing curves, given expected system parameters, we can select the better algorithms. However, during operation, parameters are likely to vary by moving the system across the equal-performing curve to the region of poor performance for the selected algorithm. To see how serious this problem is likely to be, we have chosen the average remote access time for $N = 50$, and the equal-performing point at $N = 50$, isdenoted by f_{50}. We now vary f and plot average access time for both algorithms and various values of C and D_m.

At C = 10 Mb/s, in Figure 5 there is a large difference between the two algorithms as we move away from equal performing point f_{50} of $N = 50$. For C = 100 Mb/s, the difference decreases, becoming very small as D_m decreases. That is, when the bottlenecks of network and manager are reduced, sensitivity to fault rate f at $N = 50$ is much less.

4.2 Central server versus Full-replication

Since Central server performs remarkably well relative to Migration, at low network bandwidths and large systems, we compare it to Full-replication. In Figure 6, we plot equal-performing curves with average read-to-write ratio, r, as the dependent variable. In each case, Full-replication is better above the curve and Central server is better below.

Central server performance is independent of r. As N increases, r must increase for Full-replication to remain equal to that of Central server. This is because the number of write faults increases with N. At 10 Mb/s, for $N > 70$, the network becomes the bottleneck in Central server and hence the curve levels off. At C = 100 Mb/s, this does not happen even at $N = 100$. Stumm & Zhou's model consistently underestimates the average access time of full-replication. This because of the fact that they ignored the network overhead, and queueing delays at the sequencer and the memory servers. It overestimates the performance of full-replication only for large values of Z_{fr} (typically Z_{fr} = 100 ms). This is because of the fact that they have not taken into account the local processing time between successive requests. As in the previous section, we find that with the more queueing analysis, the region of applicability of the Central server algorithm is greater than predicted by the earlier analysis.

Up to $N = 100$, we find that the utilization of the sequencer node is less than 100%. Hence, a faster sequencer node will not affect performance except possibly in much larger systems.

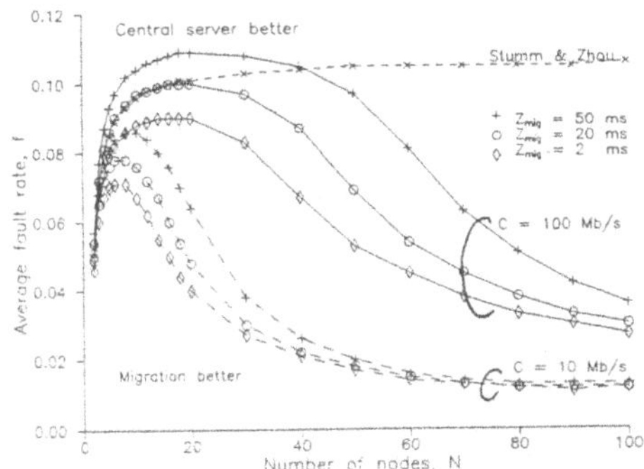

Figure 3 Equal-performing curves for Central server and Migration -- Central server is better above each curve, Migration is better below (D_m = 2 ms, Z_{cs} = 2 ms).

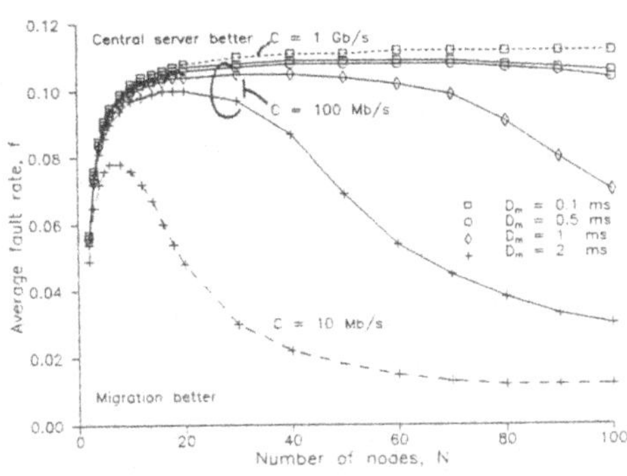

Figure 4 Effect of Manager service demand, D_m, on Central server--Migration equal-performing curves (Z_{cs} = 2 ms, Z_{mig} = 20 ms).

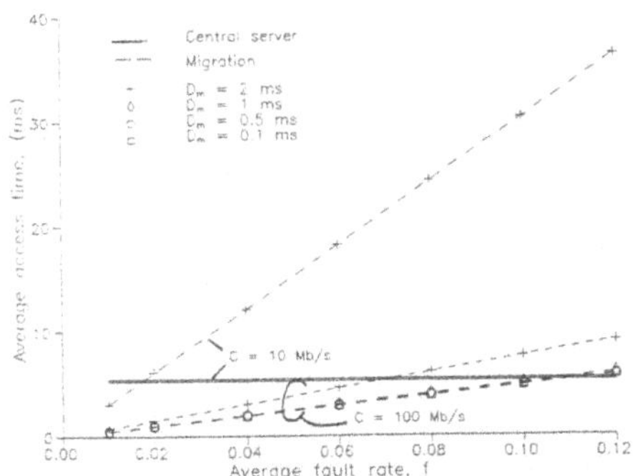

Figure 5 Sensitivity of equal-performing curves to change in average fault rate, f, Central server versus Migration. Page fault rate f_{50}, for Migration is the value that gives the equal-performing point for N = 50.

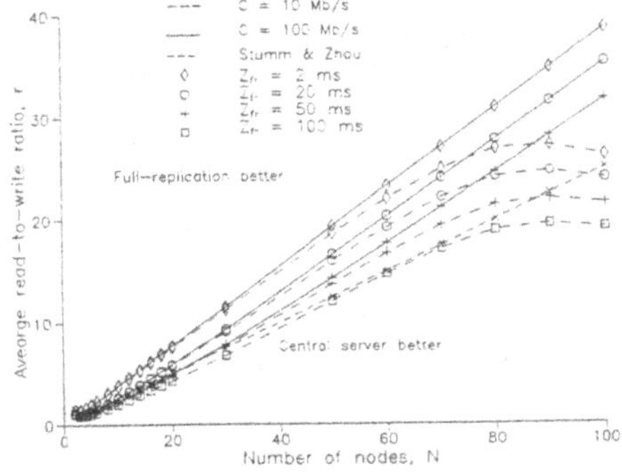

Figure 6 Equal-performing curves for Central server and Full-replication -- Central server is better below each curve, Full-replication is better above (D_{seq} = 2 ms, Z_{cs} = 2 ms).

Sensitivity Analysis

As in the previous section, in order to see the impact of moving the system across the equal-performing curve, we have chosen the average remote access time for $N = 20, 50$, and 80 for both the algorithms. The equal-performing point is denoted by r_{20} for $N = 20$. We now vary r and plot average access time for both the algorithms for different values of N and C. In Figure 7, we have chosen the Y-axis in log scale.

The curves for $C = 100$ Mb/s for both the algorithms are overlapping with $C = 10$ Mb/s curves. Hence, we have not shown them in Figure 7. For small N, the difference between the two algorithms as we move away from equal-performing point, r_{20}, is significant. The difference decreases, becoming small, as N increases. Hence, for large N, sensitivity to large r is small.

4.3 Read-replication versus Full-replication

We now compare Read-replication with Full-replication. In Figure 8, we plot equal-performing curves for various parameter values. At a low data rate of $C = 10$ Mb/s, full-replication is almost always better than read-replication. In the latter, the network is the bottleneck as the entire memory page must be transferred to each reader after a write fault. In full-replication, only the relatively small data item is sent to all the readers. That is the network cost of a write fault in read-replication is higher than in full-replication.

At high values of read/write ratio, r, in Figure 8, full-replication does better comparatively because it always maintains up to date read-only copies at all nodes thus reducing the number of read faults. For low r, frequent updating is counter-productive and read-replication performance improves.

As expected, at 10 Mb/s, in Figure 8 Stumm & Zhou's model considerably over-estimates read-replication performance because it ignores the network overhead. At higher bandwidths, the correspondence is a bit better. However, there is still considerable inaccuracy because of the other effects such as the manager.

With $C = 100$ Mb/s, in Figure 9, read-replication is seem to perform much better. For $N > 50$, the curve levels off due to the manager node in read-replication. If a faster CPU is used in the manager, we see that read-replication performs even better. As we increase the speed of the CPU at the manager the effect of $D_m < 0.5$ ms is minimal.

5 CONCLUSIONS

We have developed an efficient approximate queueing network model of several DSM algorithms. By modelling shared resources such as the network and the manager, and by considering queueing delays, these models are substantially more accurate than earlier ones.

Our results show that the Central server algorithm does surprisingly well compared to migration algorithm. This is especially true at low network data rates of 10 Mb/s. Full-replication performs well, especially at high read/write ratios.

To improve confidence in these models, we plan to perform more extensive validation by comparison with simulation. We are currently using our model to predict the performance of a DSM-Based Distributed Programming Testbed [Hariram & Gonsalves 94], and are extending it to hybrid algorithms.

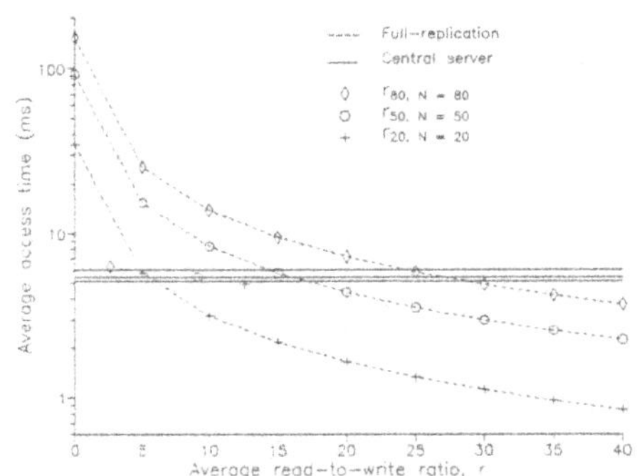

Figure 7 Sensitivity of equal-performing curves to change in read/write ratio, r, Central server versus Full-replication. Write fault rate r_{20}, for Full-replication is the value that gives the equal-performing point for $N = 20$ ($Z_{cs} = 2$ ms, $Z_{fr} = 20$, and $N = 80$, 50, and 20).

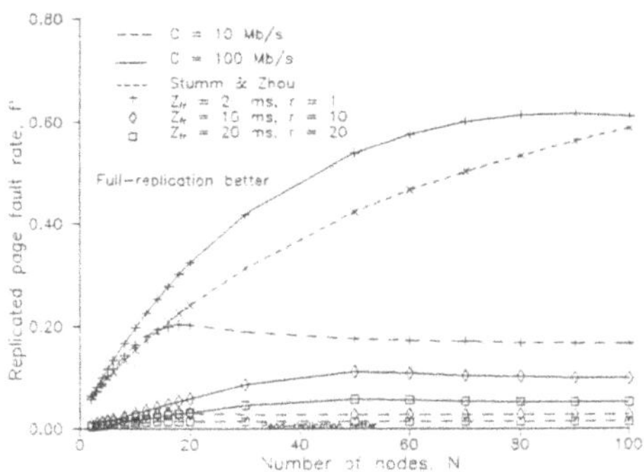

Figure 8 Equal-performing curves for Read-replication and Full-replication -- read-replication is better below each curve and full-replication is better above ($D_m = 2$ ms, and $Z_{rr} = 20$ ms).

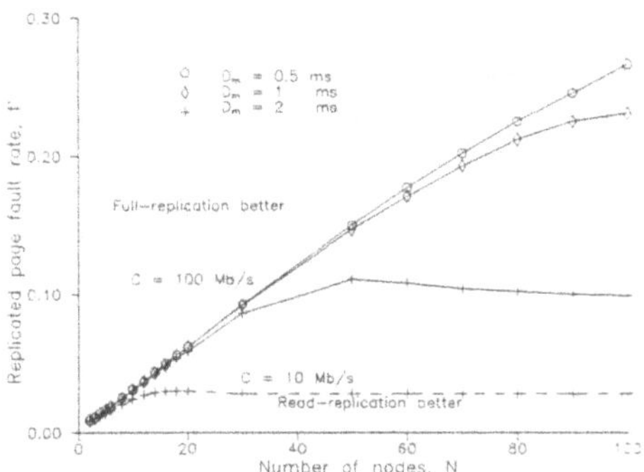

Figure 9 Effect of manager service demand, D_m, on Read-replication - Full-replication equal-performing curves ($Z_{fr} = 10$ ms, $Z_{rr} = 20$ ms , and $r = 10$).

6 REFERENCES

Agarwal, A. and Gupta, A.(1988) Memory Reference Characteristics of Multiprocessor Applications Under MACH. *Proc. ACM SIGMETRICS '88 Conf. on Meas. and Modelling of Comp. Sys.*, 16, 1, 88, pp. 215-225.

Archibald, J. and Baer, J.(1986) An Evaluation of Cache Coherence Solutions in Shared-Bus Multiprocessors. *ACM Trans. Comp. Sys.*, Vol. 4, No. 4, Nov. 86, pp. 273-298.

Bennet, J. Carter, J. and Zwaenepoel, W.(1990) Munin: Distributed Shared Memory Based on Type-Specific Memory Coherence. *Proc. 90 Conf. Principles and Practice of Parallel Programming*, ACM Press, New York, N.J., 90, pp. 168-176.

Cheriton, D.R.(1986) Problem-Oriented Shared Memory: A Decentralized Approach to Distributed System Design. *Proc. 6th Int'l Conf. Distributed Computing Sys.*, May 86, pp. 190-197.

Delp, G.S.(1988) The Architecture and Implementation of Memnet: A High-Speed Shared Memory Computer Communication Network. Ph.D thesis, Univ. of Delaware, Newark, Del., 88.

Denning, P.J. and Buzen, J.P.(1978) The Operational Analysis of Queueing Network Models. *ACM Computing Surveys*, Vol. 10, No. , Sep. 78, pp. 225-261.

Dubois, M and Wang, J.C.(1991) Shared Block Contention in a Cache Coherence Protocol. *IEEE Trans. on Computers*, Vol. 40, No. 5, May 91, pp. 640-644.

Eggers, S.J.(1991) Simplicity Versus Accuracy in a Model of Cache Coherency Overhead. *IEEE Trans. on Computers*, Vol. 40, No. 8, Aug. 91, pp. 893-906.

Fleisch, B.D.(1987) Distributed Shared Memory in a Loosely Coupled Distributed System. *Proc. of the ACM SIGCOMM' 87 Workshop*, Stowe, Vermont, Aug. 87, pp. 317-327.

Fleisch, B.D. and Popek, G.J.(1989) Mirage: A Coherent Distributed Shared Memory Design. *Proc. 12th ACM Symp. Op. Sys. Prin.*, Dec. 89, pp. 211-222.

Hariram, R.K. and Gonsalves, T.A.(1994) A DSM-Based Portable Distributed Programming Testbed. *Proc. NETWORKS '94 : An Int'l Conference on Comp. Net., Arch. and Appln.*, Madras, India, Dec. 94.

Kessler, R.E. and Livny, M.(1989) An Analysis of Distributed Shared Memory Algorithms. *Proc. 9th Int'l Conf. Distributed Computing Sys.*, CS Press, Los Alamitos, Calif., June 89, pp. 498-505.

Lazowska, E.D., Zahorjan, J., Graham, J.S. and Seivick, K.C.(1984) Quantitative System Performance-Computer System Analysis Using Queueing Network Models. Prentice-Hall Inc., Englewood Cliffs, NJ, 84.

Li, K. and Hudak, P.(1986) Memory Coherence in Shared Virtual Memory Systems. *ACM Trans. Computing Sys.*, Vol.7, No. 4, Nov. 89, pp. 229-243.

Nitzberg, B. and Lo, V.(1991) Distributed Shared Memory: A Survey of Issues and Algorithms. *IEEE Computer*, Aug. 91, pp. 52-60.

Ramachandran, U. and Khalidi, M.Y.A.(1991) An Implementation of Distributed Shared Memory. *Software-Practice and Experience*, Vol.21, No. 5, May 91, pp. 443-464.

Stumm, M. and Zhou, S.(1990) Algorithms Implementing Distributed Shared Memory. *IEEE Computer*, Vol. 23, No. 5, May 90, pp. 54-64.

Zhou, S., Stumm, M. and McInerney, T.(1990) Extending Distributed Shared Memory to Heterogeneous Environments. *Proc. 10th Int'l Conf. Distributed Computing Sys.*, Los Alamitos, Calif, May-June 90, pp. 30-37.

Zhou, S., Stumm, M., Li, K. and Wortman, D.(1992) Heterogeneous Distributed Shared Memory. *IEEE Trans. on Parallel and Distributed Systems*, Vol. 3, No. 5, Sep. 92, pp. 540-554.

7 BIOGRAPHY

D. Venkatesulu received his M.Sc. degree in Physics and M.Tech degree in Computer Science in 1985 and 1988 from Andhra University. He was in the CSE Faculty of Nagarjuna University from 1988 to 1990. He is currently pursuing his Ph.D degree at IIT Madras. His research interests include distributed systems, performance evaluation, and computer networks.

Timothy A. Gonsalves received a B.Tech.(Electronics) from I.I.T.,Madras in 1976, and M.S. (EE) from Rice University, Houston, USA in 1978, and a Ph.D.(EE) from Stanford University, USA in 1986.

From 1986 to 1988 he was an Assistant Professor in the Computer Science Department, Worcester Polytechnic Institute, USA. In January 1989, he joined the Department of Computer Science and Engineering, I.I.T.,Madras where he is currently an Associate Professor. His research interests include distributed computer systems, computer networks, and performance evaluation. He is active in technology/product development and consultancy for a number of industries. Dr. Gonsalves is a member of CSI, ACM and IEEE.

PART FIVE

Performance Studies

PERFORMANCE STUDIES

Performance evaluation studies is an integral part of any system design. From software design to total system design, performance analysis is essential for gaining insight into the operation of the system. There are five papers in this session that deals with issues such as buffer design for layered software design, efficiency of access methods, implemented experience in the design of IP routers, load patterns due to mobility in PCS and topology optimization.

The first paper is titled "Analysis of buffer management policies in layered software" authored by Raghavan and Krishnamoney. The paper focuses on different buffer management policies that is suitable for a layered communication software. The analysis is based on compaction on fragmented memory and the performance results are presented. The second paper "On the performance of the unslotted CDMA - ALOHA access protocol for finite number of users, with and without code sharing" by Anurag Kumar and Deepak Patil addresses the question of comparing the efficiency of CDMA-ALOHA and TDMA-ALOHA when they are used as bandwidth sharing mechanisms by finite number of nodes for transmitting packetized data. A simple stochastic model is proposed for channel attempts and Markov chain model is used to obtain channel capacity by varying different parameters.

The third paper "Implementation and performance of a PC based LAN/WAN router with a differential QoS feature" by Anand and Anurag Kumar discusses the implementation experiences in developing a PC based IEEE 802.3 LAN - X.25 WAN IP router. Various implementation issues that affect QoS are discussed, and it is shown how the implementation that resulted from the initial integration is improved to achieve serial lines throughput of 90% of maximum. The fourth paper "Analysis of performance and fault tolerance of data base management schemes for PCS" by Dipak Ghosal is on the topic of mobility related data management of Personal Communication Networks. The issue taken up for analysis is the signaling network load pattern in mobile communication. The fifth paper "A new approach for network topology optimization" by Srinivasa Rao, Sanjay Bose, Srivatsan and Kalyanmoy Deb focuses on how a hybrid algorithm based on link reduction algorithm and branch-X-change algorithm improves the convergence rate. Results obtained for two practical problems are presented.

16

Analysis of Buffer Management Policies in Layered Software

S.V.Raghavan and S.Krishnamoney,
Department of Computer Science and Engineering,
Indian Institute of Technology Madras, Madras 600 036, India.
{svr,kms}@iitm.ernet.in

Abstract

Use of shared memory in the transfer of protocol data units (PDUs) between layers of a network eliminates the need to physically copy the PDU, thereby saving considerable time and resources in the form of data buffers. In this paper we discuss three possible approaches to the management of shared memory between protocol layers, and analyze their relative merits and demerits. These strategies differ primarily on the compaction algorithms to cure memory fragmentation. The analysis of compaction algorithms is presented and the performance of the memory management strategies under various conditions of load are compared.

Key words

PDU (Protocol Data Unit), SDU (Service Data Unit), Shared Memory, NIU (Network Interface Unit), Compaction, Segmentation.

1 INTRODUCTION

Layering of protocol functions and engineering the related software require that information in a coded form be exchanged across layer boundaries and between peer layers [Stallings84, Tanenbaum85, Zwaenopoel85, Raghavan90]. Protocol Data Units (PDUs) can be passed across protocol layers by physically copying the data structures or by allocating these data structures in shared areas of their respective address spaces. These are referred to as the *C-scheme* and the *S-scheme* respectively in this work.

In a layered communication software, C-scheme denotes copying of PDUs across layer boundaries and S-scheme denotes the usage of *shared memory* to store PDUs and passing of pointers across layer boundaries. In C-scheme, each layer may add or remove a header (and possibly a trailer) and pass the corresponding pointer to the modified PDU to the next layer. The application layer will allocate enough buffer so that each layer simply copies its header in front of the incoming *Service Data Units* from the n + 1 layer and pass it to the n - 1 layer by giving a pointer to the PDU. The block schematic of the C-scheme and S-scheme are shown in Figure 1 and in Figure 2. *In this paper we analyze the performance of the S-scheme with special emphasis on compaction strategies.*

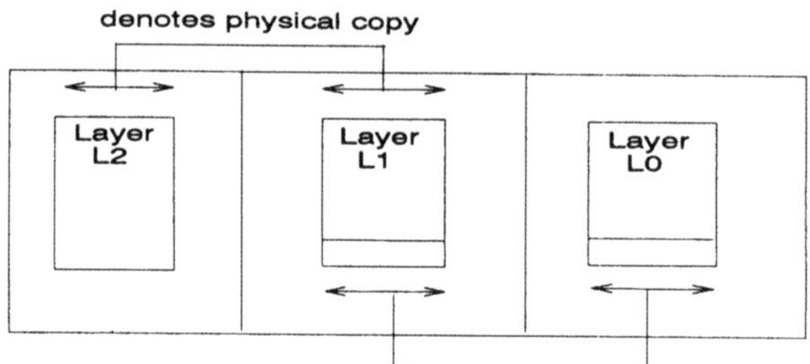

Figure 1 Block Schematic of the C-Scheme.

The rest of the paper is organized as follows: Section 2 briefly discusses related work that has already been done in this area. Section 3 states the problems of buffer management in layered software and compares the C-scheme and S-scheme. Section 4 highlights the issues in implementation, while section 5 describes the implementation itself. Section 6 presents the temporal and spatial analysis in the S-scheme leading to the identification of the optimal block size. Section 7 describes the simulation experiments that are conducted to understand the effectiveness of the compaction strategies under different load conditions and with varying shared memory parameters. Finally section 8 summarizes the work by listing the salient features of our work, the significance of the results obtained and suggests possible future course of investigation.

2 RELATED WORK

Minimization of data copying is a major issue to be addressed in a protocol implementation. It would be ideal to support data transfer across protocol layers and the NIU in just one copy, but this is not possible because of the machine architecture or due to the security requirements of the host operating system. Thus, protocol modularization

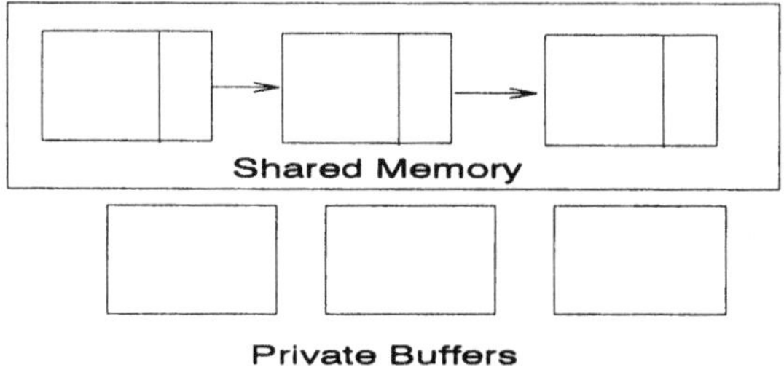

Figure 2 Block Schematic of the S-Scheme.

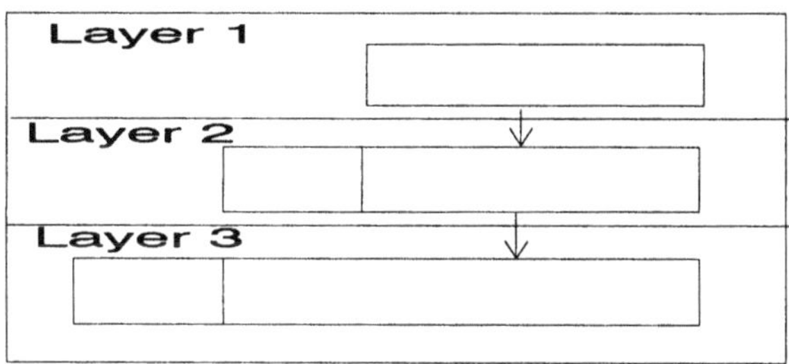

Figure 3 Transmission Phase of PDU.

requires multiple copies of the PDUs [Zwaenopoel85].

G.S. Poo et al. [G.S.Poo91] gives an efficient algorithm for passing PDUs across layers as linked lists. Their method registered a maximum improvement of 90% for bulk file transfer without software checksum. Saltzer et al. [Saltzer et al.84] discuss the implications of choosing an appropriate buffer management strategy in implementations of protocols. Buffer management strategies are also discussed in Lantz [K.A.Lanz et al.85]. Analogous to buffer management strategies are the memory management strategies and disk management strategies in Operating Systems. An overview of these strategies is presented in Peterson [Peterson et al.85]. Clark [Clark82] presents the modularity and efficiency criteria in protocol implementations, which include the issue of data copying across layers. Watson and Mamrak [Watson et al.87] also discuss this issue in detail and provide a review of previous work in the same area. ***The considerable importance given to the memory management strategies in protocol implementations and the desire to implement a clean memory management strategy that eliminates multiple copies of PDUs, especially in the context of high speed networks and high performance network interface adapters, are the motivations for this work.***

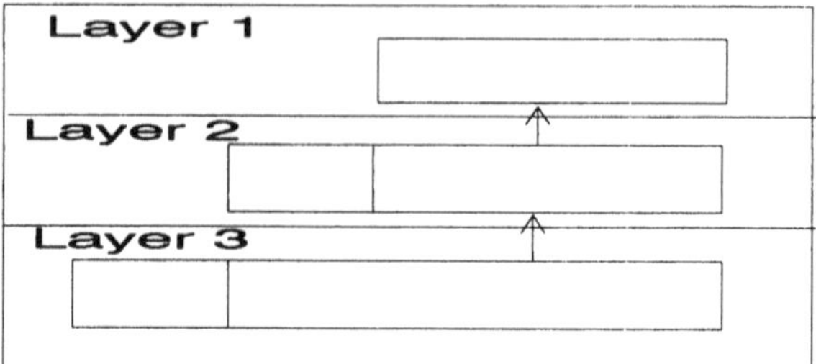

Figure 4 Reception Phase of PDU.

3 BUFFER MANAGEMENT PROBLEMS IN LAYERED SOFTWARE

In a shared memory implementation, allocation of buffers, deallocation of buffers, and management of external or internal fragmentation may be done in several ways. One simple approach is to allocate and deallocate only fixed sized buffers thereby avoiding external fragmentation and compaction. The price paid for this is extra memory due to internal fragmentation. The implemented protocol stack should handle single character packets, normal sized packets which do not need any fragmentation and large packets which do require fragmentation in an efficient manner. Therefore it becomes necessary to identify an optimal buffer management policy that is appropriate for different load conditions. In this paper we discuss both C and S schemes, compare their performance and establish through simulation, the utility and appropriateness of C and S schemes under light, medium and heavy loads.

The metric for comparison of the performance of the two schemes is *the time taken to transmit and receive one PDU across a layer. The significant operation considered in the calculation of transmit/receive time is a block copy.* The transmission phase and the reception phase are shown in Figure 3 and in Figure 4.

Let H_L be the header size of layer L in blocks.

Let D_L be the data size of layer L in blocks.

Let T be the time function of a block copy.

[Note: T is a linear function of the form $T(X) = aX$, where X is the size of the buffer and a is a constant. Therefore, $T(X_1) + T(X_2) = T(X_1 + X_2)$]

We consider a layer L_2 with user layer L_3 and service layer L_1.

Table 1 gives the ratio of C-scheme and S-scheme data copying time. This ratio establishes the S-scheme as a very attractive alternative to the C-scheme. Also it is to be noted that the gain in efficiency is directly proportional to the PDU size. However the buffer management cost has not been considered in this analysis. Processing overhead depends on the buffer management strategy used. External and internal fragmentation

Table 1 Calculation of the ratio of C-scheme to S-scheme time for data copying

Transmission time for C-scheme	
Copying time of L_2 SDU to local buffer	$T(D_{L2})$
Adding L_2header	$T(H_{L2})$
Total C-scheme transmission time	$T(D_{L2} + H_{L2})$
Reception time for C-scheme	
Copying time of L_1data to local buffer	$T(D_{L2} + H_{L2})$
Transmission and Reception time across layer L_1 in C-scheme	$2 * T(D_{L2} + H_{L2})$
Transmission time for S-scheme	
Pointer to L_2data passed from L_3 to L_2	0
Header prefixed to L_2data	$T(H_{L2})$
Pointer to L_2data passed from L_2 to L_1	0
Total S-scheme transmission time	$T(H_{L2})$
Reception time for S-scheme	
Pointer to L_1data passed from L_1 to L_2	0
Pointer to L_2data passed from L_2 to L_3	0
Total S-scheme reception time	$T(0)$
Transmission and Reception time across layer L_2 in S-scheme	$T(H_{L2})$
The ratio of the time taken by C-scheme to S-scheme $2 * T(D_{L2} + H_{L2})/T(H_{L2}) = 2 * (D_{L2} + H_{L2})/H_{L2}$	

depends on the organization of the buffer pool and the size of one block. The above mentioned factors vary for different environments. In this paper it is our aim to solve these problems and to show that buffer management cost does not affect the ratio mentioned in Table 1 significantly.

4 ISSUES IN THE IMPLEMENTATION OF S-SCHEME

1. The shared memory is treated as a *sequence of blocks* of size BLOCK_SIZE, a compile time parameter, typical values being 256 bytes, 512 bytes and 1K bytes. The blocks are treated as individual entities, the number of blocks in the shared buffer being $\lceil$ size of buffer requested / BLOCK_SIZE $\rceil$.

2. The order in which the PDUs are released from the shared buffer is not predicatable. Hence, over a period of time, such unordered freeing of blocks leads to *fragmentation* of the shared buffer, creating a spurious insufficiency of the buffer space and necessitating compaction. Memory compaction requires processing overhead.

3. Compaction involves the physical shifting of the buffer blocks and it will affect the validity of the pointers passed, thereby rendering passing of pointers impossible. To overcome this difficulty, another level of indirection ought to be introduced. This level is provided by the ***index table***. When a layer allocates buffer for an incoming PDU, an entry is allocated in the index table. This entry contains the number of contiguous blocks, a block pointer to the first of these blocks and links (next and previous) to index entries. The combination of array and list data structures in this model ensures minimum overhead in index table maintenance.

5 IMPLEMENTATION DETAILS OF S-SCHEME

Buffer pool is organized as an array of fixed sized blocks and initially free pointer points to the beginning of the buffer pool. When an allocate request comes, the buffer manager will check whether enough free space is available and if available, free pointer is advanced by the size of the allocate request and the newly allocated block is returned to the calling routine.

5.1 Data Structures

1. Buffer : Buffer of user requested size. Number of blocks allocated = $\lceil$ size/BLOCK_SIZE $\rceil$
2. Index map of buffer : An array of integers, corresponding to the blocks in the buffer pool. An Index map entry holds -1 if the corresponding block is free, else holds the index value to which it is attached.
3. Index table : A table of structures. The structure contains
   (a) Number of contiguous blocks attached to the corresponding table entry.
   (b) Block pointer to the first of these blocks.
   (c) Links to a free list. This list is required because index entries can be released in arbitrary order.
4. Current_index is the head of the free list of indices.
5. Current_buffer_block is the pointer to the next free area in the shared memory.

5.2 Initialization

1. Creation of shared buffer space.
2. Initialization of Index map.
3. Initialization of Index table. The Index table is initialized as follows : For each entry :
   (a) No. of blocks = 0.
   (b) Block_ptr = -1.
   (c) Indices are chained circularly. Initially they are in sequential order.
   (d) Current index is set to 0.

The initialized data structures are shown in Figure 5 and initial buffer state is shown in Figure 6.

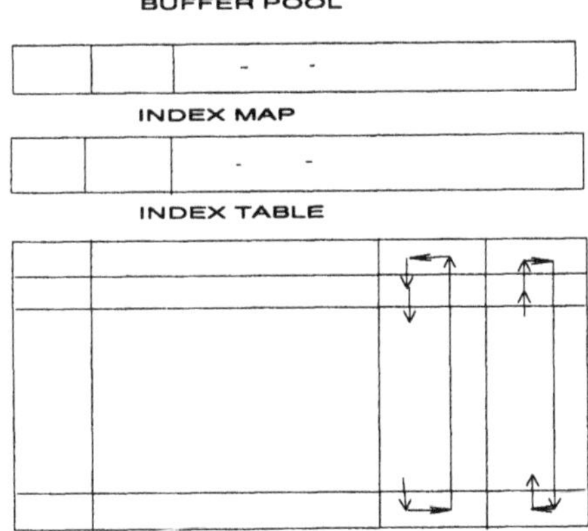

Figure 5 Initialized Data Structures.

5.3 Allocation of buffer

When a request to allocate a buffer of size SIZE is made, if the ***end run*** (the last run of free blocks) is greater than or equal to SIZE,

1. Blocks are allocated.
2. An index entry is removed from the free index list and values are entered for the number of blocks and block pointers.

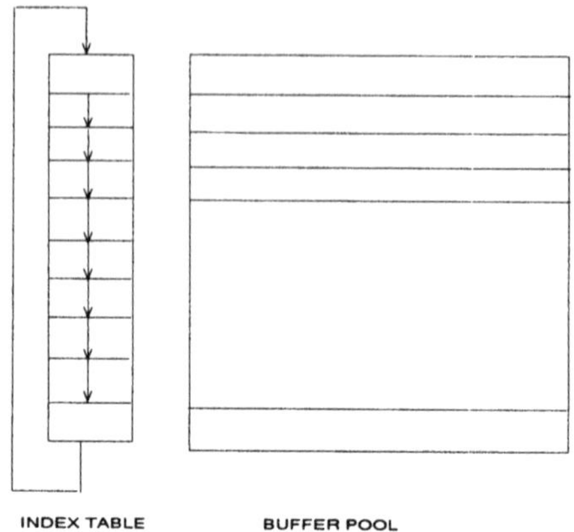

Figure 6 Initial Buffer Pool State.

3. The corresponding index map values for blocks are set to the index entry values.

4. Index entry value is returned and current index set to the next free index.

If the end run is less than SIZE and the buffer is not already compacted, the buffer is compacted and the end run is compared to SIZE again. If the end run is still less than SIZE, a REQUEST_OVERFLOW is reported. The final buffer state after some allocations and deallocations are shown in Figure 7.

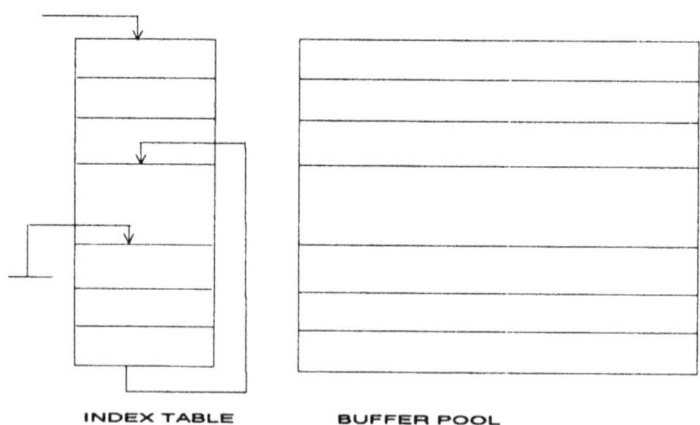

Figure 7 Final Buffer Pool State.

5.4 Freeing of buffer

To free a run of blocks, the index value is passed to the function *free_buffer*.

1. The run of buffer blocks is identified because the block pointer and the number of blocks are available.
2. The corresponding index_map entries are set to -1.
3. The index entry is added on to the free list.
4. If the run of blocks was the last occupied run, the current buffer block is set back by no_of_blocks, to increase the end run.

Because freeing blocks may occur anywhere, as per request, the buffer may be fragmented over time. This creates an artificial scarcity of space. When a request for buffer blocks cannot be granted, the buffer manager resorts to compaction. This physically moves the blocks and this is the reason for the added level of indirection. The fragmentation in this implementation is taken care of by compaction ***on demand***. This corresponds to strategy *S2* in our implementation.

6 ANALYSIS OF COMPACTION STRATEGIES

In this section, three compaction strategies are investigated and analyzed. Variations of these strategies are introduced at the end of this section. The trade-off between less space consuming strategies that require compaction and more space consuming strategies with no compaction is presented. Compaction coalesces all the free runs into one contiguous end run. This is an $O(N)$ operation where N is the number of busy blocks in the buffer pool. The different compaction strategies analyzed are:

S1 : When a sequence of blocks is deallocated, compaction is performed.

S2 : When a sequence of blocks is deallocated, no compaction is performed. When a buffer request requires a run of blocks of the size greater than the end run, then compaction is done.

S3 : When a buffer (consists of a run of blocks) is deallocated, if the length of the last run of assigned blocks is less than the length of the deallocated run, then this last run of assigned blocks is fitted into the deallocated run and the end run is increased. This operation is recursively done on the remaining run of blocks. This process is called ***stuffing***. When a buffer request requires a run of blocks of size greater than the size of the end run, a compaction is done.

C(S1)	deallocation	0
	compaction	*O(N)*
	allocation	$\sum_{i=1}^{n}[p(i) \times O(i)] + \sum_{i=n+1}^{M} p(i) \times F$
C(S2)	deallocation	0
	if size <= f allocation	$\sum_{i=1}^{f}[p(i) \times O(i)]$
	if size > f compaction	$\sum_{i=f+1}^{n}[O(B-n) + O(i)] \times p(i)$
	+ allocation	$+\sum_{i=n+1}^{M}[O(B-n) + F] \times p(i)$
C(S3)	deallocation	0
	stuffing	O(k)
	if size <= $f+k$ allocation	$\sum_{i=1}^{f+k}[O(i) \times p(i)]$
	if size > $f+k$ compaction	$\sum_{i=f+k+1}^{n}[O(i) + O(B-n)] \times p(i)]$
	+ allocation	$+\sum_{i=n+1}^{M}[O(B-n) + F] \times p(i)$
where,		
Cost of request overflow		F
Probability of the request of size i		$p(i)$
Maximum request size		M
Buffer pool size in blocks		B
Number of filled blocks after deleted run		N
Total number of free run		n
End run size		f
Number of block copies for stuffing		k

Table 2 Comparison of the three strategies.

The cost of performing a deallocation of a run of blocks followed by an allocation of a run of blocks is now evaluated for the three strategies. The significant operation in the analysis of the cost of a deallocation followed by an allocation is a block copy. We denote the cost of one deallocation followed by one allocation in scheme S by *C(S)*. Table 2 gives the cost of the buffer management overhead for the three strategies *S1, S2* and *S3*. Therefore,

$$C(S1) = \sum_{i=1}^{n}[O(i) \times p(i)] + \sum_{i=n+1}^{M}[F \times p(i)] + O(N) \quad (1)$$

$$C(S2) = \sum_{i=1}^{n}[O(i) \times p(i)] + \sum_{i=n+1}^{M}[F \times p(i)] + \sum_{i=f+1}^{M}[O(B-n) \times p(i)] \quad (2)$$

$$C(S3) = \sum_{i=1}^{n}[O(i) \times p(i)] + \sum_{i=n+1}^{M}[F \times p(i)] + \sum_{i=f+k+1}^{M}[O(B-n) \times p(i)] + O(k) \quad (3)$$

Let $\mathcal{F} = \sum_{i=1}^{n}[O(i) \times p(i)] + \sum_{i=n+1}^{M}[F \times p(i)]$. Physically, $\mathcal{F}$ is composed of two components, *allocation cost* and *Failure cost.* It may be noted that $\mathcal{F}$ is common to all the strategies.

The extra buffer management costs involved in each strategy are :

- *O(N)* in *S1* for compaction done always.
- $O(B-n)$ in *S2* for compaction done on demand.
- $O(B-n)+O(k)$ in *S3* for compaction done on demand and stuffing done always.

Therefore,

$C(S1) = \mathcal{F} + O(N)$

$C(S2) = \mathcal{F} + \sum_{i=f+1}^{M}[O(B-n) \times p(i)]$

$C(S3) = \mathcal{F} + \sum_{i=f+k+1}^{M}[O(B-n) \times p(i)] + O(k)$

We now analyze the conditions under which each strategy turns out to be the most cost effective one.

Theorem 1 *When the probability distribution, p(i) is uniform:*

1. *under light or moderate load conditions S2 is better than S1.*
2. *under light or moderate load conditions S2 is better than S3 and under heavy load conditions S3 is better.*

Proof: $C(S1) > C(S2)$ if $O(N) > \sum_{i=f+1}^{M}[O(B-n) \times p(i)]$
that is $C(S1) > C(S2)$ if

$$O(N) > O(B-n) \times (M-f)/M \tag{4}$$

Under light or moderate load, $f > M$. So *S1* is more expensive than *S2*. Under heavy load, f is comparable to M. So *S1* and *S2* are comparable, *S1* being positionally dependent on the deallocated run of blocks.

$C(S2) > C(S3)$ if

$\sum_{i=f+1}^{M}[O(B-n) \times p(i)] > O(k) + \sum_{i=f+k+1}^{M}[O(B-n) \times p(i)]$ that is

$$\sum_{i=f+1}^{f+k}[O(B-n) \times p(i)] > O(k) \tag{5}$$

that is $(B-n) > M$.

Under heavy load n is negligible. So *S2* is more expensive than *S3*. Under light or moderate load, *S3* is more expensive.

Theorem 2 ***When probability distribution, p is peaked at M/2 (Peaked bell shaped curve with no skewness):***

1. ***under light or moderate load S1 is more expensive than S2 and under heavy load, S1 is better than S2.***

2. ***under light or moderate load S3 is more expensive than S2 and under heavy load both S2 and S3 are comparable.***

Proof : Under light or moderate load, $f > M$. So *S1* is more expensive than *S2*. Under heavy load $f > M/2$. Hence *S1* is better than *S2*.

E(k) = M/4 , where E is the expectation function.

$C(S2) > C(S3)$ if

$\sum_{f+1}^{f+M/4}[(B-n) \times p(i)] > M/4$

Under light or moderate load, $f > M/2$. So *S3* is more expensive than *S2*. Under heavy load, *f* is comparable to *M*. So *S3* and *S2* are comparable.

From the above analysis, the conclusions reached are : When probability distribution is uniform, under *heavy* load, *S3* is most cost effective. Under *light* load, *S2* is most cost effective. When probability distribution is a peaked bell shaped curve with no skewness, (peaked at M/2), under *light* to *moderate* load, *S2* is most cost effective. Under *heavy* load, all strategies are comparable. These results hold for a deallocation followed by an allocation. A variant of *S3* is the modification of the *best fit* algorithm. In this algorithm, a list of fragmented free runs, ordered in descending size is maintained. When a run of blocks is deleted, the strategy *S3* is applied on the largest run in the ordered list. The remaining fragment is then inserted *in order* in the list. This reordering cost is insignificant when compared to the significant operation of block copy. We now provide a spatial and temporal analysis of the three strategies. We divide time into discrete intervals and consider p_a and p_d to be the probability of an allocate request and a deallocate request in a time interval, respectively. The probability of a deallocate request before an allocate request is :

$P = (1-p_a)^i * (1-p_d)^{(i-1)} * p_d$

Let the state of the system be denoted by the triple $<$ *work done, length of end run, total free space* $>$. Initial state of the system is $< 0, B, B >$ where B is the total shared memory space. After i releases, the state of the system corresponding to each strategy is :

S1 : $< O(n_1 + n_2 + .. + n_i), f + n_1 + .. + n_i, f + n_1 + .. + n_i >$

S2 : $< 0, f, f + n_1 + .. + n_i >$

S3 : $< O(k_1 + .. + k_i), f + k_1 + .. + k_i, f + n_1 + .. + n_i >$

where n_i denotes the ith run of deleted blocks and k_i denotes the ith run of stuffed blocks. Let the work done for allocation with end run of f and free space of n be $C(f, n)$. Then, work done for one allocation after i deallocations is,

S1 : $\sum_{j=1}^{i} O(N_j) + C(f + \sum_{j=1}^{i} n_j, f + \sum_{j=1}^{i} n_j) = C_{S1}(i)$.

S2 : $C(f, f + \sum_{j=1}^{i} n_j) = C_{S2}(i)$.

S3 : $O(k_j) + C(f + \sum_{j=1}^{i} k_j, f + \sum_{j=1}^{i} n_j) = C_{S3}(i)$.

where $C_{si}(i)$ denotes the work done for one allocation after i deallocations. The probability of the occurrence of such an event is :

$P^i * (1 - P) * C_{Sn}(i)$, where n = 1,2,3.

This is the generalized mathematical formulation of the compaction strategies. We carried out simulation studies of the three strategies under various conditions of load, mean packet size and shared memory size. We assumed a poisson arrival rate for request of allocation, an exponential distribution for the duration that a PDU is in the shared memory and an exponential distribution for packet size. We find that the simulation results generalize the results obtained from our restricted analysis of the case of a deallocation followed by an allocation. The work done in memory management due to each of the strategies is measured in terms of the number of block copies done per blocks allocated. The details of the simulation experiments are described in the next section. A strategy that totally eliminates block copies can also be implemented. In this strategy, every block has associated with it, a structure that contains the index with which the block is associated, and a pointer to the next block. When a sequence of 'n' blocks is requested, the free list is traversed down n links, and the corresponding blocks are allocated. This strategy involves an *O(B)* space overhead, where B is the size of the shared memory, and an *O(n)* time overhead in terms of pointer traversals. It however, does away with compaction. The principal advantage of splitting the shared memory into blocks is the reduction of the size of the index map and speeding up of memory management. It is based on the *principle of segmentation*, where resources of the system are divided into units, the size of which depends on the application and the particular resource. The disadvantage of this approach is a potential wastage of space in the form of internal fragmentation. A large buffer would speed up compaction and facilitate buffer management but could cause more internal fragmentation, while a very small block defeats the purpose of this scheme. We now derive the optimal block size B, for a given maximum packet request size P, and a given probability distribution p.

Theorem 3 *An optimal block size, B can be obtained form the equation* $d/dB(F(B) + C(E(w))) = 0$ *where* F(B) *is the cost of block operations for a size B,* $E(w) = (\lceil(i/B)\rceil \times B - i)/\lceil(i/B)\rceil \times B \times p(i)$, *the expected fraction of wastage and* $C(x)$ *is the cost of buffer of space* x.

Proof: If the block size B is increased then $F(B)$, the cost of block operations will decrease and $C(E(w))$, the cost of internal fragmentation will increase. The total cost is the sum of the above two factors and hence to minimize the total cost we have to solve the equation d/db $(F(B) + C(E(w))) = 0$, the cost of space of a block B be C(B). Under the conditions of uniform probability of packet request size,

$p(i) = 1/P$,

$E(w) = (1/P) \times_1^p \Sigma_{(n-1)B+1}^{nB} \Sigma(nB - i)/B$

7 SIMULATION RESULTS

The buffer management overhead variation with reference to three important simulation parameters, PDU size, buffer pool size and Residence time of the PDU in the buffer pool has been studied. Simulation was carried out in two steps. In the first step effect of varying PDU size is studied by keeping the buffer pool size constant where as in the second step, the effect of varying buffer pool size is studied by keeping the mean PDU size constant.

In Figure 8 (a) the PDU size is varied from 32 bytes to 1KB keeping the buffer pool size at 512 bytes. We assumed Poisson distribution for the arrival rate with a mean of $\lambda = 15$ and negative exponential distribution for the residence time with a mean of $\mu = 10$. From the figure we can see that out of the 3 methods *S2* has a clear advantage. When the PDU size is in the range 32 bytes to 64 bytes, there is a gradual increase in the overhead for *S2* and *S3*. Overhead of *S2* remains almost stable when the PDU size is in the range 64 bytes to 256 bytes. As the average PDU size is greater than or equal to 512 bytes, for all strategies overhead decreases dramatically well below 0.2. This is because the Buffer Manager is unable to satisfy most of the requests and hence the actual work done is quite small. Even at a low PDU size of 32 bytes there is a high failure probability of around 0.7. This indicates that even at light loads buffer pool size of 512 bytes is inadequate.

Next a series of simulation runs were conducted by varying the parameter buffer pool size to 2048 bytes, 4096 bytes, 8192 bytes, 16384 bytes and 65536 bytes. This is shown in Figure 8 (b), Figure 9 (b), Figure 10 (a), Figure 10 (b) and Figure 11 (a). When the average PDU size is 32 bytes the failure probability is as low as 0.3 (Figure 8 (a)). At this conditions overheads at a low PDU residence time are 0.47, 0.12 and 0.23 respectively for *S1*, *S2* and *S3*. Total buffer pool size is increased to 4096 bytes and the results are plotted in Figure 9 (b). When the PDU size is 32 bytes failure probability is as low as zero. In the second step of experiments the mean PDU size is kept constant and buffer pool size vs overhead is plotted for μ =15, 20, 25 and a mean arrival rate of λ = 15. (Figure 12 (a)). In the case of *S1* overhead is almost constant and well above *S2* and *S3*. In Figure 12 (a) as the buffer size = 16385 bytes for a mean PDU size of 32 bytes, overhead becomes zero. Here *S1* has a constant overhead while for *S2* and *S3* overhead decreases as buffer size increases. Comparing this results with PDU size of 128 bytes and 256 bytes, we can see that as the PDU size is small the gap between *S1* with the other two strategies become large. From Figure 12 (b) and 13 (a), we can see that all the three curves diverges from a point as we go away from origin along X axis. The conclusion drawn from the above observations are that *'Under heavy load conditions all the three strategies become comparable'*.

The simulation results confirm our conclusions reached in section 6. The conclusions drawn from the above analysis are

1. *Under light or moderate load conditions S2 is the most cost effective strategy .*
2. *Under heavy load conditions all the three strategies become comparable.*
3. *Different combinations of the values of the three simulation parameters Buffer pool size, PDU size, and Residence time has no effect on the conclusions 1 and 2 above.*

8 CONCLUSION

In this paper, various shared memory schemes have been investigated and their relative merits and demerits under various conditions of load, shared memory size and maximum packet size have been presented. The various strategies of memory management can be characterized by allocation of buffer space, deallocation of buffer space and their compaction strategies.

Under different conditions of load and shared memory size, different strategies prove to be optimal. However, since strategies *S1, S2* and *S3* differ only in their method of compaction of the fragmented shared memory, depending on the load condition, it is possible to choose the optimal one dynamically. Optimal performance may be facilitated by allowing decisions to be taken at the application level.

References

[Clark82] Clark D.D. Modularity and Efficiency in Protocol Implementation. *Internet Protocol Implementation Guide Network Information Center SRI International Menlo Park, California*, August 1982.

[G.S.Poo91] G.S. Poo, W.Ang. Cut-Through Buffer Management Technique for OSI Protocol Stack. *Computer Communications*, 14(3):166 – 177, April 1991.

[K.A.Lanz et al.85] K.A.Lanz, W.I.Nowiki, M.M.Theimer. An Empirical Study of Distributed Application Performance. *IEEE Transactions on Communications*, SE-11(10):1162 – 1173, October 1985.

[Peterson et al.85] Peterson J.L, Slibershatz. Operating System Concepts. *Addison-Wesley Publishing company*, 2nd Edition, 1985.

[Raghavan90] Raghavan S.V. *Solutions to Local Area Networks, The Indian Context.* Tata McGraw-Hill Publishing Company, 1990.

[Watson et al.87] R.W. Watson, S.A. Mamark. Gaining Efficiency in Transport Services by Appropriate Design and Implementation Choices. *ACM Transactions on Computer Systems*, 5(2):97 – 120, May 1987.

[Saltzer et al.84] Saltzer J.H., D.P.Reid, D.D.Clark. End to End Arguments in System Design. *ACM Transactions on Computing Systems*, 2(4):277 – 288, November 1984.

[Stallings84] Stallings W. Local Networks. *ACM Computing Surveys*, 16(1):3 – 42, March 1984.

[Tanenbaum85] Tanenbaum A. and R.V Reness. Distributed Operating Systems. *ACM Computing Surveys*, 17(4):479 – 490, December 1985.

[Zwaenopoel85] Zwaenopoel W. Protocol for Large Data Transfers over Local Networks. *Proceedings of the 9th Data Communication Symposium, Whistler Mountain, British Columbia, Canada, ACM, New York*, pages 22 – 32, 1985.

8 AUTHOR INFORMATION

S.V. Raghavan is on the faculty of the Department of Computer Science and Engineering, Indian Institute of Technology, Madras. He is also the Chief Investigator of the project on *Education and Research in Computer Networking* jointly sponsered by the Department of Electronics and Telecommunication Engineers. He is presently serving on the

Board of Editors of the journal of IETE for computers and control. He is also a member of the Editorial Advisory Borad for Computer Communications, Butterworth-Heinemann Ltd. His research interests are networks, protocols, multimedia systems and performance.

S Krishnamoney received his B.Tech degree in Computer Science from the Government Engineering College Trichur India in 1991 and is currently working towards his MS degree at the Department of Computer Science, Indian Institute of Technology Madras, India. His main research interests presently lie in computer networks and distributed systems.

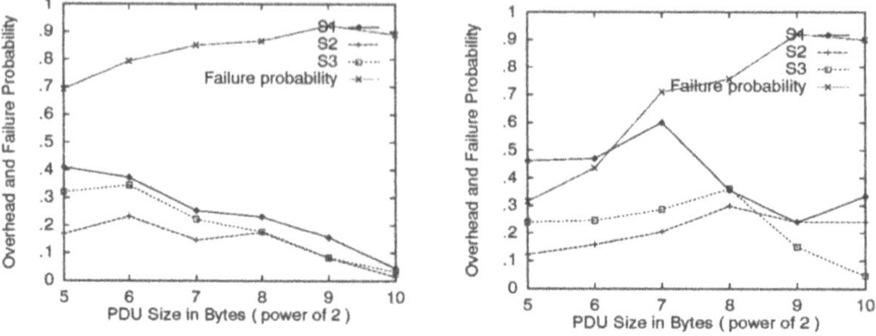

Figure 8: (a) Buffer pool size = 512 bytes $\lambda = 15$ $\mu = 10$ (b) Buffer pool size = 2048 bytes $\lambda = 15$ $\mu = 15$.

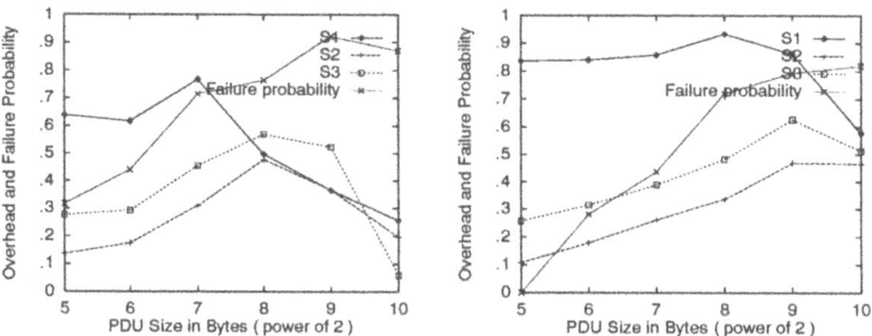

Figure 9: (a) Buffer pool size = 2048 bytes $\lambda = 15$ $\mu = 25$ (b) Buffer pool size = 4096 bytes $\lambda = 15$ $\mu = 25$.

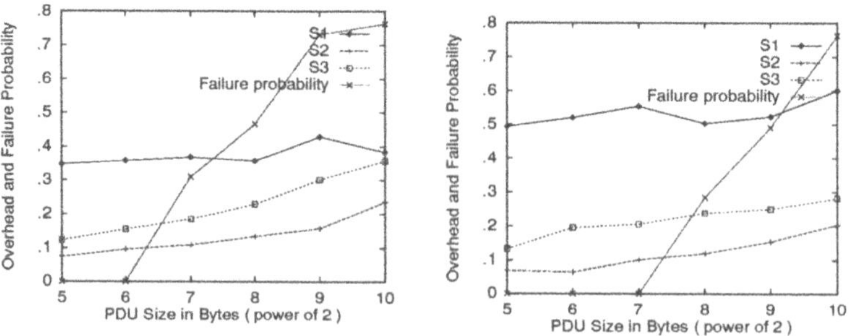

Figure 10: (a) Buffer pool size = 8192 bytes $\lambda = 15$ $\mu = 10$ (b) Buffer pool size = 16384 bytes $\lambda = 15$ $\mu = 15$.

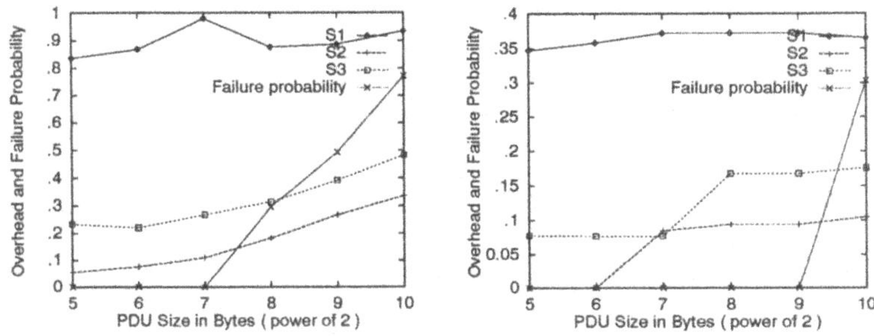

Figure 11: (a) Buffer pool size = 16384 bytes $\lambda = 15$ $\mu = 25$ (b) Buffer pool size = 65536 bytes $\lambda = 15$ $\mu = 10$.

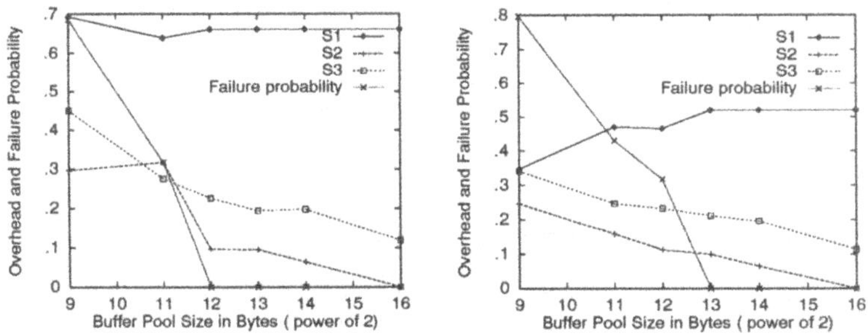

Figure 12: (a) Average PDU size = 32 bytes $\lambda = 15$ $\mu = 20$ (b) Average PDU size = 64 bytes $\lambda = 15$ $\mu = 15$.

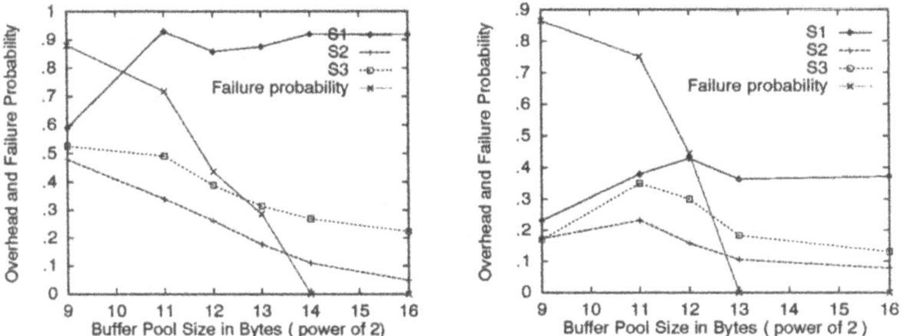

Figure 13: (a) Average PDU size = 128 bytes $\lambda = 15$ $\mu = 25$ (b)Average PDU size = 128 bytes $\lambda = 15$ $\mu = 10$.

17

On the Performance of the Unslotted CDMA-ALOHA Access Protocol for Finite Number of Users, With and Without Code Sharing

A. Kumar and D. Patil
Dept. of Electrical Communication Engineering
Indian Institute of Science
Bangalore - 560 012, INDIA
e-mail: anurag@ece.iisc.ernet.in

Abstract

We consider a system comprising a finite number of nodes, with infinite packet buffers, that use unslotted ALOHA with Code Division Multiple Access (CDMA) to share a channel for transmitting packetised data. We propose a simple model for packet transmission and retransmission at each node, and show that *saturation throughput* in this model yields a sufficient condition for the stability of the packet buffers; we interpret this as the *capacity* of the access method. We calculate and compare the capacities of CDMA-ALOHA (with and without code sharing) and TDMA-ALOHA; we also consider carrier sensing and collision detection versions of these protocols. In each case, saturation throughput can be obtained via analysis of a continuous time Markov chain. Finally, we also present some simulation results for mean packet delay.

Keywords

Optical CDMA, CDMA vs. TDMA, stability of multiple access protocols

1 INTRODUCTION

Direct Sequence Code Division Multiple Access (DS-CDMA, or just CDMA) is a technique for sharing the bandwidth of a channel among several sources of digital bit streams. In this paper we are concerned with the bandwidth efficiency, when the CDMA technique is used by several nodes to transmit packetised data. In particular, the question that we seek to answer is motivated by the following discussion.

Suppose that the sources that need to share the channel require an access rate (or burst rate) of $\frac{1}{T}$ bits per second (say, 10 Mb per sec), i.e., each information bit emanating from a source occupies T secs (e.g., 0.1 μsec). The channel to be shared has a bandwidth of B Hz. The space of signals with bandwidth B Hz, and time duration approximately T secs, has dimension approximately $2BT$ [Lee *et al.*, 1988]. In this signal space we need to choose L orthogonal signals that can be practically used for CDMA signalling. It has been found [Salehi 1989] that to do this, a signal space dimension (i.e., $2BT$) much larger than L is required. For example, $2BT = 100$ may be required for getting L=10 useful orthogonal spreading sequences. Here $2BT$ becomes the spreading factor for CDMA; thus for L=10, we need a channel bandwidth $B = \frac{100}{2T}$, and a CDMA "chip" rate of $\frac{100}{T}$ chips per sec. Now this channel of bandwidth B can be used for direct digital signalling at about $2B$ bits per second (by the Nyquist criterion, [Lee *et al.*, 1988]). In the running example we have taken, $2B = \frac{100}{T} = 1000$ Mb per sec, i.e., a direct signalling rate of 1 Gb per sec.

The question therefore arises: why not use this channel directly to implement a TDMA system at a bit rate of about $2B$ bits per sec? This paper is motivated mainly by this question, i.e., we wish to compare the performance of random multiaccess of a channel by packetised traffic, when the channel bandwidth is shared using CDMA or TDMA. As one of the advantages ascribed to CDMA is the ease of asynchronous implementation [Green, 1993], for a fair comparison, in both cases random access is assumed, i.e., we compare CDMA-ALOHA and TDMA-ALOHA. We consider *unslotted ALOHA*, without and with carrier sensing (CS) and/or collision detection (CD).

We consider TDMA-ALOHA to be a special case of CDMA-ALOHA, i.e., one in which no more than one successful transmission can exist at one time on the channel. On bandwidth limited channels we can, following the argument above, assume that the TDMA system can be operated at a bit rate equal to the CDMA chip rate. All optical CDMA ([Salehi 1989], [Prucnal *et al.*, 1986]), however, can run at a chip rate much higher than the rate at which light pulses can be modulated and demodulated by electronics. Hence, when considering all optical CDMA-ALOHA and TDMA-ALOHA, TDMA bit rate may be less than CDMA chip rate. Consequently, in our analysis we normalise all times to packet transmission times, thus eliminating the factor of bit transmission rate. The transmission rate is factored back when making comparisons between CDMA and TDMA.

We consider the situation in which a finite number of nodes, with infinite packet buffers, share the channel. A simple model for attempts and retransmissions is used. The main contribution of this paper is the saturation throughput analysis with and without *code sharing* (i.e., more than one node transmits using the same spreading code), and with and without carrier sensing and collision detection. We show that saturation throughput yields a *sufficient* stability condition for the queues. Hence saturation throughput may be considered to be the *capacity* of the multiple access system, when there is symmetric loading of all the channels. We compare the capacities of various versions of CDMA-ALOHA and TDMA-ALOHA, including versions with collision detection and carrier sensing. Finally, we also show some results for mean packet delay, ob-

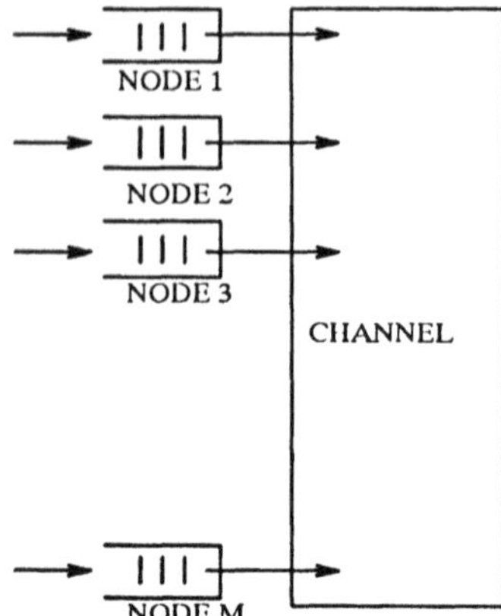

Figure 1: Model for a CDMA system with M users: The packets arriving at each user wait in a queue until they are transmitted on the channel.

tained using simulations.

There is a long history of analyses of random access systems. The work that is related to ours is the stability analysis of buffered finite user ALOHA (see [Sharma, 1990] for a list of references). The most general result for the stability of slotted ALOHA is presented in [Sharma, 1988]. See also [Szpankowski, 1990] for a survey of other related references. The result in [Sharma, 1988] basically asserts that saturation throughput yields a *sufficient* stability condition for finite user buffered slotted ALOHA. This result is easily extended to our model of unslotted CDMA-ALOHA.

In [Raychaudhari, 1981] the performance of slotted CDMA is considered. The number of nodes is finite but the queue length processes are not modelled. Throughput analysis is carried out for the case in which arrivals are blocked at backlogged nodes.

2 A MODEL FOR CDMA-ALOHA

There are M users and L CDMA codes, with each code being shared by $K = M/L$ users (it is assumed that K is an integer). The codes are assumed to be transmitter oriented, i.e., each user transmits on its own code and receives on all codes. Code-sharing (i.e., $K > 1$) reduces the complexity of the optical correlation receivers; we wish to study its impact on traffic performance. CDMA is interference limited [Lee *et al.*, 1988], and the bit error rate at the receiver deteriorates as the number of simultaneous transmissions on the channel increases. We assume that there is a desired bit error rate (e.g., the bit error rate obtainable by a TDMA system on the same channel with the same transmitter power). The desired bit error rate is achieved for a particular *interference limit* N; i.e., if more than N transmissions (involving one or more codes) take place simultaneously, they are all considered to be "bad".

The CDMA system can be very simply viewed as shown in Figure 1. Each node or user has a queue in which packets arrive and wait. The box representing "the channel" is a complex "server" that serves the head-of-the-line (HOL) packets in the queues. When a HOL packet from a user queue is transmitted, one of three things can happen to it:

1. It may be rendered "bad" owing to excessive interference if the number of active transmissions *at any time* during the transmission exceeds N.

2. It may collide with another packet transmitted by a user with the same CDMA code.

3. It may complete transmission successfully.

If case (i) or (ii) occurs, the packet is retransmitted after a random delay. The process repeats until the packet is successfully transmitted. The user then accesses the server (channel) for transmission of the next packet.

Thus transmissions from each user are either fresh transmissions or retransmissions in response to channel feedbacks. It is extremely complicated to work with exact models of the attempt, backoff and reattempt algorithms. To obtain an analysable model, we assume that when a node has a packet to send (and the packet is not already in transmission) the rate at which the HOL packet is attempted is α. There is thus a "refractory" period between successive packet attempts. This period models the channel feedback delay, and the retransmission back-off delay. Further, we assume that the refractory period is exponentially distributed with mean $1/\alpha$. All times are normalised to packet transmission time, which is assumed to be exponentially distributed with mean 1. Hence when a user has a packet to send its behaviour is modelled by the 2 state continuous time Markov chain (CTMC) shown in Figure 2; state 0 is the refractory state between attempts and state 1 is the transmitting state. Note that our model assumes that a packet arriving to an empty queue also experiences the refractory period delay (incidentally, the same phenomenon occurs in probabilistic attempt models for slotted ALOHA). While such a delay may actually occur occasionally, as the node may be waiting for a previous acknowledgement, in general it renders the model somewhat conservative at light loads. At heavier loads, however, the model is quite appropriate.

Since each receiver has correlation receivers for all the transmitting codes, it is possible to implement carrier sensing and collision detection in the following way. Each receiver monitors the output of all the correlation receivers. If the number of active outputs is N or more, then transmission from this node is deferred; this is carrier sensing. In spite of carrier sensing, owing to propagation delay, excessive interference owing to more than N active transmissions can still occur. If each node monitors the channel while transmitting, then it can detect a collision if more than N receiver correlator outputs become active. It can then abort transmission. We only consider carrier sensing (CS) and collision detection (CD) for the case $L = M$ (i.e., $K = 1$, no code-sharing).

Clearly, propagation delay must be modelled when analysing the carrier sensing and collision detection alternatives. We model the effect of propagation delay approximately as follows. When the number of active transmissions

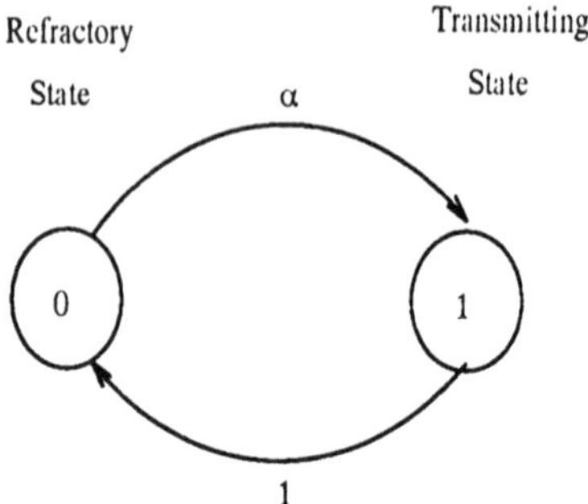

Figure 2: Channel attempt model for each user.

Table 1: Model parameters for the various access protocols; M is the number of nodes

	L	N
TDMA	1	1
CDMA, no code-sharing	M	> 1
CDMA, code-sharing	$< M$	> 1

becomes N then the system is said to be in a "vulnerable" state, as another transmission would corrupt all ongoing transmissions. We assume that after entering such a state, the system enters the safe state (in which all nodes have sensed the ongoing transmissions) after an exponentially distributed propagation delay with mean τ^{-1}; this models CS. Similarly when more than N transmissions are ongoing and hence are all corrupted, all transmissions get aborted after a propagation delay that is exponentially distributed with mean τ^{-1}; this models CD.

Note that, since we have normalised the rates to the packet transmission rate, the throughput will be obtained in packets per packet transmission time, and hence the actual transmission speeds need to be explicitly considered only when comparing TDMA and CDMA. Further, it is clear that from the point of view of our model, TDMA is just a special case of CDMA, obtained by setting $L = 1$ or $K = M$. Table 1 shows the values of L and N that yield the various access protocols that we wish to study.

3 SATURATION THROUGHPUT: THEORY AND ANALYSIS

For the queueing system of the type shown in Figure 1, saturation throughput is an important performance measure. Saturation throughput is defined as the channel throughput when the nodes always have a packet to send, i.e., the queues are *saturated.* In certain situations, saturation throughput has been shown to yield a sufficient stability condition for the queues.

The saturation condition yields the following model. Each user behaves as shown in Figure 2. When carrier sensing and collision detection is not being used then users *independently* alternate between the refractory state and the transmitting state. The saturation throughput of the system is accounted depending on the values of L and N. If during the transmission of a user, another user with the same code initiates transmission, then both users' packets are corrupted and they are counted as unsuccessful transmissions. If during the transmission of a user the total number of transmissions (of however many codes) exceeds N, then all participating transmissions are assumed to be unsuccessful. If neither of these events occur during the transmission of a packet, then the packet is counted towards channel (good) throughput. The overall throughput of good packets is then the saturation throughput. A two dimensional Markov chain can be defined for this model and its analysis yields the saturation throughput for CDMA without code-sharing, without and with CD and CS. We provide this analysis in Section 3.2. In Section 3.3 we provide a similar analysis for CDMA with code-sharing.

3.1 Significance of Saturation Throughput

We show, for CDMA-ALOHA, that, as may be expected, saturation throughput yields a sufficient stability condition for the packet buffers at the various nodes. We present detailed arguments for CDMA-ALOHA without code-sharing, or CD and CS. We use arguments similar to those used in [Sharma, 1988] and [Rao *et al.*, 1988].

We construct a "virtual attempt" process as follows. Consider M independent Poisson processes (indexed by $1 \leq j \leq M$) with rate $1 + \alpha$ and epochs $\{T_0^{(j)} = 0, T_1^{(j)}, T_2^{(j)}, \ldots,\}$. Associate with the interval $\left[T_{k-1}^{(j)}, T_k^{(j)}\right)$ $(k \geq 1)$ of the jth Poisson process the random variable $B_k^{(j)} \in \{0,1\}$. $B_0^{(j)} = 0$, and $\left\{B_k^{(j)}, k \geq 2\right\}$ is a Bernoulli sequence, $B_k^{(j)} = 0$ with probability $(1+\alpha)^{-1}$ and $B_k^{(j)} = 1$ with probability $\alpha(1+\alpha)^{-1}$, and the sequences $\left\{B_k^{(j)}, k \geq 1\right\}$, $1 \leq j \leq N$, are independent.

Define the random sequences:

$\left\{A_k^{(j)}, k \geq 0\right\}$: $A_0^{(j)} = 0$, and, for $k \geq 1$, $A_k^{(j)}$ is the number of arrivals to the jth node in the interval $[T_{k-1}, T_k)$.

$\left\{X_k^{(j)}, k \geq 0\right\}$: $X_k^{(j)}$ is the queue length in the jth queue at the epoch T_k.

$\left\{\overline{S}_k^{(j)}, 1 \leq j \leq M, k \geq 1\right\}$: $\overline{S}_k^{(j)} = 1$ if $B_k^{(j)} = 1, B_{k+1}^{(j)} = 0$, and if during the run of ones, of which $B_k^{(j)} = 1$ is the last, at most $N-1$ of the $\left\{B_k^{(i)}, i \neq j, k \geq 1\right\}$ processes were *ever* nonzero; otherwise $S_k^{(j)} = 0$

$\left\{S_k^{(j)}, 1 \leq j \leq M \quad k \geq 1\right\}$: $S_k^{(j)} = 1$ if a successful transmission attempt completes at $T_k^{(j)}$ in the actual queuing model; otherwise $S_k^{(j)} = 0$.

Now observe that owing to the assumptions that the time to attempt and the transmission time are exponentially distributed with rates α and 1, we have, for $k \geq 1$,

$$X_k^{(j)} = \left(X_{k-1}^{(j)} + A_{k-1}^{(j)} - S_k^{(j)}\right)^+$$

The only "tricky" point here is to verify that this equation is correct when an arrival joins an empty queue at a node. Suppose this arrival is to the jth queue and arrives in the interval $\left[T_{k-1}^{(j)}, T_k^{(j)}\right)$. The earliest that it can get transmitted is the interval $\left[T_k^{(j)}, T_{k+1}^{(j)}\right)$. Thus it would already have waited for an amount of time distributed as exponential $(1+\alpha)$. It is attempted in the next interval with probability $\alpha(1+\alpha)^{-1}$, and not attempted with probability $(1+\alpha)^{-1}$. Hence the time until it is first attempted is distributed as exponential α.

Observe now, from the definitions above, that $S_k^{(j)} \geq \overline{S}_k^{(j)}, \forall j, 1 \leq j \leq M$, $k \geq 1$, since, in the actual system, there is always a successful attempt that ends at an epoch at which $\overline{S}_k^{(j)} = 1$, but there can be successful attempts even when $\overline{S}_k^{(j)} = 0$. Using these bounds and $X_0^{(j)} = 0,\ 1 \leq j \leq M$ we obtain

$$X_k^{(j)} \leq \left(\max_{1 \leq \ell \leq k} \left(\sum_{i=\ell}^{k} \left(A_{i-1}^{(j)} - \overline{S}_i^{(j)}\right)\right)^+$$

It then easily follows (see arguments in [Sharma, 1988], [Borovkov, 1976, pg.12]) that if the arrival processes $\{A_k^{(j)}\}$ are asymptotically stationary, and the "virtual service" processes $\{\overline{S}_k^{(j)}\}$ are asymptotically stationary, then $\{X_k^{(j)}\}$ converge in distribution to proper random variables provided that for each j, $1 \leq j \leq M$, the arrival rate is less than $(1+\alpha)\left\{\lim_{n\to\infty} \frac{1}{N} \sum_{k=1}^{N} \overline{S}_k^{(j)}\right\}$; but this latter limit is just the saturation throughput. The asymptotic stationarity of $\{\overline{S}_k^{(j)}\}$ is clear from the Markov chain analysis in Section 3.2.

3.2 Analysis for CDMA Without Code-Sharing

In the case of CDMA-ALOHA without code-sharing, there can be no collisions, but interference can arise when the total number of ongoing transmissions exceeds N, in which case all the transmissions are considered bad. We first do not consider carrier sensing and collision detection.

Let $x(t)$ denote the total number of ongoing transmissions, and $y(t)$ denote the number of these that are good. Consider the process $(x(t), y(t))$. Observe that, owing to exponential assumptions, this is a Markov chain on a finite state space.

$$S = \{(x,y) : 0 \leq x \leq N, 0 \leq y \leq x\} \ \cup\ \{(x,0) : N+1 \leq x \leq M\}$$

For $y(t) > 0$, a transition from $(x(t), y(t))$ to $(x(t)-1, y(t)-1)$ is obtained on the completion of a good transmission, i.e., at a rate $y(t)$ (recall that packet transmission times are exponentially distributed with mean 1). Similarly, a bad transmission completion gives rise to an $(x(t), y(t))$ to $(x(t)-1, y(t))$ transition, at a rate $x(t) - y(t)$, and, if $x(t) < N$, an arrival gives rise to an $(x(t), y(t))$ to $(x(t)+1, y(t)+1)$ transition at rate $(M - x(t))\alpha$. If $x(t) = N$, an arrival will lead to a transition to $(N+1, 0)$ since all transmissions become bad. Once $x(t)$ exceeds N, good transmissions can be initiated only after the total number of transmissions eventually falls below N. If there is any transmission initiation from a state $(x(t), 0)$, where $M > x(t) > N$, then it will result in a transition to state $(x(t)+1,0)$. This transition will occur at a rate $(M-x(t))\alpha$. Transitions from $(x(t),0)$ to $((x(t)-1),0)$ occur at the rate $x(t)$.

For $\alpha > 0$, this is an irreducible finite state CTMC. Let π denote its stationary distribution. The throughput of successful packets in this case is given by:

$$g_{CDMA-ALOHA}(L,M,N,\alpha) = \sum_{1 \leq i \leq N, 1 < j \leq i} j\pi(i,j) \tag{1}$$

where j is the transition rate from state (i,j) to state $(i-1,j-1)$.

With collision detection, each transmitting user can detect that the number of transmissions has exceeded N, after a propagation delay distributed as $\exp(\tau)$ and it then aborts its own transmission. The packet is then reattempted after a delay distributed as $\exp(\alpha)$. The Markov chain is the same as in case 1 with the addition being that whenever $x(t)$ exceeds N, the system can go back to the state $(0,0)$ at the rate τ. The state space is the same as for case 1.

$$S = \{(x,y) : 0 \leq x \leq N, 0 \leq y \leq x\} \cup \{(x,0) : N+1 \leq x \leq M\}$$

The throughput of successful packets in this case is given by:

$$g_{CDMA-ALOHA-CD}(L,M,N,\alpha,\tau) = \sum_{1 \leq i \leq N, 1 < j \leq i} j\pi(i,j) \tag{2}$$

where j is the transition rate from state (i,j) to state $(i-1,j-1)$

We can further include carrier sensing in this analysis. Now every node senses the channel before transmitting. If there are N users already transmitting, it refrains from transmitting, since otherwise, all transmissions will be rendered bad. The interference limit N may still be exceeded since a node may not perceive N ongoing transmissions, due to the finite propagation delay.

Now there are 2 sets of states corresponding to $x(t) = N$, denoted by (N,y) and (N',y). A state (N,y) is an "unsafe" or "vulnerable" state, and is entered from $x(t) = N-1$ upon a transmission initiation. After a propagation delay distributed as $\exp(\tau)$, it is assumed that all users are able to sense that the maximum limit has been reached and the remaining nontransmitting users refrain from transmitting; this is a transition from the unsafe state (N,y) to the corresponding "safe" state (N',y). This transition occurs at the rate τ. From the safe states there are only transmission completions. Any transmission initiation while in the unsafe state will, however, result in causing all the good ongoing transmissions to also become bad. The state space of the Markov chain $(x(t),y(t))$ is now given by

$$S = \{(x,y) : 0 \leq x \leq N, 0 \leq y \leq x\} \cup \{(N',y) : 0 \leq y \leq N\} \cup \{(x,0) : N+1 \leq x \leq M\}$$

The throughput of successful packets in this case is given by:

$$g_{CDMA-ALOHA-CD/CS}(L,M,N,\alpha,\tau) = \sum_{1 \leq i \leq N, 1 < j \leq i} j\pi(i,j) + \sum_{1 < j \leq N} j\pi(N',j) \tag{3}$$

where j is the transition rate from state (i,j) to state $(i-1,j-1)$.

The saturation throughput results for the above three cases are given in the next section

Returning to the point of comparison with TDMA, an interesting observation can be made regarding the propagation delay. Let a be the propagation delay (between the furthest nodes in the network) normalised to the mean packet transmission time. Considering the running example in the Introduction, in which the CDMA bit rate is .01 times the TDMA bit rate, it is clear that for TDMA the factor a is 100 times the value of a for CDMA. With the numbers in that example in mind, while comparing throughput, we will compare the TDMA case with $a = 1$, with the CDMA case with $a = .01$. Recall that the propagation delay, though fixed in practical situations, is assumed to be distributed as exponential(τ), for simplification in the analysis. Note that a higher a will make the carrier sense and collision detection techniques less efficient.

3.3 A Reduced State Space Model for CDMA with Code-Sharing

In order to extend the CDMA analysis, to the case of CDMA with code-sharing, it appears necessary to keep track of the number of active users on each code. This will result in a Markov chain of higher dimensions and hence will complicate the analysis. For CDMA-ALOHA (without CS and CD) it is possible to work with a Markov chain that keeps track only of the transmissions of one code. This works since the problem is symmetric in the various codes.

Fix a particular code, and consider the Markov Chain $(x(t), y(t))$, where $x(t)$ is the total number of ongoing transmissions, while $y(t)$ is the number of ongoing transmissions of *that particular code.* A value $y(t) = 1$ indicates a good transmission, while a negative value $y(t) = -j$ indicates j bad ongoing transmissions using this code.

The state space is given by :

$$S = \{(x,1) : 1 \leq x \leq N\} \cup \{(x,y) : 0 \geq y \geq -K, -y \leq x \leq M - K - y\}$$

Note that only one transmission on each code can be good; as soon as there is another transmission on the same code by any of the remaining $K-1$ users, both become bad, i.e., a transition from $(x(t), 1)$ to $(x(t)+1, -2)$ occurs at a rate of $(K-1)\alpha$. If the transmission on this code is completed successfully, a transition from state $(x(t), 1)$ to state $(x(t)-1, 0)$ is obtained at a rate of 1 (since rates have been normalised to packet transmission time). If $y(t) = 1$ and there is a transmission completion on any other code, then a transition occurs to state $(x(t)-1, 1)$ at the rate $x(t)-1$. If $x(t) < N, y(t) = 1$, then a transmission initiation on any other code leads to state $(x(t)+1, 1)$ at the rate $(M-K-x(t)-1)\alpha$. If $x(t) = N, y(t) = 1$ then a transmission initiation on any other code leads to state $(N+1, -1)$ at the rate $(M-K-N-1)\alpha$. When the system is in a state $(x(t), y(t))$ such that $y(t) < 0$, four kinds of transitions can occur:

1. A transition to state $(x(t)+1, y(t))$ at a rate of $(M-K-x(t)-y(t))\alpha$ corresponding to a transmission initiation by a user on any other code.

2. A transition to state $(x(t)+1, y(t)-1)$ at a rate of $(K+y(t))\alpha$ corresponding to a transmission initiation by a user on this code.
3. A transition to state $(x(t)-1, y(t))$ at a rate of $x(t)+y(t)$ corresponding to the end of transmission of a user on any other code.
4. A transition to state $(x(t)-1, y(t)+1)$ at a rate of $-y(t)$ corresponding to the end of transmission of a user on this code.

When the system is in a state $(x(t), 0)$ with $x(t) < N$, a transmission initiation will cause a transition to state $(x(t)+1, 1)$ if it is on this particular code, else a transition to state $(x(t)+1, 0)$ is obtained. However, if $x(t)$ were more than N, then a transmission initiation on this code would give rise to a transition to state $(x(t)+1, -1)$.

The above chain also correctly captures the interdependence among the codes, while separating the model for each code. The channel throughput is given by :

$$g_{CDMA-ALOHA-Code-Sharing}(L, M, N, \alpha) = L \sum_{i=1}^{N} \pi(i, 1)$$

where $\pi(i, j)$ is the stationary probability vector, and the term involving the summation is the throughput per code.

Note that for $L = 1$, the above model will correspond to TDMA-ALOHA. Also, with $L = M$, it corresponds to CDMA ALOHA without code-sharing. It is expected that the results in this case should match exactly with the results obtained in Section 3.2. This is indeed true, as shown by the results in the next section.

4 SATURATION THROUGHPUT: NUMERICAL RESULTS

The results of the saturation throughput analysis are presented in Figures 3 through 8.

Each Markov chain is seen to be irreducible and is defined over a finite state space. Hence it is positive recurrent and there is a stationary probability distribution $\pi(i, j)$. The saturation throughputs have been obtained using the equations given in Sections 3.2 and 3.3. The computations were carried out with various values of $\alpha, M, N,$ and L.

4.1 CDMA Without Code-Sharing

Figure 3 shows the saturation throughput results for CDMA without code-sharing, for $N = 1$, and various values of M, the number of users. Recall that $N = 1$ corresponds to pure TDMA-ALOHA. Note that α is the attempt rate per node, and the curves are plotted against $M\alpha$ the aggregate attempt rate when all the nodes are in the refractory state, to enable comparison with the classical infinite user ALOHA model. Observe the peak of about 0.2 near $M\alpha = 0.5$. Note, however, that in this model $M\alpha$ is not the attempt rate on the channel.

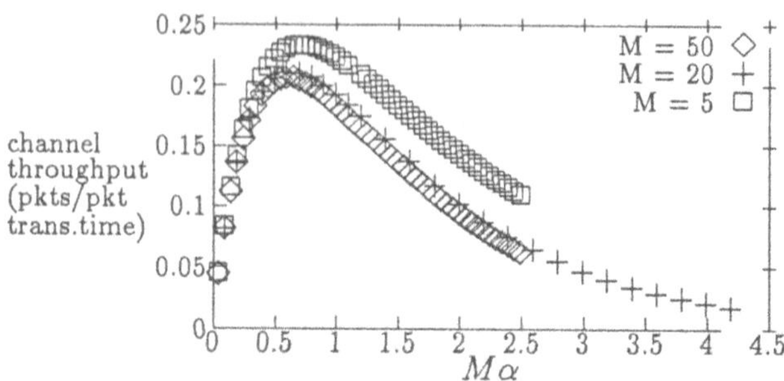

Figure 3: Saturation throughput analysis of unslotted finite user TDMA-ALOHA. Graph of total channel throughput (in packets per packet time) vs. $M\alpha$, for various values of M, the number of users.

Figure 4 shows saturation throughput results for CDMA without code-sharing, but with its collision detection and carrier sensing enhancements, for $M = 20, N = 5$, with $\tau = 200$ (normalised to packet transmission time). It is observed that the throughput increases progressively as collision detection and carrier sensing features are added. Although not shown in the curves, as α increases further, the CD and CS curves decrease owing to excessive collisions.

In Section 3.1 we have seen that saturation throughput yields a sufficient stability condition for the packet buffers. We interpret this saturation throughput as channel capacity. But then it follows that channel capacity depends on α, a parameter that models the backoff algorithm used by the nodes. In a real system, if adaptive backoff is used, it may be expected that α will be adapted on-line till throughput is maximised. The throughput will then correspond to the peak of the corresponding saturation throughput curve. It is therefore meaningful to compare the peaks of the saturation throughput curves.

From Figures 3 and 4 observe that for $M = 20$ (and $K = 1$) without CD or CS, TDMA-ALOHA has a peak normalised throughput of about 0.2, whereas CDMA-ALOHA has a peak normalised throughput of about 2 for $N = 5$, and 5 for $N = 10$ (see Figure 5). If CDMA chip rate is the same as the TDMA bit rate then for a spreading factor of 100 (say), TDMA-ALOHA has a channel capacity of 20 packets per CDMA packet transmission time, compared with, for CDMA, 2 for $N = 5$ and 5 for $N = 10$. On the other hand if CDMA chip rate can be higher than the maximum achievable TDMA bit rate (as may be the case with optical CDMA), then the channel capacity with CDMA-ALOHA can exceed that with TDMA-ALOHA.

Table 2 shows how the peak saturation throughput varies with M for the TDMA case with $\tau = 1, a = 1$ and $N = 1$, and for the CDMA case with $\tau = 100, a = .01$ and $N = 10$, both with CD and CS. For $M = 20$, the maximum throughput for CDMA ALOHA is about 8 packets per packet transmission time, while for TDMA ALOHA the maximum throughput is about 0.35 packets per packet transmission time. For TDMA bit rate equal to CDMA chip rate, and a CDMA spreading factor of 100, it follows that the ratio of CDMA throughput

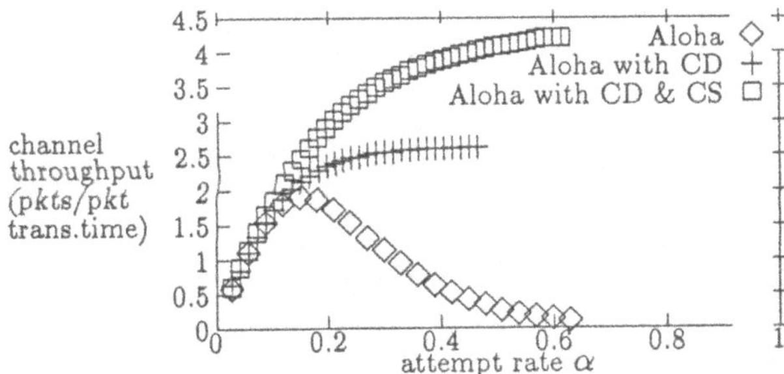

Figure 4: Saturation throughput analysis of unslotted finite user CDMA-ALOHA, CDMA-ALOHA with collision detection, and CDMA-ALOHA with collision detection and carrier sensing, for CDMA system without code-sharing. Graph of total channel throughput (in packets per packet time) vs. α, for $M = 20$ users, with interference limit $N = 5$, and $\tau = 200$ (normalised to packet transmission time).

to TDMA throughput is $\frac{8}{100 \times 0.35} \approx \frac{1}{4.5}$.

It may have seemed at the outset that since there are 10 codes, after taking into consideration the bandwidth expansion factor, CDMA would yield an aggregate throughput that is $(1/10)^{th}$ of the TDMA throughput; owing to the different impacts of ALOHA overheads, however, it turns out that the CDMA throughput is (1/4.5) of the TDMA throughput.

4.2 CDMA with code-sharing

Figure 5 shows the saturation throughput results for CDMA-ALOHA without code-sharing with $M = 20, N = 10$, obtained from the analysis in Section 3.2, and the total throughput for all codes for CDMA with code-sharing for $M = 20, N = 10, K = 1$, obtained from the analysis in Section 3.3. The throughput values match exactly thus showing the consistency of the two analyses.

Figure 6 shows the throughput per code versus the attempt rate α, for various values of K, the number of users sharing each code. Figure 7 shows the total throughput of the system versus the attempt rate α, for various values of K. The total throughput has been obtained simply by multiplying the throughput per code (as plotted in Figure 6), by the number of codes (L), since the system is symmetric in the various codes. The nature of the curve is observed to be similar to the usual ALOHA throughput curve. At low values of α the throughput increases with α; it peaks and then reduces with increasing attempt rate since the interference and collisions increase. For $K = 1$, the reduction in the throughput per code with α is solely due to the interference limit set by N. If there was no such limit, then the throughput would increase to 1 and saturate. At higher K, increasing α reduces throughput due to an increase in collisions, in addition to an increase in the probability of exceeding the interference limit N.

Table 2: Variation of maximum throughput with M, for TDMA with $\tau = 1$ & $N = 1$, and for CDMA with $\tau = 100$ & $N = 10$, considering the case of ALOHA with CS & CD

M	Maximum throughput	
	TDMA	CDMA
10	0.3581	10
11	0.3563	8.99
12	0.3555	8.72
13	0.3541	8.54
14	0.3537	8.40
15	0.3524	8.28
16	0.3519	8.19
17	0.3516	8.125
18	0.3510	8.06
19	0.3497	8.01
20	0.349	7.97

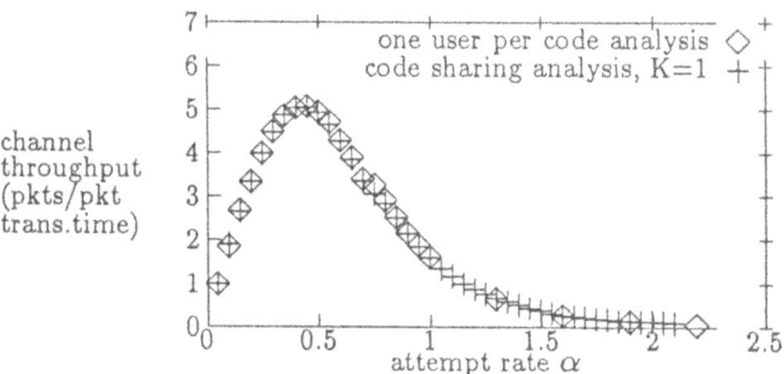

Figure 5: Saturation throughput analysis of unslotted CDMA-ALOHA without code-sharing. Comparison of results obtained from the analysis of the two Markov chain models, for $M = 20 and N = 10$.

Figure 7 shows that the value of α at which the curves peak reduces as K increases, i.e., the maximum permissible attempt rate per user reduces as the level of code-sharing increases. It is also observed that the system throughput decreases for any α as the level of code-sharing increases.

Finally Figure 8 again shows more explicitly the decrease in the total throughput when the level of code-sharing is increased.

5 PACKET DELAY: SIMULATION STUDY

The system was modelled as M stations connected to each other via L links, made to represent codes. A station can transmit on only one link (its own code) but can receive on all the links (codes) simultaneously. Each code is shared by $K = M/L$ users.

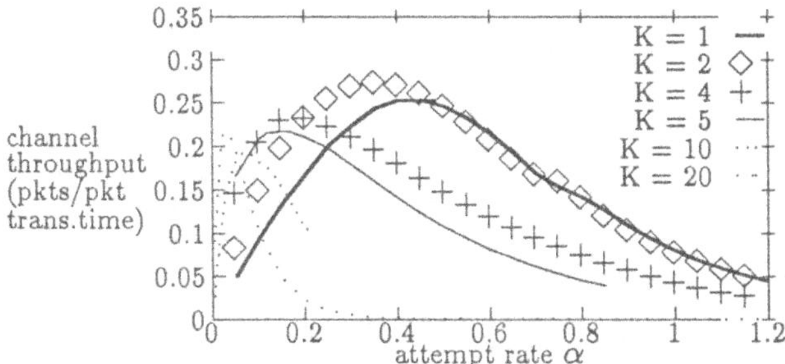

Figure 6: Saturation throughput analysis of unslotted CDMA-ALOHA with code-sharing(i.e. $K > 1$). Graph of channel saturation throughput per code(in packets per packet time) vs. α, for $M = 20$ users and the interference limit $N = 10$, for various values of K, the number of users per code.

Each node operates in accordance with the model presented in Section 2 and shown in Figure 1 and Figure 2. We assume Poisson arrival processes in our simulation. The value of α for each user was set to the value that gave the peak saturation throughput for the particular access protocol with the corresponding values of $L, M,$ and N.

The results are shown in Figure 9. The results assume that TDMA bit rate is the same as CDMA chip rate. The arrival rate has been normalised to the number of packets arriving per packet transmission time of CDMA. Hence the curves take into consideration the fact that packet transmission time for CDMA is 100 times the packet transmission time for TDMA.

Figure 9 contains the plots for the delay versus the normalised arrival rate. The results obtained are as expected. The delay increases to a large value as the arrival rate is increased to the saturation throughput. This stable region is upto 22 packets per CDMA packet transmission time for TDMA-ALOHA (see Figure 3), 5 for CDMA-ALOHA without code-sharing, and 2.5 for CDMA-ALOHA with code-sharing and $K = 2$ (see Figure 7). The mean delay increases when code-sharing is done.

6 CONCLUSION

This study was motivated by the question of the comparative efficiency of CDMA-ALOHA and TDMA-ALOHA when they are used as bandwidth sharing mechanisms by a finite number of nodes for transmitting packetised data. We have proposed a simple stochastic model for channel attempts by the nodes, we have shown that saturation throughput in this model yields a sufficient stability condition for the packet buffers, and we have used Markov chain models to obtain saturation throughputs for several versions of these medium access protocols; i.e., CDMA-ALOHA with or without code-charing, and with or without collision detection or carrier sensing.

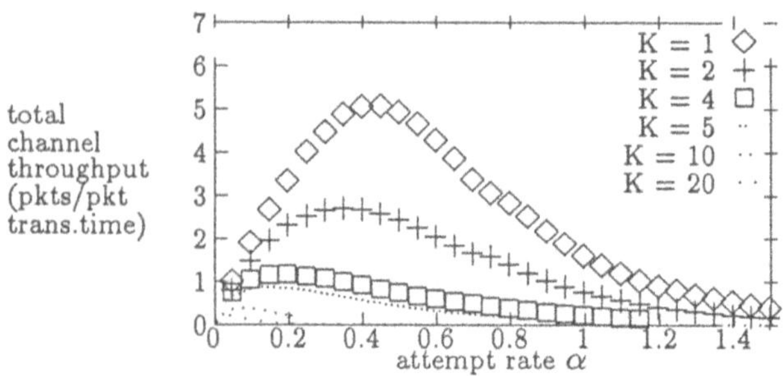

Figure 7: Saturation throughput analysis of unslotted CDMA-ALOHA with code-sharing(i.e. $K > 1$). Graph of total channel saturation throughput (in packets per packet time) vs. α, for $M = 20$ users and the interference limit $N = 10$, for various values of K, the number of users per code.

If the best that can be done for CDMA-ALOHA is that its chip rate be the same as the maximum achievable TDMA-ALOHA bit rate, then the channel utilisation efficiency is quite poor with CDMA-ALOHA, though not as poor as it may seem from a simple minded calculation based on the spreading factor and the interference limit N. In CDMA-ALOHA, code sharing further reduces the bandwidth efficiency but makes for a cheaper implementation of the correlation receivers. If CDMA-ALOHA chip rate can substantially exceed the achievable TDMA-ALOHA bit rate, as may be possible in optical CDMA, then CDMA-ALOHA will have the combined advantages of high efficiency, less sensitivity to propagation delays (owing to the lower transmission rate from the nodes), and ease of implementation.

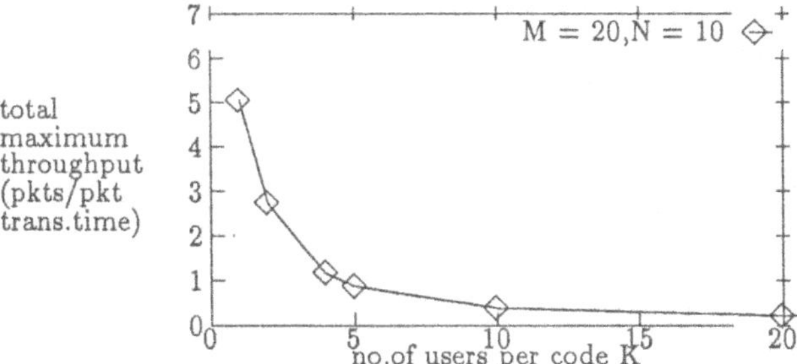

Figure 8: Saturation throughput analysis of unslotted CDMA-ALOHA with code-sharing. Graph of maximum total channel saturation throughput (in packets per packet time) vs. number of users sharing a code (K). The number of users $M = 20$ and interference limit $N = 10$.

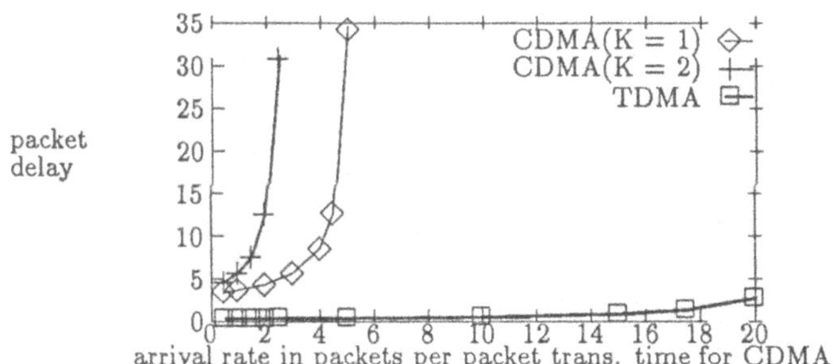

Figure 9: Simulation analysis of packet delay for unslotted finite user TDMA-ALOHA and CDMA-ALOHA, with and without code-sharing. Graph showing the mean packet delay, in number of packet transmission times for CDMA, vs. normalised arrival rate. The number of users $M = 20$ and the interference limit $N = 10$.

ACKNOWLEDGEMENTS

This problem was motivated during a discussion with Prof. A. Selvarajan, who is himself developing an optical CDMA network. We are thankful to Drs. Utpal Mukherji, Vinod Sharma, and Vivek Srinivas for discussions and exchange of ideas that have benefited this work.

REFERENCES

[Borovkov, 1976] A.A. Borovkov. *Stochastic Processes in Queuing Theory* . Springer-Verlag, New York, 1976.

[Rao *et al.*, 1988] Ramesh R. Rao and A.Ephremides. "On the Stability of Interacting Queues in a Multiple Access Environment". *IEEE Tran. on Inf. Thy.*, Vol.34, No.5, pp 918-930, Sept.1988.

[Lee *et al.*, 1988] E.Lee and D.Messerschmitt. "*Digital Communication*". Kluwer Academic Publishers, Boston, 1988.

[Salehi 1989] J.A. Salehi. "CDMA Techniques in Optical Fibre Networks-Part 1 & 2". *IEEE Transactions on Communications*, Vol.37, No.8, pp. 824-842, rAug. 1989.

[Prucnal *et al.*, 1986] P.R.Prucnal and M.A.Santoro. "Spread Spectrum Fibre-Optic Local Area Network using Optical Processing". *IEEE Journal of Lightwave technology*, Vol.LT-4, No.5, May 1986.

[Raychaudhari, 1981] D.Raychaudhary. "Performance Analysis of Random Access Packet Switched CDMA systems". *IEEE Transactions on Communications*, vol.29, No.6, pp. 895-901, June 1981.

[Green, 1993] P.E.Green. "*Fibre Optic Networks*". Prentice Hall, New Jersey, 1993.

[Sharma, 1988] V. Sharma. "Stability and Continuity for Slotted ALOHA with Stationary Non-independent Input Traffic". *Journal of Appl. Prob.*, Vol.25, pp 797-807, 1988.

[Sharma, 1990] V.Sharma. "Stability of Multiple Access Packet-Switched CDMA Systems". *Sadhana*, Proceedings of the Indian Academy of Sciences, V.15, Parts 4 and 5, pp 365-380, Dec 1990.

[Szpankowski, 1990] W. Szpankowski. "Towards Computable Stability Criteria for Some Multidimensional Stochastic Processes". *Stochastic Analysis of Computer and Communication Systems*, Hideaki Takagi (ed.), Elsevier, North-Holland, pp.131-172, 1990.

BIOGRAPHIES

Anurag Kumar has a B.Tech. in E.E. from the Indian Institute of Technology at Kanpur, and a PhD from Cornell University. He was a Member of Technical Staff at AT&T Bell Labs, Holmdel for over 6 years. Since 1988 he has been with the Indian Institute of Science (IISc), Bangalore, in the Dept. of Electrical Communication Engineering, where he is now Associate Professor. He is also the Coordinator at IISc of the Education and Research Network Project, which has set up a country-wide computer network for academic and research institutions, and conducts R&D in the area of communication networks. His own research interests are in the area of modelling, analysis, control and optimisation problems arising in communication networks and distributed systems.

Deepak Patil has a B.Tech. in Electronics and Telecommunication Engg. from V.J.T.I., Bombay, and an M.E. in Electrical Communication Engg. from the Indian Institute of Science. He is currently a Senior Engineer software R&D in Wipro Infotech, Bangalore.

18

Implementation and Performance Enhancement of a PC Based LAN/WAN Router with a Differential QOS Feature

S.V.R. Anand and A. Kumar
Department of Electrical Communication Engineering
Indian Institute of Science
Bangalore-560012, INDIA
e-mail : anand, anurag@ece.iisc.ernet.in

Abstract

In this paper we describe our approach to, and experiences in, developing a PC based IEEE 802.3 LAN - X.25 WAN IP router. The basic router functionality is achieved by integrating a commercially available synchronous serial port card for the PC-AT bus, and our enhancements of public domain or licensed pre-production source code for the communication protocols. The router implements a strategy developed in [Kumar *et al.*, 92] for providing differential QOS to bulk transfer and interactive traffic on WAN links. Various link level policies are studied to further improve the QOS. Several other performance issues that we encountered while developing the router are discussed. We study various task handling strategies for the router software to arrive at the ideal combination that gives the peak performance. The issues, and their resolutions for physical interface handling are detailed. Experimentation done to compare the various alternatives for resolving these issues are reported.

Keywords

PC based router, CLNS over CONS internetworking, X.25, RFC 877

1 INTRODUCTION

In a typical LAN-WAN interconnection with slow WAN links (less than 64kbps), to support network traffic with different performance requirements, the router should have the capability of providing differential QOS on the slow WAN links. The absence of this capability can be felt when we try doing telnet from a LAN node while ftp is in progress from another node on the same LAN, both using the same WAN link to reach a host on a remote LAN. The response times seen by the telnet user can be quite annoying. A detailed study of this issue can be found in [Kumar *et al.*, 92], wherein various internetworking strategies have been considered. From this paper we learn that a CONS based WAN can be effectively used to support services with varied network demands. The study was made in the context of LANs running a CLNS network protocol. Preliminary experimental results were given that validated the theoretical results.

In this paper we present our approach to, and experiences in developing a PC based IEEE 802.3 LAN-X.25 WAN router, that implements the strategies discussed in [Kumar *et al.*, 92].

At the time of this report, we have implemented and successfully tested our router software in the lab environment. The system is stable and has been deployed in four places; further installations are planned.

With the commercially available routers (including Cisco's router) it has been observed that the response of an interactive application is very poor in the presence of file transfers, both using the slow WAN(X.25) link. In those implementations we found that all the IP traffic is treated alike, and the packets are queued in a single FCFS queue after routing packets onto the WAN link. This in turn giving rise to the problem mentioned. In this aspect our implementation is different from the existing routers. We make use of the X.25 connection management effectively to support differential QOS required by the IP traffic.

In section 2, we give some of the implementation details. These include the platform on which the router was built, the software development, and the main features of the software package. In section 3, we discuss the performance issues that were considered regarding support for differential QOS and our approach to resolve these. We also report the experiments done to further improve QOS. Section 4 pertains to performance issues such as task scheduling and physical interface handling. We present the approach we have taken to resolve these issues by looking at the performance bottlenecks in the router software. In the light of this, a description of various strategies considered is presented. Then we describe in detail the experiments conducted, and evaluate the strategies and discuss the results. We conclude the paper by summarizing the achievements of the work done and outline the future work underway.

2 IMPLEMENTATION

Our router software has been built on an i386 based PC-AT running DOS. Presently the router supports IP and X.25 protocols. The two physical interfaces we handle are an Ethernet interface,and a synchronous serial interface. The hardware for the Ethernet interface is a Western Digital (WD8003) compatible add-on card. For the serial interface we have used a commercially available i82530 SCC based card that can support link speeds upto 19200 bps. This card does not have an on-board memory and hence relies on the host's services.

The software is derived by integrating two modules: Retix's X.25 ([Retix, 88]) and CMU's PC/IP. The initial work involved in integrating these two software packages was to change the model of the PC/IP code from small to large. This was due to the large memory requirements of the router. The design and the implementation of PC/IP follows the now popularly known thread mechanism([Tanenbaum, 92]). PC/IP realises a multi-tasking environment using this mechanism. The tasks are non-preemptive. The tasks are scheduled in a round robin fashion. There are no priorities associated with the tasks. We have customised the scheduler to some extent so that a priority based scheduling is possible.

The main advantage with the thread approach is that one can create various tasks based on some functionality, and these will be executing independently using their own stacks. This independent stack makes task switching possible at any point of execution. Resumption of the task execution continues from where it was left. For the non-preemptive multi-tasking environment that we have, based on an event, a task can be programmed to yield to another task, may be high priority one, at any point of the execution. Since modularity is inherent, the addition and deletion of tasks can be done without intervening in other tasks. Thus it is easy to add or remove functionality without affecting the overall system. It is possible to build a customised scheduler to schedule tasks according to some performance criteria. Since it is possible to make a task sleep, a better CPU utilisation is achieved. From the developer's view point debugging becomes very simple and stack overflow problem can be localised and corrected just by looking at one task.

Considering the advantages with this approach, especially for a router environment, we have changed the main Retix functional modules to threads. A function can be changed to a thread by giving it a stack frame, so that the function executes as an independent entity. Realising that the CMU package was designed for a host, and not for a router, additional functionality had to be introduced for routing purposes. This required an attachment of an additional logical link interface at IP, apart from the existing Ethernet interface.

We have included the TCP task along with other tasks to support telnet service at the router. The independent buffer and timer managements of both the pieces of software have been retained for the simplicity of integration.

For the PC/IP - X.25 interface (SNDCF in OSI terminology) we have implemented the *RFC 0877* specification [RFC877, 83]. The mapping of destination IP addresses to X.25 DTE addresses has been done at this interface. We have given a provision for establishing an X.25 connection directly to a remote router more than a hop away. This is useful because it avoids IP packets requiring processing at the IP layer of the intermediate routers, thus improving switching efficiency.

Interoperability tests with the Cisco(MGS) router have been done successfully. In fact Cisco router has throughout been a part of our experimental test-bed, on which all our performance related experiments were conducted. So far as X.25 connection establishment is concerned, we have used flow control negotiation facilities for packet size and window size. These parameters are useful not only to support differential QOS, but also to determine the router performance.

Finally, many performance related issues have been experimentally studied to enhance the performance. From these, we have incorporated the strategies that gave the best results. We have also attempted to explain the experimental observations.

2.1 Main Features

- Our implementation supports differential QOS required for bulk transfer vs. interactive traffic. The window flow control that is required for this purpose is configurable. These window sizes are directly related to the QOS given to each service class. The details of this issue will be discussed later.

- On the router console we provide the utilities ping and telnet. These are useful in managing the network. For one thing, we can use ping to check for reachability. We can alter the link level parameters by telnetting to other router(s), among the other things.

- The software comes with a customisation package to configure various parameters pertaining to each layer. The channel management menu enables one to set parameters that govern QOS. Some typical configurable parameters are: window sizes, timeouts, and number of retransmissions for X.25 and LAPB layers. Of course the serial card can be configured to have an external or an internal clock. The serial line bit rates can be chosen upto a maximum of 19200 bps.

- The routing customisation can be done in a simple way by specifying the mapping information. We can also make use of X.25 switching thus avoiding IP processing if the intermediate routers to a destination support X.25 switching.

- Debugging facilities can be enabled to observe the data movement at different layers of the router stack. The information provided is related to packet and frame information and the state information at LAPB and X.25 layers.

- We have developed a program that can be used to remotely monitor the health of the router. e.g., one can find the buffer occupancy on both LAN and WAN interfaces. We have extended this feature to control router parameters from a remote host.

- We also support SNMP agent on our router based on CMU's implementation.

3 DIFFERENTIAL QOS FOR IP TRAFFIC

In a LAN-WAN environment like ours where the WAN link speeds are less than 64 kbps (typically 9600 bps), and the serial line quality is poor requiring an ARQ protocol, the delays experienced by a LAN user who is remotely logged-in to a host on a remote LAN can be quite annoying in the presence of file transfers between other nodes on the same two LANs. The reasons for this are the following:

(i) Packet sizes for the file transfers are large (552 bytes typical) when compared to interactive traffic where typical packet sizes are 64 bytes (we assume that the user is typing characters on the terminal). Each file transfer packet can give rise to several link packets depending on the Maximum Transmission Unit (MTU) of the link.

(ii) The packet rate into the router for file transfer is much higher, because TCP window sizes can be quite large and disk access times can be ignored due to the parallelism we get with the low speed WAN.

Because of these reasons we find that the link window is hogged by the file transfer packets, thus causing intolerable response times for the interactive user. Hence we have an unfair sharing of the link, in a QOS sense, i.e., two services are sharing a facility, each with different QOS requirements, and one gets relatively better QOS than the other.

Note that simple priority before the link level window will not be enough to help the interactive traffic if the link window is large. Since interactive traffic arrives slowly, with a high probability, interactive packets find the link window full of file transfer packets. Thus even if this arriving packet is given priority before entry into the link window, it will still be delayed by a window full of file transfer packets. For a link packet size of 512 bytes, and link window of 7, this can be as much as 1.5 secs. at a link speed of 2400 bps. Inserting the arriving interactive packet at the head of the transmitter's queue is infeasible as then all sequence numbers of packets admitted into link window will have to be altered.

If the serial link card has buffering and runs the ARQ protocol, then provision of priority will require modification of the card. For analysis and simulation of this issue see [Kumar *et al.*, 92].

An alternative is to limit the number of file transfer packets that can be outstanding on the link. This can be achieved by segregating the file transfer traffic into a window controlled connection thus limiting the number of file transfer packets that can occupy the link window. This is easily done with a standard protocol, i.e., X.25. Note that simple priority (to telnet) before link insertion is not adequate since the telnet packets will have to wait for a link window full of ftp link packets anyway; and the link window can be large to support a range of link packet sizes. If the link window is small, say 2, then priority before insertion into link window should be enough.

There is another issue, however, that our approach addresses; the provision of fair bandwidth sharing between, for example two ftps one originating from a LAN and the other from a slow serial link attached to the router, both being multiplexed onto the slow outgoing link (see [Kumar *et al.*, 92]). Further, the X.25 interface permits us to connect the router to a Cisco and to the public X.25 network.

There are differences between our work and the existing work on QoS control in high speed networks. The latter work assumes that quantifiable, hard QoS guarantees are required for the various traffic classes; hence very fine grained control strategies are proposed and studied in this context ([Demers *et al.*, 90]). Implementation of these strategies requires complex nonstandard procedures, and may even require hardware support. The services that we are concerned with (namely, ftp and telnet) do not have any hard QoS requirements. The

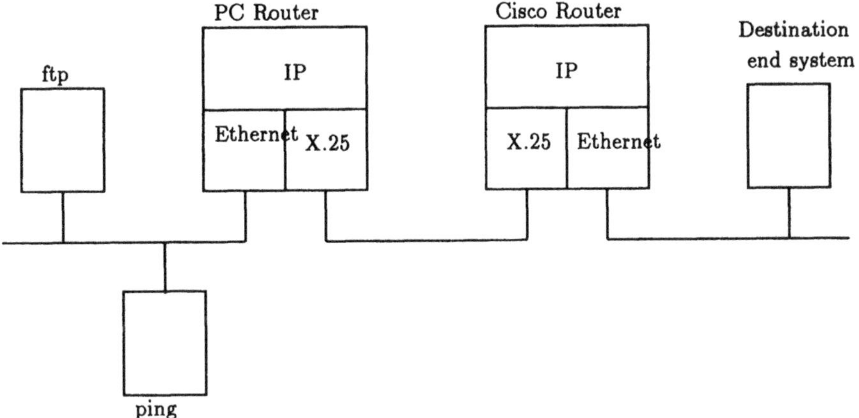

Figure 1: Setup for the QOS experiment

objective is simply to provide reasonably low response times for telnet, high throughput for ftp, fairness among services of the same class, and all this while efficiently utilising link bandwidth. We have realised this objective by utilising the existing protocol support in routers.

Since predominant number of hosts that provide Internet services do not make use of Type Of Service (TOS) field in the IP header to indicate the QOS class of the IP packet, we cannot make use of this field to identify the service class and provide the required QOS. Hence to circumvent this problem and identify the service class of the Internet service, we have used TCP port numbers which are globally unique for most of the popular services. At the router, apart from parsing IP header for routing purposes we also parse the TCP header. Once we identify the end-application, and thereby service class, we open an X.25 connection with an appropriate window size negotiation facility. Thus, for example, we open one connection for ftp packets, and another connection for telnet/rlogin packets, and so on. We find that the response time of an interactive application decreases dramatically with little decrease in ftp throughput. This corroborates the analysis in [Kumar *et al.*, 92]. To appreciate this better, quantitative measures have been shown here.

In the current implementation we classify three classes of traffic, namely, ftp, telnet/rlogin and others, and thus we have three X.25 connections with corresponding flow control parameters.

We now present the results obtained from some experiments. We have used an i386 machine running at 33 Mhz clock to run our router software. The testbed is shown in Figure 1. Since ping delays can be measured, we use ping to emulate a telnet like session. ping packets are 64 bytes long including the IP header. The serial link speed is set to 19200 bps. The ftp packet size is 552 bytes. At the serial interface with no fragmentation done at X.25, the lengths of ping and ftp packets become 69 and 557 bytes respectively with the inclusion of X.25 and LAPB headers.

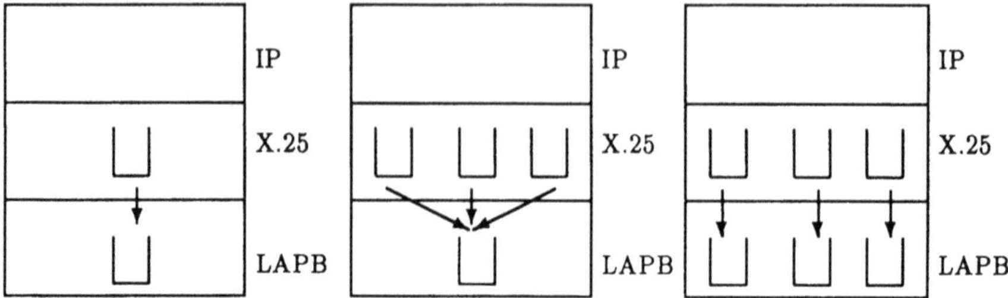

Figure 2: Traffic streams at Link level and X.25

Single X.25 connection:

To start with, we have taken throughput and delay measurements independently for ftp and ping with X.25 window size 2, and LAPB window size 7. The following are the results.

ftp throughput (in bytes/sec) : 1970
ping average response time (in millisecs.) : 270

Next, we have done a similar experiment with both ftp and ping initiated simultaneously.

ftp throughput (in bytes/sec) : 1910
ping average response time (in millisecs.) : 1300

We can clearly see the high ping response times once a bursty file transfer is in progress.

Multiple X.25 connections:

We now see ping delays by opening multiple X.25 connections with X.25 windows for ftp, and ping being 2 and 7 respectively. LAPB window size is 7 as in the earlier case. The following results were obtained.

ftp throughput (in byte/sec) : 1828
ping average response time (in millisecs.) : 518

It can be seen that even in the presence of ftp traffic, the response times for ping are quite low thus encouraging one to go for X.25 based WAN and have service based network connections. Note that we are not hurting the ftp traffic as can be seen from the less than 5% fall in ftp throughputs.

3.1 Link Level Scheduling

To attempt to enhance the QOS further, we assigned priorities to the multiple data streams obtained from the various X.25 connections. We thus provided priority based queues. When the link protocol gets credits, it serves these queues on a priority basis with the highest priority queue served first. The number of elements served will be equal to the number of credits. We have given the highest priority to the interactive traffic, and the lowest to the ftp traffic. The rest of the traffic such as UDP is assigned a medium priority.

We have implemented FCFS, Round-Robin and Exhaustive policies. In the FCFS policy, we serve the data streams in the order of their arrival to the priority queues. In other words this amounts to serving a single queue and treating the traffic alike. For Round-Robin policy, we serve the queues by taking one element from each queue. If there are no elements in a particular queue, and there are credits available, the server then serves the remaining queues in the order of priority within the remaining queues. For the Exhaustive policy, the server serves all the elements in the queue it is currently serving before serving the next queue. Segregation of traffic streams is shown in Figure 2. The first part of the figure shows the case of a single X.25 connection with a single queue at the link level. The data stream cannot be segregated at the link level since the order of packet transmission has already been decided at X.25 which assigns the sequence numbers for the packets. The second part of the same figure shows multiple data streams being put into a single link level queue. However, in this part traffic segregation has been done at the X.25 level. Both these parts depict the FCFS policy at the link level. The third part shows the traffic on multiple X.25 connections, and also the segregation done at the link level based on the priority given to the data streams.

In this context it should be mentioned that the policy that was employed by Retix's X.25 implementation was FCFS. During this experiment we also observe the delay and throughputs at the LAN end systems for various X.25 window sizes for the connection corresponding to ftp. Table 1 shows the experiments and the results obtained. From Table 1 we note that once the traffic is segregated into different channels, we get all the obtainable QOS improvement. The link level scheduling policy did not give any further appreciable improvement.

Table 1 Effect of Link Level Scheduling on QOS

Link Scheduling Policy	X.25 Window for ftp	ftp Throughput (bytes/sec)	Ping Delay (milli secs)
FCFS	2	1828	516
	4	1828	515
	7	1828	522
Round Robin	2	1828	524
	4	1828	518
	7	1828	518
Exhaustive	2	1828	514
	4	1828	523
	7	1818	521

4 TASK SCHEDULING IN THE ROUTER SOFTWARE

The software of any router comprises parts that handle the transmission and reception of packets at the physical interface, and parts that handle the movement of packets between these physical interfaces. Let us go through the software structure and see how various tasks have been handled. In doing so we identify the performance bottlenecks in the system. We experiment with various strategies for task scheduling to improve performance at these bottlenecks.

4.1 Software Structure

The software design used in building the router is inherited from CMU's PC/IP and is based on the well known concept of threads. This concept has helped us to create a cooperative multi-tasking environment under DOS. The implementation involved integration of CMU's PC/IP, and Retix's X.25 packages. Of these two main modules, PC/IP inherently follows the thread approach. We have introduced a task structure into Retix code to comply with the overall design philosophy. The main tasks created in the overall system are: Ethernet packet reception, Ethernet packet processing, serial interface reception, serial interface processing, and timer recovery. Ethernet packet processing include routing, and forwarding for packets coming from LAN. These functions are achieved through function calls from within the same task. The frame and packet level processing for the serial interface are done in the similar way from the corresponding task.

We address some of the performance issues considered for the transmission and reception mechanisms on the interfaces. We give their resolutions that enhance the router performance.

As is generally the case, there are different mechanisms for transmission and reception: blocked mode polling, unblocked mode polling, and using interrupt mode. In order to describe these, let us consider a transmission or a reception on a physical interface to be an event. In blocked mode polling, after initiating an event, CPU polls a status register till the status of the completed event is obtained. Thus the task from which the event was initiated blocks all the other tasks. In unblocked mode polling, the task which had generated the event polls a status register at regular time intervals to obtain the status of the completed event. In between the polling intervals the scheduler can schedule other tasks. For the interrupt mode, the status of the completed event is posted by an interrupt from the physical interface responsible for the event. Interrupt mode is unblocked and polling is completely avoided.

4.2 Ethernet Interface

Transmission:

In our implementation, the transmission discipline on the Ethernet interface is blocked polling. After initiating a transmission on the Ethernet interface, the CPU will be polling for an event that indicates the status of the completed transmission. Considering the non-preemptive multi-tasking structure, this essentially results in a blocked task. However, the blocking delay can be ignored since we are dealing with the Ethernet speeds and a moderate load on our Ethernet LAN. Further, this delay is negligible compared to the slower serial interface. Hence, blocked mode transmission at the Ethernet interface cannot be the bottleneck for the system performance.

Reception:

The reception process is inherently asynchronous, and hence it is natural to prefer interrupt mode. The Ethernet adapter is programmed to post the event and interrupt the CPU. The pending reception event can also be known by polling a register of the Ethernet adapter. This method is adopted to take care of the possible missed interrupts at the CPU. The polling interval is suitably chosen so as not to overly tax the host.

On a heavily loaded network, incoming Ethernet packets arrive back to back. This results in a condition wherein one interrupt can correspond to multiple messages buffered at the receiver. On an interrupt, we do a batch transfer of messages from the Ethernet adapter to the host buffers. Thus we prevent receiver overflow and the consequential dropping of packets.

4.3 Serial Interface

It is clear that the performance bottleneck in LAN-WAN interconnection over low speed links is the serial link itself [Kumar *et al.*, 92]. It is therefore imperative in the design of the router software that the link is utilised as best as possible. Thus, for example, no time should be wasted between the completion of one frame transmission on the serial link and the initiation of the next one. On the other hand since the link is slow, and it takes long to transmit a packet, and the software should not block while a frame is being transmitted but should proceed with other activities. These rather obvious requirements need special attention, however, in the DOS environment in which multiprocessing does not come naturally. In fact, since we integrated two existing pieces of software to build our router we were initially confronted with decisions made by the original software modules that were not necessarily optimal for our application.

In order to ascertain the best combination of strategies we experimented with several. Before moving on to the experiments, and the results obtained we mention the strategies considered. In our implementation, the end of transmission or reception will be notified through an interrupt from the serial device. We use this event to wake up the respective tasks. Based on our discussion on the Ethernet interface, we can think of similar service modes for serial line handling.

Reception on the serial line is through DMA, and after the reception of a frame, the CPU will be notified by an interrupt. Interrupt routine then wakes up the task designated to do further frame processing. Hence this reception process is asynchronous. There are no significant delays involved while receiving incoming frames. We now look at the possible variations for handling serial transmission.

1. Blocked mode transmission:

A call to transmit routine gets blocked within the routine till all the frames pending get transmitted. This method hurts the performance, especially on slow links as other activities get locked out while transmission on the serial link is proceeding. We also see from our experimental results the conditions in which this mode does not hurt the performance seriously.

2. Non-Blocked mode transmission:

As the term implies, the transmitter initiates transmission but does note wait for completion of transmission. We have introduced a task which is woken up once the frame gets transmitted. This high priority task will then initiate further transmissions. This is to minimise the delay between the link access times. Let us call this the transmit_task. Though a better method than the earlier one in terms of CPU utilisation, there is still some finite delay between frame transmissions. The scheduler can schedule the transmit_task only when the currently running task yields to the scheduler. Thus there will be times when the serial interface will be idling. This in turn gives rise to the under utilisation of the serial interface.

3. Asynchronous initiation of transmission from within the Interrupt Service Routine:

Here, instead of just posting an event for the scheduler, as in the earlier case, the end of transmission interrupt routine will trigger further transmission of frames from the LAPB transmit queue. Hence we no longer need the additional transmit_task just for frame transmissions. Certainly this should be the best mechanism compared to the other two options. From the software engineering point of view, we take care of the possible race condition at the shared transmit queue. LAPB accesses this queue, to insert outbound frames, as does the asynchronous occurrence of an interrupt, which causes an element to be removed from the same queue. We use the semaphore mechanism to prevent simultaneous access.

In section 4.4 we present the experiments that were carried out for comparing the above mentioned strategies.

4.4 Test-bed for Studying Serial Line Handling Issues

For these set of experiments we have used i286 based PC-AT to run our router software as our main i386 platform was out of order during this experimentation. We have used ftp throughputs as the performance measure. We measured throughputs by doing file transfers between two end systems on LANS connected by routers as shown in Figure 1 (ignoring the machine labelled "ping"). We have conducted two sets of experiments. The first two options mentioned for the serial line will be studied in our first test. The results have been tabulated in Table 2 and Table 3.

4.5 Discussion

Description of various columns of Tables 2 and 3 is the following. The first column shows the service strategy for the incoming frame processing on the serial line. It indicates whether the task that does frame processing services all the frames (exhaustive) or processes one frame before yielding to other tasks. The second column is the mode of frame transmission on the serial line. The Ethernet packet receive strategy is shown in the third column. The interpretation of the notation given for column 1, is also applicable here with respect to the Ethernet packet processing. The last column gives the ftp throughputs for various combination of strategies. Rows 1 to 4 of the Tables 2 and 3 give the measurements obtained for the blocked mode transmission on the serial link. Rows 5 to 8 of these tables give the measurements for non-blocked frame transmission on the serial link. The Ethernet packet transmission delays are ignored because of the moderate load on our Ethernet LAN.

Table 2 ftp throughput for various reception and transmission strategies at the interfaces.
Configuration : LAPB window 7, X.25 window 7, Baudrate 19200 bps

Sr.No.	Frame Reception on Serial	Frame Transmission on Serial	Ethernet Packet Reception	ftp thruput in bytes/sec
1.	Single	Blocked	Single	1360
2.	Single	Blocked	Exhaustive	1458
3.	Exhaustive	Blocked	Single	1758
4.	Exhaustive	Blocked	Exhaustive	1458
5.	Single	Non-Blocked	Single	1734
6.	Single	Non-Blocked	Exhaustive	1734
7.	Exhaustive	Non-Blocked	Single	1734
8.	Exhaustive	Non-Blocked	Exhaustive	1734

Table 3 ftp throughput for various reception and transmission strategies at the interfaces
Configuration : LAPB window 2, X.25 window 7, Baudrate 19200 bps

Sr.No.	Frame Reception on Serial	Frame Transmission on Serial	Ethernet Packet Reception	ftp thruput in bytes/sec
1.	Single	Blocked	Single	1834
2.	Single	Blocked	Exhaustive	1834
3.	Exhaustive	Blocked	Single	1808
4.	Exhaustive	Blocked	Exhaustive	1808
5.	Single	Non-Blocked	Single	1734
6.	Single	Non-Blocked	Exhaustive	1734
7.	Exhaustive	Non-Blocked	Single	1734
8.	Exhaustive	Non-Blocked	Exhaustive	1734

From the tables we see that for a large LAPB window size (i.e. 7 in Table 2), non-blocked frame transmission on the serial line gives a better performance irrespective of the kind of reception strategies followed for the serial and Ethernet interfaces. For small LAPB window size (i.e. 2 in Table 3), non-blocking transmission mode hurts the performance compared to the blocking mode, since with non blocked transmission the transmit_task keeps yielding to other tasks more frequently after the initiation of the transmission. Because of the non-preemptive nature of the task scheduler even the high priority transmit_task has to wait till the currently executing task gives up control to the scheduler. This causes link access delays and therefore we get low ftp throughput.

In our next experiment we try the option wherein the serial link transmission is asynchronously triggered from the interrupt service routine, and there is no blocking while transmission is in progress. This option eliminates the CPU idling or the starvation of the serial link, and the link access delays are minimised. Hence we expect encouraging results from this experiment. Considering the results obtained from the previous experiment we have fixed the servicing strategies for the incoming message processing on the two physical interfaces to be exhaustive. Apart from studying the influence of serial line handling we also see how X.25 window affects the performance. The results are shown in Table 4.

We compare the best ftp throughput obtained in Tables 2 and 3 with that obtained in Table 4. We see a remarkable improvement in the performance. LAPB window size does not matter since we always get back the acknowledgement by the time the LAPB frame gets transmitted. This causes one more frame to enter the transmit queue, which will be immediately transmitted. A larger X.25 size improves the performance further because of the parallelism that is obtained at the X.25 protocol processing while the serial line is busy sending frames.

Table 4 Configuration : Baudrate 19200 bps
Servicing incoming messages on the interfaces: exhaustive

LAPB Window	X.25 Window	ftp thruput bytes/sec
2	2	1970
2	7	2019
7	2	1970
7	7	2019

5 FINAL RESULT

From the previous section we have observed that the version of the router software incorporating asynchronous initiation of frame transmission on serial link with frame transmissions triggered from within interrupt routine has given the best performance. Recall that the performance tuning was done on a i286 PC/AT. We have used this tuned version of the software on i386 m/c and measured ftp throughput again. We now present the measurement thus obtained for i386 m/c as the final result. The X.25 window is set to 7, and LAPB window is set to 2 in accordance with our previous discussion. The serial link speed is configured to 19200 bps (i.e. 2400 bytes/sec).

ftp throughput (in bytes/sec) : 2170

Compare this result with that obtained from our earlier version prior to the performance enhancement (see section 3). We clearly see an improvement of about 10% (from 1970 bytes/sec to 2170 bytes/sec) in the ftp throughput from our latest version.

6 CONCLUSIONS & FUTURE WORK

We have proposed a feasible method for building LAN/WAN router that provide differential QOS using the well-known port numbers for Internet applications, and the channel management introduced at SNDCF. Our implementation showed an appreciable improvement in the response times for interactive traffic in the presence of bursty ftp traffic. From the measurements we can also conclude that, once the channeling of the IP traffic is done, the nature of the link level scheduling policy for the serial interface does not change the performance. Of the serial line handling mechanisms considered, and tested, non-blocking with asynchronous activation of the transmitter has given the best performance. We have seen further enhancement in the performance by choosing a larger X.25 window. Also, the servicing strategy at the Ethernet interface for incoming Ethernet packets, played no role in improving the performance.

We are planning to introduce more serial interfaces to the router to interconnect more subnets. Presently we use static routing information. The expansion of the router demands that we use dynamic routing protocols. There is a thought in this direction to incorporate at least one of the existing dynamic routing protocols.

References

[Kumar *et al.*, 92] Anurag Kumar, T.V.J.Ganesh Babu and S.V.R.Anand, "Comparative Performance of Queueing Strategies for LAN-WAN Router in Packet Data Networks", *Proceedings of the International Conference on Computer Networks, Architectures and Applications, Networks 92* .

[Demers *et al.*, 90] A. Demers, S. Keshav, S. Shenker, "Analysis and Simulation of Fair Queuing Algorithms", *Journal of Internetworking Research and Experience*, September 1990, pp. 3-26.

[Retix, 88] Retix Manual on "X.25 Packet Level Protocol", Retix Inc., Publication Number 1080057-00-B, March 7, 1988.

[Tanenbaum, 92] Andrew S. Tanenbaum "Modern Operating Systems", Prentice Hall 1992.

[RFC877, 83] J.T. Korb, "A Standard for the Transmission of IP Datagrams Over Public Data Networks", Internet RFC 0877, September 1983.

BIOGRAPHIES

S.V.R. Anand has a BSC(Hons) from Osmania University, and a B.E. in Electrical Communication Engg. from the Indian Insititute of Science (IISc). He is currently working in Education and Research Network project (ERNET) at IISc.

Anurag Kumar has a B.Tech. in E.E. from the Indian Institute of Technology at Kanpur, and a PhD from Cornell University. He was a Member of Technical Staff at AT&T Bell Labs, Holmdel for over 6 years. Since 1988 he has been with the Indian Institute of Science (IISc), Bangalore, in the Dept. of Electrical Communication Engineering, where he is now Associate Professor. He is also the Coordinator at IISc of the Education and Research Network Project, which has set up a country-wide computer network for academic and research institutions, and conducts R&D in the area of communication networks. His own research interests are in the area of modelling, analysis, control and optimisation problems arising in communication networks and distributed systems.

19

Analysis of Performance and Fault-Tolerance of Database Management Schemes for PCS

Dipak Ghosal,
Bellcore,
331 Newman Springs Road,
Red Bank, NJ 07701-7020.

Abstract

This paper presents the performance and fault-tolerance analysis of four different placement schemes for the HLR (Home Location Register) and the VLR (Visitor Location Register) databases in the signalling network. Based on a mean value analysis it has been shown that distributing the HLR and the VLR database may be required for high degree of penetration of PCS usage. A simple distributed scheme where the HLR is maintained in the Service Control Point (SCP) and VLR is co-located (physically or logically) in the SP (Signalling Point) can significantly reduce the impact of PCS signalling on the existing signalling network. Performance of distributed architectures with replication depends largely on how the databases are replicated. The results show that performance gains are achievable when the replication is based on the roaming characteristics of the mobile station. Fault-tolerance of the different placement strategies is based on a simple failure and recovery model of the databases. The results show that the system can be made more robust against database failures with a distributed and replicated database placement strategy if all the distributed replicated copies are always consistent with each other. This may incur additional performance penalties. The analysis is based on a number of simplifying assumptions and is only intended for a comparative study of different database placement schemes.

1 INTRODUCTION

Personal Communication Services (PCS) require significant intelligence in the signalling network to keep track of the mobile subscribers and appropriately route the calls. This study focuses on the performance and fault-tolerance issues of the databases that are required to provide PCS in a high density and highly mobile subscriber community.

In this study we consider a network supporting PCS providers offering cellular services to end customers. The mobile network considered in this study consists of wireless mobile terminals/stations (MS) which have radio access capability with base stations (BS). The type of terminal equipment and its radio access capability depends largely on the mobility of the user [Cox, 1990; Steel, 1990]. The service area consists of cells, each of which is serviced by a base station. A number of base stations is served by a controller such as a Mobile Switching Center (MSC) which is assumed to be co-located in the SP and provides access to the Public Switched Telephone Network (PSTN) [Kathy et al., 1992; CCITT, 1988]. The call control and routing are based on information maintained in two databases, namely the Home Location Register (HLR) and the Visitor Location Register (VLR). These are databases which store user profiles related to the nature of the service to which the user has subscribed. The HLR holds the permanent subscriber parameters while the VLR contains the database of subscriber parameters within a particular location area. In this study, we are interested in the placement strategies of these databases in the signalling network and to determine their impact on the performance and fault-tolerance of the PCS.

The signalling network considered in this study is the Common Channel Signalling (CCS) network based on the Signalling System Number 7 protocol [Bellcore, 1989]. The CCS network consists of the Signalling Points (SPs) interconnected via Signalling Transfer Points (STPs). The network database is provided in the Service Control Points (SCPs) which are connected to the STPs. SCPs are centralized databases which receive queries from the Signalling Points (SPs) and return the appropriate information. The 800 services and Calling Card are examples of two services that are provided by the SCP [Homa et al., 1992; Skoog et al., 1990].

The issue of SCP query and update rates for PCS is also studied in [Lo et al., 1992] in which the authors have considered support of end-to-end PCS with the subscriber data co-located in the regional SCP. Based on a Monte-Carlo simulation it has been shown that the mean real-time query rate can be up to an order of magnitude higher than the estimated load on the SCP due to 800 services. Furthermore, the real-time update rate is comparable to the query volume.

Both the above results show that providing region wide end-to-end PCS service solely based on the performance level of current SCP technology may impose limitations on the degree of penetration and mobility of the mobile terminal users. The effect of relocating databases to the SPs has been addressed in [Kathy et al., 1991]. In [Kathy et al., 1991] the authors have quantified the SS7 traffic load associated with the terminal mobility when Pan European standard GSM is used to provide PCS. Based on a detailed analysis of the SS7 signalling messages required for GSM, the authors have determined the impact of GSM PCN and GSM cellular with respect to normal POTS traffic. The results show that the placement of the VLR has impact on the additional traffic that must be handled by the SS7 network due to PCS.

This paper extends the work reported in [Kathy et al., 1992] in a number of ways. First it provides a more detailed analysis of four different placement strategies which involve both the HLR and the VLR. In [Kathy et al., 1992] only the placement of the VLR is considered while the HLR is assumed to be always outside the MSC. Second, in this paper we consider a more accurate characterization of the traffic mixes and environments. Finally, this study also provides analysis of the fault-tolerance of the system under different database placement strategies.

In this paper we investigate the following four placement strategies for HLR and VLR.

1. HLR and VLR in SCP.
2. HLR in SCP and VLR in SP.
3. Master copy of HLR in SCP and cached copy in SP. VLR in SP.
4. Master copy of HLR in SCP and cached copies replicated in SPs. VLR in SP.

Clearly these strategies provide different performance and fault-tolerance capability for personal communication services. When HLR and VLR are co-located in the SCP, many call set-up/tear-down messages and handovers and location update messages require queries to the SCP. This will require a large number of messages to be exchanged across the signalling network. Furthermore, since the SCP is a centralized database from the services and protocol perspectives, any congestion and/or isolation of the SCP can severely affect the personal communication services to a large user community. Moving the VLR to the SP can improve performance but not the fault-tolerance of the system [Gray et al., 1993]. Maintaining master copy of HLR in SCP and cached copies replicated in one or more SPs along with the VLR can improve fault-tolerance but only at the cost of performance. Thus there is a trade-off between the performance and fault tolerance for the different placement strategies. This study is an attempt to quantify this trade-off.

This paper is organized as follows. Section 2 gives a detailed description of the system analyzed in this paper. The various messages that are required for mobile call set-up, handovers and location updates are discussed in this section. The assumptions made in the analysis are discussed in Section 3. One of the key parameters of the system is the model of the MS and characterization of their mobility. This is also discussed in Section 3. A summary of the various parameters of the system is also tabulated in this section.

Section 4 discusses the four placement strategies and their performance impact on the signalling network. The fault-tolerance of the system under different placement strategies is discussed in Section 5. Finally Section 6 concludes this paper and points out some future research directions.

2 SYSTEM DESCRIPTIONS

This study is based on the network shown in Figure 1. We consider a total service area A_{sa} which is divided into L equal sized location areas of size A_{la}. Each location area is serviced by a controller referred to as the Mobile Switching Centre (MSC) which is co-located in the Signalling Point (SP) which also provides wire-line access services.

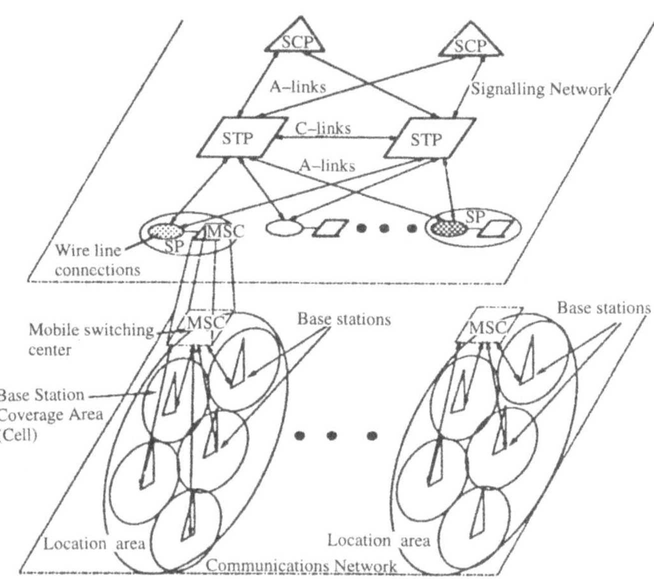

Figure 1: Internetworking between the wireless and the signalling networks.

Each location area is divided into equal sized cells each of area A_{cell}. Each cell is served by a base station which provides the radio interface to the wireless mobile terminals/stations (MS) within the coverage area of the base station. The signalling network is the Common Channel Signalling (CCS) network based on the SS7 protocol [Bellcore, 1989; Skoog et al., 1990]. It consists of SPs connected via A-links to a Signalling Transfer Point (STP) pair which in turn is connected by C-links. The STP is a packet switch that routes SS7 signalling messages between SPs. The SS7 messages are call control messages for call set-up and disconnect and network control messages for maintaining the correct and stable operation of the signalling network. The network database is provided in the Signalling Control Points (SCPs) which is connected to the STP pair through A-links.

As mentioned in the introduction, call control and routing in the mobile network is performed based on two databases, namely, the Home Location Register (HLR) and the Visitor Location Register (VLR) [CCITT, 1988]. The HLR contains the permanent and temporary data about each subscriber. The permanent data includes the current services the subscriber receives and temporary data includes a pointer to the current VLR which services the subscriber. There is one VLR for each location area and it contains a list of MS stations that are currently registered in that location area. For each MS, the VLR also contains a copy of the permanent data stored in the HLR. In this study we will assume that each location area is served by one VLR and one VLR serves only one location area.

Because the MS can roam from one location to another both the VLR and the HLR need to be updated with the most recent location of the MS. This study focuses on the placement strategies of these databases in the signalling network and their impact on the performance and fault-tolerance of PCS. In order to perform a detailed analysis of database location strategies it is necessary to understand the various messages that are needed for call set-up/tear-down, location update and handovers. These are described in the following sub-sections with respect to the IS-41 protocol [Homa et al., 1992].

2.1 Call Set-up/Tear-down

Call set-up with or from a mobile station, in general, involves three major tasks: *terminal control, routing* and *call control.* Terminal control includes set-up functions between the mobile station and the controlling MSC. These include set-up radio channels and authentication and validation procedures and may require access of the VLR in which the mobile station is currently registered. Routing tasks are required to find the current location of the mobile station so that the call can be appropriately routed. This information is obtained from the HLR of the mobile station. Finally, call control functions relate to the messages that are exchanged between the controlling SPs to set-up the call.

Figure 2 shows the messages that are exchanged for a call set-up originating at a MS and terminating at a wire-line equipment. The originating SP sends a call set-up request to the controlling VLR which invokes application processes to manage the call set-up with the mobile station. The primary function of these processes involves authenticating the originating mobile station and providing communication privacy. This requires messages

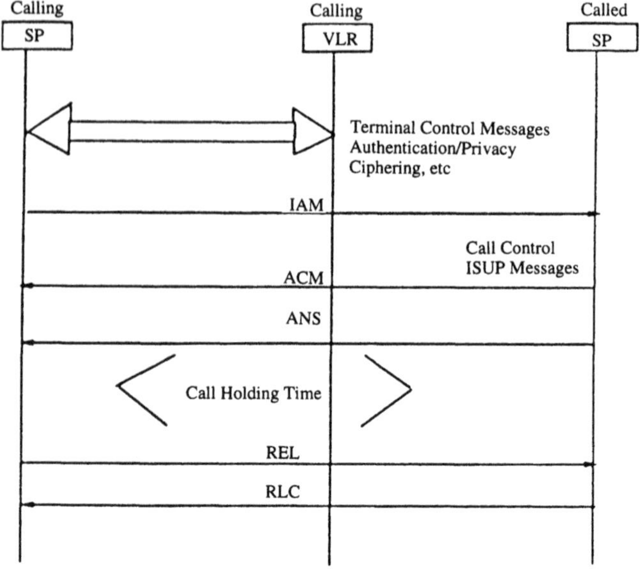

Figure 2: Messages exchanged for call set-up originating at a MS.

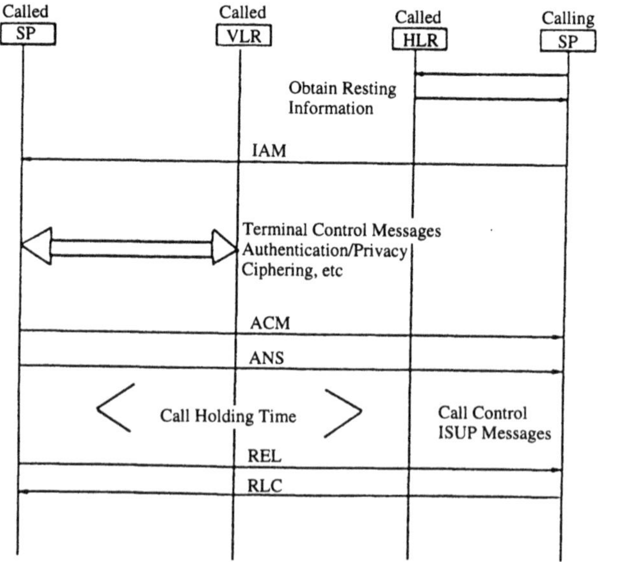

Figure 3: Messages exchanged for call-set terminating at a MS.

to be exchanged between the VLR and the SP. The number of such messages depends on the algorithms used for session key agreement and terminal authentication. In this study we assume that 4 messages are required in each direction to achieve terminal control functions. Note that since this is only a comparative study, an exact enumeration of various messages will not significantly alter the conclusions of this study.

The call control functions are achieved by the standard SS7 call set-up messages, namely Initial Address Message (IAM), Address Complete Message (ACM) and Answer Message (ANS). The call is cleared by the Release(REL) and Release Clear (RLC) messages. If the called number is also a mobile station, then the routing function must be invoked to determine the controlling VLR of the terminating mobile station. This is done by querying the HLR which contains a pointer to the VLR where the terminating mobile station is registered. This is achieved by the **loqreq** and **locreq acknowledge** messages as shown in Figure 3. The remainder of the call set-up involves terminal control functions for the terminating mobile station and call control functions between the controlling SPs.

2.2 Location update

Location update is required whenever the mobile station crosses location areas. Location update involves three main functions; 1) updating the location information maintained in the HLR (i.e., the pointer to the VLR), 2) registering the mobile station with the VLR which controls the new location area, and 3) cancelling the registration in the previous VLR. Figure 4 shows the different messages which achieve these tasks. Note that the new VLR obtains the subscriber parameters and authorization parameters from the HLR using the *profreq* and *qualreq* messages respectively.

2.3 Handovers

There are three types of handovers[CCITT, 1988];

1. Handovers between radio channels of the same radio port. This arises primarily due to interference or disturbances in the radio channels;
2. Handovers between base radio ports within the same SP; and
3. Handovers between radio ports of different SPs.

In this study we will assume that handovers of types 1 and 2 will be handled by the controlling SP and do not require any access to the signalling network. Inter-SP handovers, i.e., handovers of type 3, is similar to a call set-up and provides a through connection between the SPs involved in the handover. Inter-MSC handovers based on the IS-41 MAP use dedicated facilities between MSCs and thus do not require ISUP capabilities. The messages are exchanged between the MSCs involved in the handover and consists of 3 invoke and 2 response messages. These are ***hand-off measure request, facilities directive*** and ***mobile channel*** invoke messages and ***hand-off measure request*** and ***facilities directive*** response messages. In this study we will assume that inter-SP handovers require 4 inter-SP messages in each direction and one query to the VLR database.

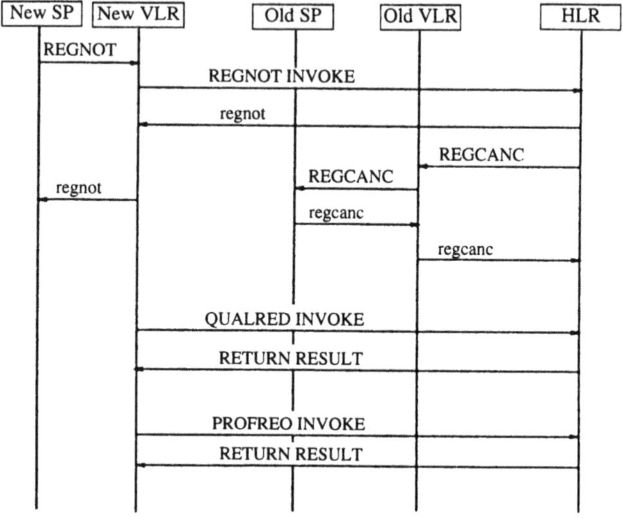

Figure 4: Messages exchanged during location update.

3 MODEL AND ASSUMPTIONS

In order to simplify the analysis we will make the following assumptions regarding the network;

Assumption 1 : We will assume that Figure 1 represents the entire network, i.e., all calls originate and terminate with the subscribers connected to the network.

Assumption 2 : We will assume that each location area is served by one VLR and one VLR serves only one location area.

Assumption 3 : We will assume that the cells in a location area are identical in size.

Let A_{la} denote the size of the location area in square kilometers and N denote the number of subscribers in the location area. Thus, N also represents the size (in terms of number of lines) of the SP serving the location area.

3.1 Traffic mixes and environments

Let Λ denote the number of calls per hour serviced by an SP in an existing POTS (Plain Old Telephone System). These calls can be approximately[1] divided into three components: 1) Outgoing calls which constitute 40% of the overall calls, 2) Incoming calls which also constitute 40% of the overall calls and 3) Intra-SP calls which originate and terminate in the SP constitute 20% of the calls. We will assume the following three types of SPs:

METRO : An end-office in a major metropolitan area with 50,000 to 70,000 lines serving 3.59 calls per line per hour in a HDBH (High Day Busy Hour) period.

Single System City : An end-office in a medium size town with 20,000 to 50,000 lines serving 2.9 calls per line per hour during a HDBH period.

Suburban : An end-office residing in the outskirts of a metropolitan area with 10,000 to 30,000 lines generating 1.9 calls per line per hour in a HDBH period.

Let λ_w (λ_m) denote the number of calls per hour made by a wireline (mobile) subscriber. Also, let t_{hw} (t_{hm}) denote the mean call holding time for a wireline (mobile) call.

3.2 Performance measure

In the CCS network, A-links connect the SPs with the STP pair. In this study we assume that there is one 56 Kbps link in each direction from an SP to an STP. We will use the utilization of the A-link in the outgoing direction as our performance measure. Note that one performance objective is to maintain the A-link utilization below 40%. The A-link utilization, U, assuming only POTS call and 30 bytes per message, is given by

$$\begin{aligned} U &= \Lambda \times \textit{fraction of o/g POTS calls per sec} \\ &\quad \times \textit{messages per call} \times \textit{bits per message} \\ &= \frac{\Lambda \times 0.4 \times 30 \times 8 \times 5}{2 \times 36 \times 56 \times 10^5} \end{aligned} \tag{1}$$

where Λ denotes the calls per hour serviced by the SP. From the above relation we can obtain the A-link utilization for different SP calling rates. In the following analysis, this will be used as the base case for comparing the various database placement schemes.

3.3 Mobility model

Let ξ denote the penetration of PCS usage. This implies that in a location area of N users, ξN are mobile stations and $(1-\xi)N$ are wireline subscribers. Furthermore, for Λ

[1] For exact numbers it is necessary to account for the mis-dials and calls which terminate due to a busy signal

Table 1: Number of outgoing messages per call type for HLR and VLR both in SCP. WL : Wireline, MS : Mobile station

Call types	Outgoing calls	Intra-SP calls
WL $\rightarrow$ WL	5	0
WL $\rightarrow$ MS	10	5
MS $\rightarrow$ WL	9	4
MS $\rightarrow$ MS	14	9

Table 2: Number of outgoing messages per call type for HLR and VLR both in SCP.

Event types	Handover	Location Update
Outgoing messages	9	2

calls per hour served by an SP, the proportions of different call types assuming that a person calls any other person with equal probability are as follows;

$$\begin{aligned} MS \rightarrow MS &= \xi^2\Lambda \\ MS \rightarrow Wireline &= (1-\xi)\xi\Lambda \\ Wireline \rightarrow wireline &= (1-\xi)^2\Lambda \\ Wireline \rightarrow MS &= (1-\xi)\xi\Lambda \end{aligned} \tag{2}$$

In order to evaluate the impact due to PCS signalling, it is necessary to determine the number of handovers and location updates. Let λ_{ho} and λ_{lu} denote the number of handovers and location updates per hour. These are obtained based on the simple flow model described in [Thomas et al., 1988; Kathy et al., 1991]. Based on the assumption that the MSs are uniformly distributed on the surface of the cell and the direction of the movement is uniformly distributed between $[0, 2\pi]$, it can be shown that the number of boundary crossings per hour, $R_{cross}(area)$, is given by

$$R_{cross}(area) = V\rho L/\pi \tag{3}$$

where V is the mean velocity, ρ is the density of mobile station in the area and L is the length of the boundary of the area.

Now, the inter-SP handovers are location area crossings while the mobile station is active, i.e., there is a call in progress. Thus

$$\lambda_{ho} = R_{cross}(A_{la}) \times s \tag{4}$$

where s is equal to the fraction of an hour a mobile station is active. The latter depends on the call holding time and the number of calls per hour made by the mobile station and is given by

$$s = \frac{\lambda_m \times t_{mw}}{3600} \tag{5}$$

Note that the above equation gives the number of handovers out of the area. By the fluid flow approximation it is also equal to the number of handovers into the area. Based on the same idea, the number of location updates per hour is given by

$$\lambda_{lu} = R_{across}(A_{la}) \times (1 - s) \tag{6}$$

This is because the number of location updates due to location area departures is equal to the number of location area crossings while the terminal is idle.

4 DATABASE PLACEMENT STRATEGIES

In this section we consider the various database placement strategies and analyze their performance impact on the signalling network.

4.1 VLR and HLR in SCP

In this scheme the SCP maintains both the HLR for each subscriber and the VLR for each location area. As a result all terminal control and routing messages which require access to the VLR or the HLR must be sent to the SCP. Note that in this scheme even for intra-SP calls in which at least one mobile station is involved will require access to the SCP. In order to determine the link utilization for this placement strategy, it is necessary to quantify the number of messages in the outgoing direction. This is shown in Table 1.

The above table is obtained from the figures in Section 2 which illustrate the messages that are exchanged during call set-up. For example the number of outgoing messages for an outgoing call from a wireline to a mobile station consists of 1 message to the HLR to obtain the SP of the terminating mobile station and 5 POTS message to set up the call. Note that the proportion of different call types can be found using Equation (2).

The other factor which will affect the link utilization is the number of messages required to perform handovers and location updates. From the call flows described in Sections 2.2 and 2.3 and making the fluid flow assumption that the number of handovers (location updates) due to mobile stations roaming out of the location area is equal to the number of handovers (location updates) due to mobile stations roaming into the location area, it is easy to obtain Table 2 which shows the number of outgoing messages for every handover and location update.

Let $\lambda_{o/g}$ and λ_{in} denote the number of outgoing and intra-SP calls per hour, respectively. Using Table 1 and Table 2 and assuming a mean message length of 30 bytes it can be shown that the link utilization U is given by

Table 3: Number of outgoing messages per call type for HLR in SCP and VLR in SP

Call types	Outgoing calls	Intra-SP calls
WL → WL	5	0
WL → MS	6	1
MS → WL	5	0
MS → MS	6	1

Table 4: Number of outgoing messages for handover and location updates for HLR in SCP and VLR in SP.

Event types	Handover	Location Update
Outgoing messages	8	4

$$
\begin{aligned}
U = & \frac{\lambda_{o/g} \times 30 \times 8}{36 \times 56 \times 10^5}[5(1-\xi)^2 + 15(1-\xi)\xi + 10\xi^2] \\
& + \frac{\lambda_{in} \times 30 \times 8}{36 \times 56 \times 10^5}[9(1-\xi)\xi + 9\xi^2] \\
& \frac{\lambda_{ho} \times 30 \times 8 \times 9}{36 \times 56 \times 10^5} + \frac{\lambda_{lu} \times 30 \times 8 \times 2}{36 \times 56 \times 10^5}
\end{aligned} \tag{7}
$$

where $\lambda_{o/g} = \frac{\Lambda \times 0.4}{2}$ and $\lambda_{in} = \frac{\Lambda \times 0.2}{2}$. Since PCS signalling messages will have higher average message length, assuming a lower number of bytes per message gives advantage to this placement scheme.

4.2 HLR in SCP and VLR in SP

In this scheme the HLR for each subscriber is stored in the SCP and the VLR for each location area is stored in the SP. We assume that the VLR contains a copy of the subscriber profile and can perform the authentication, validation and ciphering procedures. The HLR for each subscriber contains the permanent copy of the subscriber data and a pointer to the current VLR in which the subscriber is currently registered. This placement strategy significantly reduces the number of messages for call set-up as shown in Table 3.

Note that the intra-SP calls that terminate in a mobile station require accesses to the HLR to obtain the current VLR of the terminating mobile station. This is an assumption and may not be true under certain implementations. In this study we assume that when the SP determines the called number to that of a mobile station it obtains its current location from the HLR. Searching the local VLR can save an access to the HLR if the mobile station is currently registered in the same VLR.

The number of handovers and location update messages are shown in Table 4. The

Table 5: Number of outgoing messages per call type for HLR replicated in SCP and SP and VLR in SP

Call types	Outgoing calls	Intra-SP calls
WL → WL	5	0
WL → MS	7	p_1
MS → WL	5	0
MS → MS	6	p_1

number of handover messages decreases because the VLR is co-located in the SP. The number of location update messages increases because the messages between the two VLRs must be sent over the signalling network. Following the same approach as before, the link utilization can be written down as follows:

$$U = \frac{\lambda_{o/g}\times 30\times 8}{36\times 56\times 10^5}\,[5(1-\xi)^2 + 11(1-\xi)\xi + 6\xi^2] + \frac{\lambda_{in}\times 30\times 8}{36\times 56\times 10^5}\,[(1-\xi)\xi + \xi^2] \frac{\lambda_{ho}\times 30\times 8\times 8}{36\times 56\times 10^5} + \frac{\lambda_{lu}\times 30\times 8\times 4}{36\times 56\times 10^5} \quad (8)$$

where $\lambda_{o/g}$ and λ_{in} are defined as before.

4.3 Master copy of HLR in SCP and cached copy in SP. VLR in SP

In this scheme each mobile station is assigned to a *home location area* and the permanent subscriber parameters of all subscribers with the same home location area are kept in a HLR database co-located in the SP. The HLR databases in each SP are referred to as the cached copies of the master HLR database which is maintained in the SCP. The VLR for each location area is also maintained in the SP of that location area. The home location area for a subscriber can be chosen to be the location area in which the mobile station is most often registered. The mobile station identifier of mobile stations with the same home location area has a unique identifier that can be recoginzed by all SPs for the purpose of routing the calls.

When the called number belongs to a mobile station, the calling SP queries the HLR in the home location area of the called mobile station to obtain the location of the controlling VLR. When the mobile station roams out of the home location area, both the cached HLR in the home location area and the master copy of HLR in the SCP are updated. However, note that HLR is not used in obtaining the routing information and thus can be updated at a much slower rate than the cached HLR and this rate will depend on the consistency that needs to be maintained between the master copy and the cached copy.

Table 6: Number of outgoing messages for handover and location updates for HLR replicated in SCP and VLR in SP.

Event types	Outgoing messages
Handover	8
Updating cached HLR	5.33
Updating master HLR	δ

The number of outgoing messages for intra-SP and outgoing calls are shown in Table 5. Note that the number of messages for outgoing calls terminating at a mobile station is greater than the previous case. The simple way to figure this out is to realize that in this scheme obtaining the routing information involves two SPs while in the previous case only the calling SP and the SCP is involved. In this case most intra-SP calls will be completed without requiring access to the signalling network. The only exception will be the intra-SP calls which terminate at mobile stations that belong to a different home location area. For these calls the routing information needs to be obtained from the home HLR which must be queried via the signalling network. The fraction of these calls is denoted by p_1.

The number of outgoing messages for location update and handover are shown in Table 6. The number of location update messages depends on the roaming characteristics of the mobile station with respect to the home location area. If the mobile station roams in two location areas which are different from the home location area then 8 outgoing messages are required per location update. This assumes that for every location area crossing in a particular direction by a subscriber there is another location area crossing by a similar subscriber in the opposite direction. On the other hand of one of the location areas is the home location area then the number of outgoing messages is 4. We assume that a mobile station roams randomly and uniformly in three location areas of which one is the home location area. Based on this assumption the mean number of location update messages is 5.33. The rate at which the master copy of the HLR is updated will be denoted by δ per hour.

Based on the above tables we derive the link utilization which is given by

$$
\begin{aligned}
U = & \frac{\lambda_{o/g}\times 30\times 8}{36\times 56\times 10^5}\,[5(1-\xi)^2+12(1-\xi)\xi+7\xi^2] \\
& +\frac{\lambda_{in}\times 30\times 8}{36\times 56\times 10^5}\,[p_1(1-\xi)\xi+p_1\xi^2] \\
& \frac{\lambda_{ho}\times 30\times 8\times 8}{36\times 56\times 10^5}+\frac{(\lambda_{lu}+\delta)\times 30\times 8\times 5.33}{36\times 56\times 10^5}
\end{aligned} \qquad (9)
$$

Table 7: Number of outgoing messages for handover and location updates for HLR replicated in SCP and neighboring SPs and VLR in SP.

Event types	Outgoing messages
Handover	9
Updating 2 HLRs	3.33
Updating 3 HLRs	2.0
Updating master HLR	δ

4.4 Master copy of HLR in SCP and cached copies of replicated in SPs. VLR in SP

This scheme is the same as before except that the HLR is cached into a number of nearby location areas. The VLR is still co-located in the SP. In this scheme the location update involves updating all the cached copies and updating the master copy at a certain rate. Note that if the HLR is properly replicated then the probability of a call to a MS with a different home location area, denoted by p_2, can be reduced significantly. The number of location update messages also depends on the number of replications. Assuming as before that the mobile station roams in three location areas, the number of outgoing location update messages depends on the number of replication as shown in Table 7.

The link utilization is given by

$$U = \frac{\lambda_{o/g}\times 30\times 8}{36\times 56\times 10^5}\,[5(1-\xi)^2 + 12(1-\xi)\xi + 7\xi^2] + \frac{\lambda_{in}\times 30\times 8}{36\times 10^5}\,[p_2(1-\xi)\xi + p_2\xi^2$$

$$\frac{\lambda_{ho}\times 30\times 8\times 8}{36\times 56\times 10^5} + \frac{(\lambda_{lu}+\delta)\times 30\times 8\times 3.33|2.0}{36\times 56\times 10^5} \quad (10)$$

4.5 Results and discussion

Table 8 lists the link utilization as a function of the call serving rate for different database placement strategies. The various parameters are shown below the table.

The velocity of the mobile station is based on the characteristics of a "student pedestrain" and is obtained from [kathy et al., 1992]. The size of the location area is assumed to be 2.25 square kilometers. This would correspond to a high density metro area. In this experiment we assume a switch with 70000 subscribers handling x (O+I) calls per hour per subscriber. The different calling rates were obtained by varying x. When the HLR and the VLR are co-located in the SCP, the link utilization is almost double that of the existing POTS network. As a result, the link utilization becomes greater than the required 40% limit for almost half the call handling rate that can be supported in the existing of penetration of mobile users. Let x denote the probability that the originating call is an intra-SP call. From the above one can easily establish the following four different call types and their respective probability of occurrence.

$$\begin{aligned} MS \rightarrow MS(Intra - SP) &= \xi x \\ MS \rightarrow Wireline(Intra - SP) &= (1 - \xi)x \\ MS \rightarrow wireline(Inter - SP) &= (1 - \xi)1 - x \\ MS \rightarrow MS(Inter - SP) &= (1 - x)\xi \end{aligned} \tag{11}$$

Note that based on the statistics of the various traffic mixes it can be easily established that x is equal to 0.3333.

Since the SCPs operate in a replicated mode, for the SCPs to be unavailable, both SCPs must be in the failed state. Based on a simple analysis it can be shown that the probability that at least one SCP is in the active state, i.e., availability of the SCPs, η_{scp} is given by

$$\eta_{scp} = 1 - \frac{\gamma^2}{\gamma^2+\gamma\beta+\beta^2} \tag{12}$$

If both the HLR and the VLR are co-located in the SCP, then the probability that a mobile station can initiate a call is equal to the probability that at least one SCP is in the active state. Thus the availability for this case is given by

$$\eta_{case1} = 1 - \frac{\gamma^2}{\gamma^2+\gamma\beta+\beta^2} \tag{13}$$

Next, consider Case 2 in which the HLR is in the SCP and the VLR is in the SP. We make the assumption that if the VLR has failed then call control and routing functions can be performed by the HLR database. Note that this will require extra messages to be exchanged which has not been modeled here. Based on this assumption it can be shown that the probability that an MS station can initate a call can be written as

$$\begin{aligned} \eta_{case2} = \xi(1 - \frac{\gamma^2}{\gamma^2+\gamma\beta+\beta^2}) + (1 - \xi) \\ [1 - (\frac{\gamma^2}{\gamma^2+\gamma\beta+\beta^2}\frac{\gamma}{\gamma+\beta})] \end{aligned} \tag{14}$$

The above equation is dervied by considering all the possible call types and noting that $\eta_{vlr} = \frac{\gamma}{\gamma+\beta}$ and η_{scp} is given by Equation(12).

Similarly, when the HLR and the VLR are cached in the SP, it can be shown that

$$\begin{aligned} \eta_{case3} = x[1 - (\frac{\gamma^2}{\gamma^2+\gamma\beta+\beta^2}\frac{\gamma}{\gamma+\beta})] \\ +(1 - x)(1 - \xi)[1 - (\frac{\gamma^2}{\gamma^2+\gamma\beta+\beta^2}\frac{\gamma}{\gamma+\beta})] \\ (1 - x)\xi(1 - \frac{\gamma^2}{\gamma^2+\gamma\beta+\beta^2}) \end{aligned} \tag{15}$$

Table 8: Utilization as a function of the calling rate for different placement strategies. Parameters: $\xi = 0.25, N = 70000, A_{la} = 2.25, V = 5.17, L = 1.5, t_{hm} = 180, p_1 = 0.3, p_2 = 0.005, \delta = 1/hr$

Calling rate	70000	126000	182000	238000
POTS	0.083	0.150	0.217	0.283
HLR and VLR in SCP	0.166	0.281	0.396	0.511
HLR in SCP, VLR in SP	0.137	0.211	0.285	0.358
HLR cached in SP	0.155	0.230	0.305	0.380
HLR cached in 2 SPs	0.132	0.208	0.283	0.360
HLR cached in 3 SPs	0.118	0.194	0.270	0.346

POTS network. Moving the VLR into the SP results in a 25% to 30% reduction in the link utilization for the same calling rate. However, caching the HLR in the SP does not result in any improvement. In fact, there is a marginal increase in the A-link utilization. The main problem with this scheme is that the number of messages for outgoing calls increases without a compensating decrease in the messages for the intra-SP calls. Finally, replicating the HLR in one or more SPs results in a decrease in utilization for the same calling rate. Replication reduces the number of location update messages and also messages for intra-SP calls terminating at a mobile station. Note that in order to reduce the number of location update messages the replication must be correlated with the roaming characteristics of the mobile station.

The effect of different degrees of penetration of PCS usage is shown in Figure 5. From the figure it is clear that a higher degree of penetration will make distributed and/or replicated architectures a more attractive alternative. The results also show that a simple distributed scheme where the VLR is co-located in the SP and the HLR is maintained in the SCP can perform as well as more complex distributed and replicated architectures. Also note that the gains due to replication increase when the databases are replicated based on the roaming characteristics of the mobile station.

Finally, the effect of mobility is shown in Table 9 which shows the link utilization for different velocity in kilometers per hour (kph). The increase in velocity results in a higher number of crossings and hence more location updates and hence higher gains due to replication.

5 FAULT-TOLERANCE ANALYSIS

We consider a very simple model to study the fault-tolerance of the various database placement strategies. We make the following assumptions;

Table 9: A-link utilization as a function of the mobility for different placement strategies. Parameters: $\xi = 0.40, N = 70000, A_{la} = 9.00, \lambda_m = 3.0, L = 3.0, t_{hm} = 180, p_1 = 0.3, p_2 = 0.005, \delta = 1/hr$

Velocity (V kph)	5	11	17	23	29
POTS	0.25	0.25	0.25	0.25	0.25
HLR and VLR in SCP	0.55	0.58	0.61	0.64	0.68
HLR in SCP, VLR in SP	0.32	0.37	0.42	0.47	0.52
HLR cached in SP	0.34	0.40	0.46	0.53	0.59
HLR cached in 2 SPs	0.33	0.37	0.41	0.45	0.50
HLR cached in 3 SPs	0.32	0.35	0.38	0.41	0.44

- We consider only the failure of the databases. We assume that the inter-failure time and the recovery time are (negative) exponentially distributed random variables with mean rates γ and β, respectively.
- We assume that the two SCPs are operated in the replicated mode (as opposed to shared mode). This implies that both SCPs have the same information.

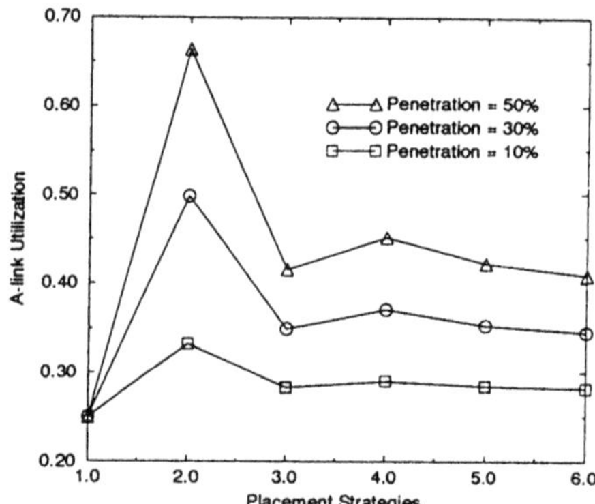

Figure 5: The link utilization for different placement strategies plotted for different degrees of penetration. X-axis : 1 → POTS, 2 → HLR and VLR in SCP, 3 → HLR in SCP and VLR in SPs, 4 → HLR cached in SPs, 5 → HLR cached in 2 SPs, 6 → HLR cached in 3 SPs. Parameters : $\xi = 0.40$, $N = 70000$, $A_{la} = 9.00$, $\lambda_m = 3.0$, $L = 3.0$, $t_{hm} = 180$, $p_1 = 0.3$, $p_2 = 0.005$, $\delta = 1/hr$.

We consider fault-tolerance only with respect to PCS. In view of this we define *availability* to be the probability that an MS can initiate a call. As before ξ denotes the degree

Table 10: Unavailability (1 - Availability) as a function of the failure rates for different placement strategies. Parameters : $\xi = 0.25, \beta = 1, x = 0.3333$.

γ	Unavailability		
	HLR and VLR in SCP	HLR in SCP, VLR in SP	HLR Cached in SP
5.0e-01	1.4e-01	0.72e-01	0.635e-01
5.0e-02	2.4e-03	0.7e-03	0.49e-03
5.0e-03	2.5e-05	0.63e-05	0.43e-05
5.0e-04	2.5e-07	0.6e-07	0.42e-07
5.0e-05	2.5e-09	0.6e-09	0.4e-09

Table 10 shows the availability of the three cases for different failure rates with $\beta = 1$ and $\xi = 0.25$. From the results it is clear that the availability of case 3 is the highest. This is not very surprising since replication in general results in higher availability. Based on the same analysis it can also be shown that if the HLR is replicated in the neighbouring SPs, then the availability can be further improved.

One of the key assumptions which leads to the above results is that the call control and routing functions can be performed by the HLR in case the VLR has failed. This may not be true for multiple reasons, e.g., the HLR and VLR may not be consistent with each other. If this assumption is not valid then it can be shown using a very similar analysis that availability when both the HLR and VLR is kept in the SCP is the highest. This suggest that in order to achieve high fault-tolerance it is necessary to maintain tight synchronization between the HLR and VLR databases. This implies higher performance penalties for higher degrees of replication.

6 CONCLUDING REMARKS

In this paper we have addressed performance and fault-tolerance of database management schemes to provide PCS in the existing signalling network. Specifically, we consider the placement of the VLR and th HLR databases in the existing signalling network. Four different schemes have been studied reflecting different degrees of distribution and replication. Based on a mean value analysis it has been shown that distributing the HLR and the VLR databases may be required for high degree of penetration of PCS usage. Simple distributed schemes where the VLR is co-located in the SP and the HLR is maintained in the SCP can significantly reduce the impact of PCS signalling on the existing signalling network. Performance of distributed architectures with replication depends largely on how the databases are replicated. The results show that performance gains are achievable when the replication is based on the roaming characteristics of the mobile station.

Fault tolerance of the different placement strategies was based on a simple failure and recovery model of the databases. The results show that the claim that the network can be made more robust against database failures with a distributed and replicated database placement strategy is valid only if all the distributed replicated copies are always consistent with each other. This may incur additional performance penalties.

There are a number of important future research directions. First, the mobility model assumed in this study is very simplistic and may not hold for different service areas. Realistic models are needed to quantify the number of location updates more accurately. Second, this study is based on a homogeneous signalling network. The database issues addressed in this paper should be analyzed for the case when the signalling network provides accesses to all different classes of PCS providers. Finally, this study is based on a mean value analysis or in other words, the "sunny day" scenario. It is important to study the performance and fault-tolerance of different database management schemes under failure and/or congestion conditions with temporal and spatial correlations.

7 REFERENCES

[Awerbuch et al., 1989] Baruch Awerbuch and David Peleg, "Online Tracking of Mobile Users", MIT Technical Report, October 1989.

[Bellcore, 1989] "Bell Communication Research Specification of Signalling System No 7", TR-NPL-000246, Issue 1,1985, reissued in June 1991.

[Cox, 1990] D. C. Cox, "Personal Communications – Viewpoint", IEEE Communications Magazine, November 1990.

[CCITT, 1988] "Public Land Mobile Network : Interworking with ISDN and PSTN", CCITT Recommendations Q.1000 – Q.1032, November 1988.

[Homa et al., 1992] Jonathan Homa and Steve Harris, "Intelligent Network Requirements for Personal Communication Services", IEEE Communications Magazine, February 1992, pp. 70–76.

[Jabbari, 1992] Bijan Jabbari, "Intelligent Network Concepts in Mobile Communications", IEEE Communications Magazine, February 1992, pp. 64–69.

[Kathy et al., 1992] K. S. Meier-Hellstern, E. Alonso and D. R. Neil, "The Use of SS7 and GSM to Support High Density Personal Communications", Winlab Workshop, April 1992.

[Kathy et al., 1991] K. S. Meier-Hellstern and E. Alonso, "Signalling System No 7 Messaging in GSM", Winlab Technical Report No. 25, December 1991.

[Lo et al., 1992] C. N. Lo, R. S. Wolff and R. C. Berhardt, "Expected Network Database Transaction Volume to Support Personal Communications Services", UPT Conference in Italy, January 1992.

[Minhas, 1987] H. Minhas, "GSM Signalling on the Radio Path : A Dimensioning Study", Proceedings International Conference on Digital Land Mobile Radio Communications, Venice 1987.

[Raymond, 1991] Paul-Andre Raymond, "Performance Analysis of Cellular Networks", IEEE Transactions on Communications, Vol. 39, No. 12, December 1991, pp. 1787–1793.

Skoog et al., 1990] A. R. Modarressi and R. A. Skoog, "Signalling System No. 7 : A Tutorial", IEEE Communications Magazine, July 1990, pp 19–35.

[Steel, 1990] R. Steel, "Deploying Personal Communications Networks", IEEE Communications Magazine, pp. 12–15, September 1990.

[Thomas et al., 1988] R. Thomas, H. Gilbert and G. Mazziotto, "Influence of the Movement of the Mobile Station on the Performance of a Radio Mobile Cellular Network", Proc. of 3rd Nordic Seminar, paper 9.4, Copenhagen, September 1988.

[Gray et al., 1993] Jim Gray and Andreas Reuter, "Transaction Processing : Concepts and Techniques", Morgan Kaufman Publishers, 1993.

20

A New Approach for Network Topology Optimization

T.Srinivasa Rao, S.K.Bose, K.R.Srivathsan
Dept. of Elec. Engg. I.I.T. Kanpur - 208016
INDIA.
Kalyanmoy Deb
Dept. of Mech. Engg. I.I.T. Kanpur - 208016
INDIA.

Abstract

We consider the problems of designing netwrok topologies with minimum cost. In the first problem, basic optimum topology which guarnatees only one path from one to the other node, is considered. This problem is extended in the second part to include one redundant path between two nodes. These two problems are solved in this work in three stages. In the first stage, modified version of random feasible topology generation algorithm is used to generate initial feasible solutions. In the second and third stages, Link Reduction Algorithm and Branch-X-Change algorithms are used sequentially to get a locally optimal solution. These three stages are required to repeat with different orderings of the nodes to get a global optimal solution. We tested the above approach with two problems.

1 Introduction

The central aim of network topology optimization is to minimise the total cost of the links subject to constraints on link utilization, redundant connections between any pair of nodes and the number of links emanating at each of the node. In a general optimization problem, the design variables are link decision variables, $\{l_{mn}\}$, defined as follows:

$$l_{mn} = \begin{cases} 1 & \text{if there is link between m and n} \\ 0 & \text{otherwise} \end{cases}$$

Here, the link decision variables, l_{mn} take an integer value (strictly either 0 or 1) and the capacity of the link is a continuous variable. This optimization problem is one of Mixed Integer Programming. In the objective function, the cost of a given link may depend linearly either on the link capacity or on the distance between the two nodes. If the objective function is also a linear cost function with respect to the design variables, then the optimization would be one of Mixed Integer Linear Programming.

Apart from communication system engineers, people from various engineering disciplines like transportation, production and computer systems engineering are also interested in this problem. The exact solution to this problem is not known as it is established that general network optimization is an NP- complete problem (Steiglitz, 1969). In the literature, various heuristic and approximate algorithms are proposed for solving such optimization problems. Unfortunately, most of these techniques converge to locally optimum solutions (Gersht,1990 and Magnanti,1984). Therefore one is required to solve this problem with different initial solutions to get locally optimum solutions for each case; we then select one among them as a globally optimum solution. Since it involves considerable amount of computational effort, this approach may not be practical for large sized networks.

In this report, we first look for a basic optimum topology. This means a minimum-cost topology which guarantees at least one path from one node to every other node. In the second part of the report, we concentrate on the minimum-cost topology which guarantees atleast two paths from one node to every other node in the network; this is so that the failure of any link will not make a node unreachable from other nodes.

To solve the network topology optimization problem, we use heuristic procedures in three stages. In the first stage, a feasible but not optimal topology is generated. In the second stage of the problem, we use the link reduction algorithm on the initial feasible topology to obtain a feasible topology with reduced cost. In the third and final stage, we use the Branch-X-Change algorithm to get a local optimum solution.

In Section 2, the formulation of the basic topology and its extension to a topology with one redundant path is discussed. The methodology to solve the above problem is presented in Section 3. In Section 4, the implementation details are given. The results are presented in Section 5. Fianlly in Section 6, Conclusions and future extensions are discused.

2 Problem Formulation

We consider a communication network with N nodes, with each node transacting various types of packet traffic. In this design, however, we only consider one type of traffic. The other types of traffic will be considered implicitely by constraints on the link utilisation for this traffic. The nodes are proposed to be interconnected by voice grade telephone lines. The cost C_{ij} of link l_{ij} between node i and node j is the sum of a fixed installation cost A and a variable quantity which is proportional to the distance d_{ij} between two nodes. In our approach, we need to suggest a basic topology with minimum cost, subject to the following constraints:

1. Every node is reachable from every other node.
2. One of the nodes (say a given node K), should be connected to every other node such that the shortest path for this has 4 hops or less.
3. The total number of links at some given nodes J, K and L, should not be more than 4 and at all other nodes the number of links should be either 1 or 2.
4. The link utilization factor should be in the range $(0, u_f)$.

In the second part of the work, this problem is extended to get a topology with redundant links. For this part, the following constraint is added to the constraint set:

5. There should be at least one redundant(arc-disjoint) path from one to every other node.

2.1 Optimization Problem:

The optimization problem considered here is formulated as follows–
Given:

$$\text{Requirement Matrix, R} = \begin{bmatrix} 0 & r_{12} & r_{13} & \dots & r_{1N} \\ r_{21} & 0 & r_{23} & \dots & r_{2N} \\ \dots & \dots & \dots & \dots & \dots \\ r_{N1} & r_{N2} & r_{N3} & \dots & 0 \end{bmatrix}$$

where r_{ij} = Traffic flow (Bytes/hr) between node i and node j.

$$\text{Distance Matrix, D} = \begin{bmatrix} 0 & d_{12} & d_{13} & \dots & d_{1N} \\ d_{21} & 0 & d_{23} & \dots & d_{2N} \\ \dots & \dots & \dots & \dots & \dots \\ d_{N1} & d_{N2} & d_{N3} & \dots & 0 \end{bmatrix}$$

where d_{ij} = Distanace between node i and node j in units of 100 kilometers.

Find :

$$\text{Link Matrix} = \begin{bmatrix} 0 & l_{12} & l_{13} & \dots & l_{1N} \\ l_{21} & 0 & l_{23} & \dots & l_{2N} \\ \dots & \dots & \dots & \dots & \dots \\ l_{N1} & l_{N2} & l_{N3} & \dots & 0 \end{bmatrix}$$

Such that :

$$Minimize \quad \bar{C} = \textstyle\sum_{i,j} l_{ij}(A + d_{ij}B)$$

Subject to :

$$(a) \quad 0 \le \tfrac{f_{ij}}{2400} \le u_f \le 1$$

where f_{ij} = traffic flow on link l_{ij} in bits per sec.

$$(b) \quad 1 \le \textstyle\sum_j l_{ij} \le 4 \quad \text{if i = j, K or L}$$

$$1 \le \textstyle\sum_j l_{ij} \le 2 \quad otherwise$$

(d) The no. of hops in the shortest path from node K to every other node is also constrained to be in the range of (1,4).

The constraint (d) also implicitely ensures that every node is reachable from every other node. For the extended problem, one more constrain is added. This states that–

(e) There should be at least one redundant path between one node and every other node.

3 Heuristic Approach

As mentioned earlier, most heuristic algorithms for solving the topology problems, result in a locally optimum solution for a given initial solution. We need to use these heuristics repeatedly for different initial feasible solutions to obtain different local optimum topologies. In this section, we discuss an algorithm to generate an initial feasible topology. This algorithm is a version of the Randomised routin proposed in (Steiglitz,1969), which has been modified to suit our problem.

3.1 Randomised Starting Routine

The algorithm is proposed in (Steiglitz,1969) to generate initial feasible solution randomly in the absence of any constraint other than the redundancy paths. i.e. there should be at least s_{ij} paths between node i and node j. To generate an initial topology which satisfies the above constraint, the following steps are followed–

1: Order the nodes randomly.

2: A new link is added with the highest degree of requirement and a node with next highest degree of requirement, where–

Degree of requirement of node i = maximum of s_{ij} over all j.

Of all those nodes with the next highest degree of requirement, the node with least distance is chosen.

3. The degree of requirement of the nodes between which a link added is reduced by one for each.

4. If the degree of requirement of all the nodes are zero, the procedure terminates. Otherwise, repeat from (2).

This algorithm will generate different initial feasible topologies with different orderings of the nodes.

3.2 Modified Starting Routine

The above algorithm is modified to suit our problem. The modified definition of degree of requirement is the maximum number of links emanating at the node. The terminating condition, Step 4, is also modified as follows:

4. If the highest and next highest degrees of requirement are greater than zero, then go to step 2. Otherwise terminate.

The topology so obtained from the above modified algorithm, will satisfy only the constraint on the nuber links at each node. Hence before accepting the topology as feasible, the algorithm is extended as follows with the addition of the following steps:

5. For every node i other than node K, find shortest distance to node K. If the shortest path from any node is more than 4 (as specified in the constraint), go to Step (8). Otherwise go to Step (6).

6. From every node i to every other node j, find the shortest path. If the shortest path is equal to zero, go to Step (8). Otherwise update the flow f_{mn} for every link l_{mn} in the shortest path between node i and node j by r_{ij}.

7. For every i and j, calculate the ratio of the flow f_{ij} and capacity of the link l_{ij}. If for any link, the ratio is more than u_f (as given in the constraints), go to Step (8). Otherwise terminate.

8. Permute the ordering of the nodes and rearrange distance matrix d_{ij} and flow requirement matrix r_{ij} accordingly. The repeat from Step (2).

3.3 Link Reduction Algorithm

This algorithm is proposed in (Gersht,1990) for improving (reducing) the cost of the topology.

1. Let the initial topology be T^0 and set v = 1. Calculate the cost C^0 of the topology T^0.

2. Pre-elimination of link l_{mn}:

Generate topology T^v_{mn} by eliminating a new link l_{mn} from T^{v-1}.

3. check T^v_{mn} to see whether the constraints are satisfied or not. If not, go to step 2. Otherwise include T^v_{mn} in T^v, the set of retained topologies for iteration v.

4. Compute $\Delta C^v_{mn} = C^{v-1} - C^v_{mn} \rightarrow \Delta C^v$.

If every feasible link is reduced then go to step 4. Otherwise go to step 2.

5. Get cost set ΔC^v for topologies T_v.

Find $\max_{k,l} \Delta C^v_{kl} \rightarrow (\Delta C^v_{mn}, T^v_{mn})$.
Then $T^v = T^v_{mn}$; $C^v = C^v_{mn}$.

6. If $C^{v-1} - C^v > 0$, then $v = v + 1$, and go to step 2. Otherwise, terminate.

3.4 Branch-X-Change Algorithm

The topology obtained from Link Reduction Algorithm is used as the initial topology in Branch-X-Change algorithm (Steiglitz,1969). This algorithm is another popular algorithm for network optimization. The basic idea in this algorithm is that in the topology T, link l_{ij} and link l_{mn} are replaced by links l_{in} and link l_{mj}, if the topology T' so obtained, is feasible and of lower cost compared to that topology T.

The Algorithm is:

1. Let the initial topology be T_0 with cost C_0. And set $v = 1$.

2. select links l_{ij} and l_{mn}. Remove links l_{ij} and l_{mn} add links l_{in} and l_{mj} to get a topology T_1' .

3. Calculate the cost C_1' of T_1'. If $C_1' < C_0$, go to Step 4. otherwise go to Step 5.

4. Check for feasibility of T_1'. If feasible, $T_1 = T_1'$ and $C_1 = C_1'$. Othewise $T_1 = T_0$ and $C_1 = C_0$.

5. If all links are checked terminate. Otherwise, set $v = v + 1$, go to Step 2.

The topology so obtained is a locally optimum solution. Since there is no guarantee that this topology would be globally optimum, we need to compute different locally optimum solutions by starting with different initial solutions. As discussed above, we can get different initial feasible topologies from the modified starting routine by varrying the ordering of the nodes every time.

4 Implementation

The above discussed algorithm was implemented in C for solving the network topology problems presented in Section 2. The module level flow chart is given in Fig. 1.

The code is developed in an obeject-specific modular fashion so that the same code can be used for different kinds of network topologiy problems with little changes like addition and deletion of particular mudules. In this work, we have solved both the problems presented in Section 2 using essentially same code. This has been done by appropriately adding submodules in order to take care of additional constraints in the second problem.

5 Results and Discussion

Most of the topology optimization algorithms proposed in the literature are based on the searching of the feasible topology space. The efficiency of any such search algorithm depends on the range of the search space. For a given initial topology, this varies from algorithm to algorithm. In the Link Reduction algorithm, the search space is such that the cost of the feasible topologies reduces monotonically. The disdvantage of this is that it searches for an optimal topology in only one narrow direction. In other words, in the initial feasible topology, if there is no link between node i and node j, there will not be a link between node i and node j in the locally optimum topology obtained from the Link Reduction Algorithm. Hence, the range of search space for this algorithm is very small. This drawback is partly overcome in our approach by using the Branch-X-Change algorithm. As explained earlier, in this algorithm, the link l_{ij} and link l_{mn} are replaced by links l_{in} and l_{mj}, if the topology so obtained is feasible and at improved cost. In other words, the Branch - X - Change algorithm searches for a better solution in neighborhood regions and hence results in an increase in the range of the search space.

We have tested the code on an HP 9000/300 workstation with two network topology problems. As discussed earlier, like most of the search algorithms, the above approach also converges to a locally optimum topology which varies from one initial topology to another. It may not be practical from a computational point of view, to generate locally optimum solutions for every initial feasible solution; the code is tested for the generation of fixed but not exhaustive number of locally optimum solutions. In this report, we have tested the code for generating thirty locally optimum solutions for each instance of the porblem. The results are given below:

5.1 Problem 1

Here we have a 6- node network, whose distance Matrix D is given by,

$$D = \begin{bmatrix} 0.0 & 3.8 & 8.0 & 10.3 & 14.2 & 13.7 \\ 3.8 & 0.0 & 11.2 & 10.6 & 16.0 & 16.3 \\ 8.0 & 11.2 & 0.0 & 14.2 & 13.6 & 11.7 \\ 10.3 & 10.6 & 14.2 & 0.0 & 7.7 & 9.6 \\ 14.2 & 16.0 & 13.6 & 7.7 & 0.0 & 2.7 \\ 13.7 & 16.3 & 11.7 & 9.6 & 2.7 & 0.0 \end{bmatrix}$$

Flow requirement Matrix R, is given by,

$$R = \begin{bmatrix} 0 & 1885 & 1071 & 13097 & 1402 & 870 \\ 6736 & 0 & 2425 & 25547 & 3252 & 1919 \\ 2146 & 2399 & 0 & 18581 & 1870 & 1070 \\ 64892 & 61328 & 50346 & 0 & 74692 & 35490 \\ 2403 & 3116 & 1759 & 37391 & 0 & 1455 \\ 2264 & 2770 & 1551 & 27185 & 2047 & 0 \end{bmatrix}$$

Also $J = 2, K = 4, L = 5$ and $u_f = 0.5$.

The installation cost of link = Rs. 40,000 (fixed) The incremental cost of link = Rs. 75,000 per unit distance.

5.1.1 Basic Optimal Topology

For one particular ordering of the nodes, the initial feasible topology is shown in Fig. 2. Its total cost is Rs. 54,37,500. After Link Reduction Algorithm, the improved topology is given in Fig. 3. Its cost is Rs. 41,97,500. After Branch - X - Change Algorithm, the local optimum topology is as shown in Fig. 4. The cost of this topolgy is Rs. 35,97,500. We have run this program and generated thirty locally optimum topologies (among these, some may be identical). Among these topologies, the topology with the lowest cost is considered as global optimal topology. The global optimal topology is as shown in Fig. 5 and its cost is Rs. 26,60,000. The execution time of the code for obtaining these thirty locall optimal topologies is 0.16 seconds.

5.1.2 Optimal Topology with One Redundant Path

The global optimal topology (best of thirty locally optimal topologies) is shown in Fig. 6 and its cost is Rs. 35,77,500. To generate these thirty topologies, the execution time required is 0.23 seconds.

We have also obtained optimum topologies for different value of u_f (i.e. $u_f = 0.4$, 0.3, 0.2). It is observed that for the given traffic flow requirements, the optimal topology remained unchanged if u_f is as low as 0.3.

5.2 Problem 2

In this problem we considered an 8 - node network whose distance matrix is given by D given below,

$$D = \begin{bmatrix} 0.0 & 3.8 & 8.0 & 10.3 & 14.2 & 13.7 & 9.3 & 8.5 \\ 3.8 & 0.0 & 11.2 & 10.6 & 16.0 & 16.3 & 11.2 & 12.0 \\ 8.0 & 11.2 & 0.0 & 14.2 & 13.6 & 11.7 & 9.7 & 1.0 \\ 10.3 & 10.6 & 14.2 & 0.0 & 7.7 & 9.6 & 5.5 & 15.4 \\ 14.2 & 16.0 & 13.6 & 7.7 & 0.0 & 2.7 & 4.7 & 14.4 \\ 13.7 & 16.3 & 11.7 & 9.6 & 2.7 & 0.0 & 5.0 & 12.6 \\ 9.3 & 11.2 & 9.7 & 5.5 & 4.7 & 5.0 & 0.0 & 11.0 \\ 8.5 & 12.0 & 1.0 & 15.4 & 14.4 & 12.6 & 11.0 & 0.0 \end{bmatrix}$$

The flow requirement matrix for this problem is chosen as,

$$R = \begin{bmatrix} 0 & 1885 & 1071 & 13097 & 1402 & 870 & 870 & 1071 \\ 6736 & 0 & 2425 & 25547 & 3252 & 1919 & 1919 & 2425 \\ 2146 & 2399 & 0 & 18581 & 1870 & 1070 & 1070 & 3000 \\ 63892 & 61328 & 50346 & 0 & 74692 & 35490 & 35490 & 50346 \\ 2403 & 3116 & 1759 & 37391 & 0 & 1455 & 1455 & 1759 \\ 2264 & 2770 & 1551 & 27185 & 2047 & 0 & 3000 & 1551 \\ 2264 & 2270 & 1551 & 27185 & 2047 & 5000 & 0 & 2500 \\ 2146 & 2399 & 5000 & 18581 & 1870 & 1070 & 2000 & 0 \end{bmatrix}$$

Also $J = 2, K = 4, L = 5$ and $u_f = 0.5$.

5.2.1 Basic Optimal Topology

The global optimal topology obtained from our program is shown in Fig. 7. Its cost is Rs. 32,65,000. The time for execution to generate thirty local optimal topologies is 0.33 seconds.

5.2.2 Optimal Topology with One Redundant Path

The global optimal topology resulting from our program is given in Fig. 8 and its cost is Rs. 39,57,500. The time for the generation of the thrity optimal topologies is 0.35 seconds.

6 Conclusions and Future Extensions

In this work, we studied some heuristic algorithms for netwrok topology optimisation. The first is for generating an initial feasible topology. This is a modified version of the algorithm presented in (Steiglitz,1969). The other two algorithms are for optimising these initial feasible topologies. They are Link Reduction Algorithm (Gersht,1990) and Branch - X - Change Algorithm (Steiglitz,1969). In this work, we observed that the topology obtained from Link Reduction Algorithm may not always be optimum. This is because of the fact that the link reduction algorithm searches only in the direction of the initial feasible topology. Hence improvement in the solution was observed when we execute the Branch - X - Change Algorithm with the topology obtained from Link Reduction Algorithm, as i.e. the initial feasible topology for the Branch - X - Change Algorithgm. We have developed the code for these algorithms in C and tested it with two different problems. An increase in range of the search space, due to the application of Branch - X - Change Algorithm preceded by Link Reduction Algorithm, is observed. The range of search space can be improved further by applying both Link Reduction Algorithm and Branch - X - Change Algorithm in alternative iterations. The initial starting routine can modified so that the ordering of the nodes is generated randomly.

7 References

Gerla,M. and Kleinrock,L.(1977) *On the Topological Design of Distributed Computer Networks*, IEEE Trans. on Communications, Vol. COM-25, No. 1, Jan 1977

Gersht,A. and Weihmayer,R.(1990) *Joint Optimization of Data Network Design and Facility Selection*, IEEE Jl. on S.A.C., Vol. 8, No. 9, Dec 1990

Magnanti,T.L. and Wong,R.T.(1984) *Network Design and Transportation Planning: Models and Algorithms*, Transport Science, Vol. 18, No. 1, Feb 1984

Steiglitz,K., Weiner,P. and Klietman,D.J.(1969) *The Design of Minimum-cost Survivable Networks*, IEEE Trans. on Ciruits Theory, Vol. CT-16, No. 4, Nov 1969

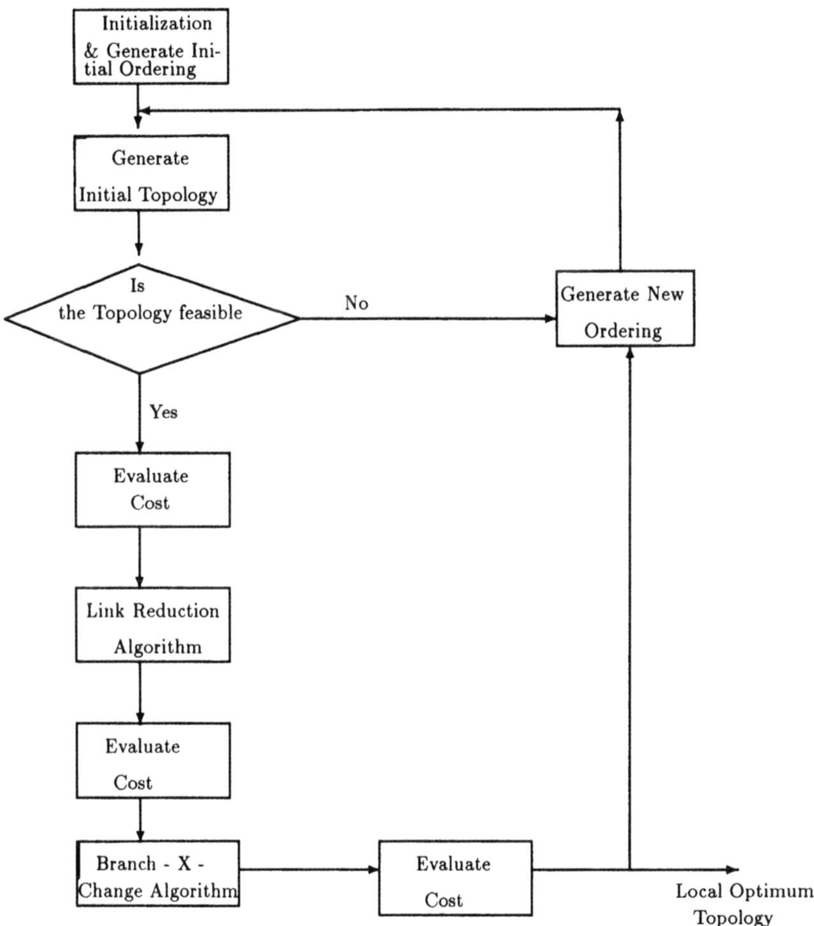

Fig. 1 Flow Chart

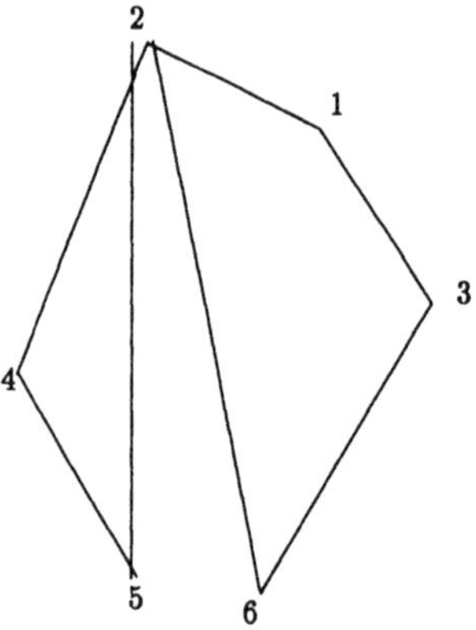

Fig. 2 Initial Feasible Topology

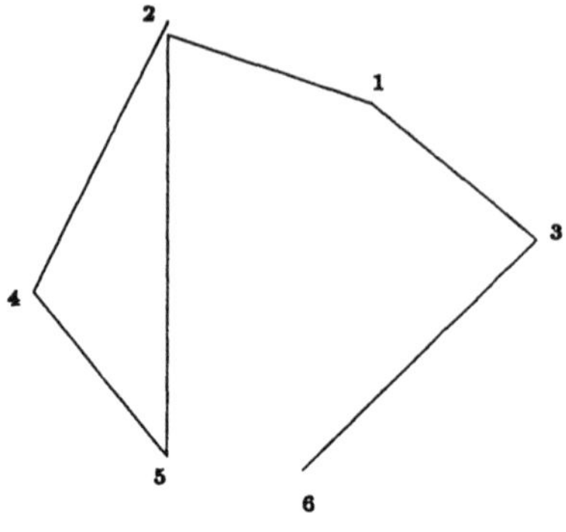

Fig. 3 Reduced Cost Topology, from Link Reduction Algorithm

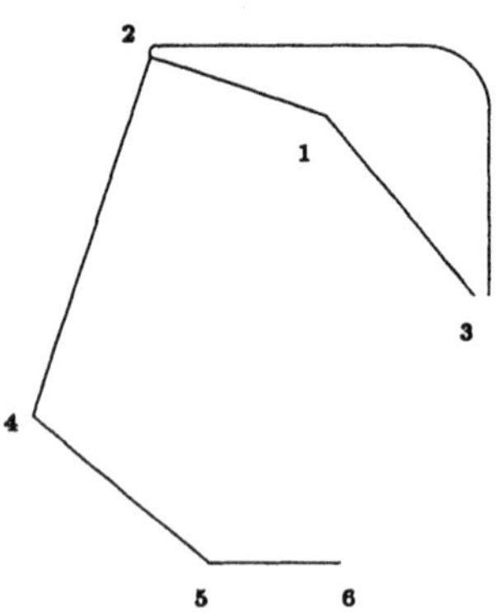

Fig. 4 Locally Optimum Topology

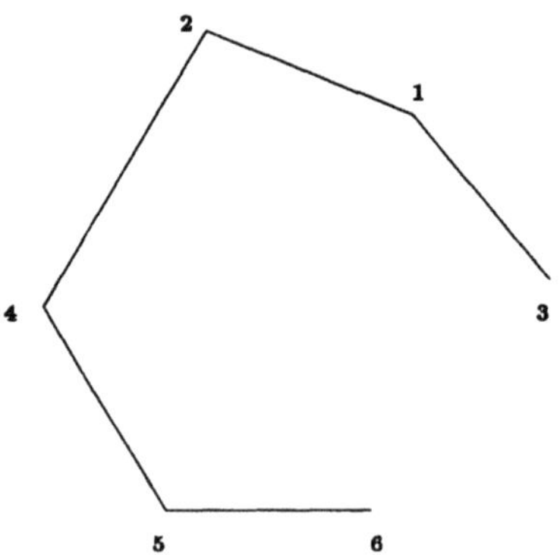

Fig. 5 Global Optimal Topology

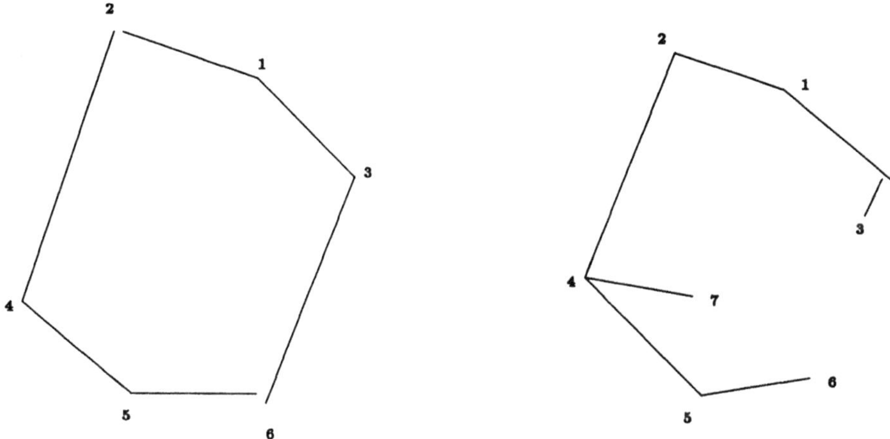

Fig. 6 Optimal Topology With One Redundant Path

Fig. 7 Basic Global Optimal Topology

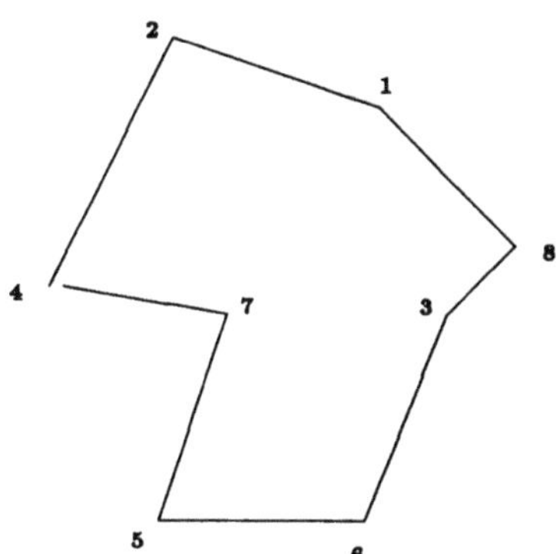

Fig. 8 Global Topology With One Redundant Path

INDEX OF CONTRIBUTORS

KEYWORD INDEX

www.ingramcontent.com/pod-product-compliance
Ingram Content Group UK Ltd.
Pitfield, Milton Keynes, MK11 3LW, UK
UKHW020058200726
13856UKWH00002B/273